Advances in Tea Agronomy

Tea is big business. After water, tea is believed to be the most widely consumed beverage in the world. And yet, as productivity increases, the real price of tea declines while labour costs continue to rise. Tea remains a labour-intensive industry.

This book is intended to assist all those involved in tea production to become creative thinkers and to question accepted practices. International in content, it will appeal to practitioners and students from tea-growing countries worldwide. In addition to providing a comprehensive review of the principal tea-growing regions in terms of structure, productivity and constraints, it examines the associated experimental evidence needed to support current and future crop management practices.

Mike Carr is Emeritus Professor of Agricultural Water Management at Cranfield University, UK. He has over 50 years of experience in the management and delivery of international research, education, training and consultancy in agriculture and natural resource management and was previously Executive Director of the Tea Research Institute of Tanzania. He is the author of *Advances in Irrigation Agronomy: Plantation Crops* (Cambridge University Press, 2012), *Advances in Irrigation Agronomy: Fruit Crops* (Cambridge University Press, 2014) and former editor-in-chief of the Cambridge University Press journal *Experimental Agriculture*.

Advances in Tea Agronomy

M. K. V. CARR

Cranfield University

With written contributions from:

T. C. E. CONGDON

Fellow of the Royal Geographical Society and of the Royal
Entomological Society

R. H. V. CORLEY

Independent Consultant on Plantation Crops Research

G. K. TUWEI

Research and Development Manager, Unilever Tea Kenya Limited

C. J. FLOWERS

Group Managing Director, Eastern Produce Kenya and Kakuzi Ltd

M. UPSON

M. PEREZ ORTOLA

CAMBRIDGE
UNIVERSITY PRESS

CAMBRIDGE
UNIVERSITY PRESS

University Printing House, Cambridge CB2 8BS, United Kingdom

One Liberty Plaza, 20th Floor, New York, NY 10006, USA

477 Williamstown Road, Port Melbourne, VIC 3207, Australia

314–321, 3rd Floor, Plot 3, Splendor Forum, Jasola District Centre, New Delhi – 110025, India

79 Anson Road, #06–04/06, Singapore 079906

Cambridge University Press is part of the University of Cambridge.

It furthers the University's mission by disseminating knowledge in the pursuit of education, learning and research at the highest international levels of excellence.

www.cambridge.org
Information on this title: www.cambridge.org/9781107095816
DOI: 10.1017/9781316155714

First published 2018

Printed in the United Kingdom by TJ International Ltd, Padstow, Cornwall

A catalogue record for this publication is available from the British Library.

Library of Congress Cataloging-in-Publication Data
Names: Carr, M. K. V., author.
Title: Advances in tea agronomy / M.K.V. Carr ; with written contributions
from: Timothy Colin Elphinston Congdon, Richard Hereward Vanner Corley,
Gabriel Kiplangat Tuwei, C.J. Flowers, Mathew Upson, Marta Perez Ortola.
Description: New York, NY : Cambridge University Press, 2017. | Includes
bibliographical references and index.
Identifiers: LCCN 2017048713 | ISBN 9781107095816
Subjects: LCSH: Tea.
Classification: LCC SB271 .C37 2017 | DDC 633.7/2–dc23
LC record available at https://lccn.loc.gov/2017048713

ISBN 978-1-107-09581-6 Hardback

This book is dedicated to my grandchildren:
Tabitha Dolton, James Carr and Gracia Carr.
One day they will wonder what Grandpa did for a living . . .

Contents

Colour plates are to be found between pp. 242 and 243.

Foreword

Professor Michael K. V. Carr's long journey through the 'Tea World' commenced as early as 1966 at the Tea Research Institute of East Africa, Kericho, Kenya. In 1998 he deservedly became the founder Executive Director of the Tea Research Institute of Tanzania. Having occupied senior positions in different research institutes on varied crops and undertaken professorial assignments in several universities, Professor Carr has been a visiting professor, a professor emeritus and an adviser/consultant to national and international academia and multinational commercial organisations on tea and other plantation crops. In connection with his professional role he has travelled extensively to over fifty countries. He continues to serve as a consultant and/or adviser to research and commercial organisations.

Professor Carr has published numerous scientific articles and four books on tea, irrigation and related subjects. He edited the prestigious journal *Experimental Agriculture* for seven years. With such a distinguished career and rich experience in diverse crops, he is eminently qualified to provide an intelligent discourse on tea agronomy. This book, *Advances in Tea Agronomy*, is not a routine planters' guide. The author has attempted to identify and evaluate the research findings that form the basis for current cultivation practices and also those that often defy explanation and interpretation.

The book commences with a 'world view' of tea and weaves its way sequentially through basic physiology, breeding techniques (including that unique to tea, clonal selection), establishment of young tea plants in the field, their training into strong and healthy bushes, shoot growth and development of shoots, and harvesting techniques. The discussion on the difficulty in explaining the varying, if not conflicting, results from fertiliser trials and the chapter on irrigation are particularly engrossing. Tree species that play an important role in tea cultivation are highlighted.

The chapters that stand out most conspicuously are two towards the end of the book on ethics in the tea industry and sustainability. They are likely to induce some controversy, stimulate intelligent argument and wake up the 'generals' of the industry!

V. S. Sharma, MSc, PhD
Previously: Director UPASI Tea Research Institute and
Editor-in-Chief, **International Journal of Tea Science**

Preface

It All Depends

After graduating from Nottingham University (UK) in 1966 as an ambitious 22-year-old aspiring to see the world, my first job was with the Tea Research Institute of East Africa as it was then known. Its headquarters were in Kericho, Kenya, but with substations in Tanzania and Uganda. Each of these neighbouring countries has since gone its separate way, but the tea is still there.

My challenge was to establish a small research station in the southern highlands of Tanzania, where there was an extended annual dry season. It was anticipated that this would be a good site to undertake some fundamental studies of the water relations of tea, and the irrigation requirements of the tea crop, without interference by rain.

On arrival in Kericho, I was taken to meet the Chairman of the Governing Body; he told me to come back in 10 years, by which time I should know something about tea. I refrained from telling him that I was only on a two-year contract and therefore was unlikely to be of much use. It is now 50 years since that meeting, and perhaps I am now qualified to write about tea. Fifty years' experience could mean one year's experience multiplied by 50 or, at best, it could represent 50 years of diverse activities throughout the tea world.

My experience in tea agronomy has been focused on research in eastern Africa, but I have also been involved with a range of other projects in the tea industry. These include management training courses, consultancy services for commercial companies, supervision of PhD projects, and formal education programmes. From my UK base, mainly at Cranfield University, I have been privileged to visit and study tea cultivation in many countries in Africa – including Kenya, Tanzania, Uganda, Malawi, Zambia, Zimbabwe in eastern and southern Africa plus two in West Africa – Nigeria and Cameroon – and in Asia – India, Sri Lanka, Bangladesh, Indonesia and Japan. I have also witnessed tea growing in the USA, Australia and England. Tea is grown in some of the most beautiful parts of the world, with idyllic scenery and equable climates, and it is cultivated by nice people. What more can one ask for?

Much has changed over the past 50 years, while much too has remained the same. In 1966 (the year to remember, as it was when England won the football World Cup), a yield of 1000 kg made tea per hectare was considered to be good in East Africa, and 2000 kg ha^{-1} was exceptional. Now, the same bushes are yielding 3500–4000 kg ha^{-1}, while the target yields on the best estates are 5000–6000 kg ha^{-1} with the best fields achieving 7000–8000 kg ha^{-1}. The world record still seems to be the 10,000 kg ha^{-1} obtained in Kericho in 1980.

Research can, justifiably, claim some credit for these increases in productivity, not least through the selection and propagation of new cultivars or clones, although many of the earlier successful selections owe as much to chance as to good science. Perhaps the biggest change is in people's higher expectations of achievable yields. The smartest way is to identify and adopt the best agronomic practices to raise the good yields to those achieved by the best growers, and to bring the poor yields up to today's average.

Many changes that occurred in productivity are the result of incremental improvements in agronomic practices rather than any single major breakthrough (an exception being the introduction of chemical weed control in the 1960s). The challenge for the immediate future is how best, and when, to replant tea that is fast approaching the decline of its productive life, the so-called *'geriatric tea'*.

The aim of this book is to identify – based on 50 years of hard-learnt experience – the questions that need to be asked by an inquisitive farmer/manager. It is not aimed at necessarily providing answers, because quite often there is no single answer. The standard response of 'it all depends' is annoying to the person asking the question, but invariably it is the correct answer! This book attempts to explain 'on what does it all depend?' And, can we make allowances for these other factors? If you want a recipe book for tea growing, this is not the book for you. There is no shortage of books on tea husbandry targeted at local conditions.

Acknowledgements

Since the start of my tea career, my principal 'tea mentor' has been Colin Congdon, who, as an estate manager and later estates director, always questioned established practices – often to the annoyance of his 'superiors', who wanted 'standing orders' imposed from afar to be followed. Colin has contributed his views as a commentary at appropriate places in the text. A younger mentor is Chris Flowers, who despite his relative youth (being much younger than Colin) also asks probing questions (and attempts to answer them). Chris has contributed to the chapter on mechanical harvesting as well as reviewing individual chapters. Other reviewers include Philip Owuor, Tom Brazier, Rob Lockwood and Peter Rowland. Special thanks go to Hereward Corley, who read and critically reviewed nearly every chapter, and to William Stephens, who read and commented on the final script.

Hereward together with G. K. Tuwei wrote Chapter 5 on tea breeding. Marta Pérez Ortola and Mathew Upson collated the weather data presented in Chapter 3 and prepared diagrams used elsewhere in the text. George Woodall helped to prepare the photographs for publication, as did my son Julian. I thank them all.

I acknowledge too with thanks the directors of the Tea Research Institute of Tanzania, the Tea Research Institute of Sri Lanka, the Bangladesh Tea Research Institute, the Tea Research Foundation of Kenya, UPASI Tea Research Institute and Tocklai Tea Research Institute for granting me permission to use their weather records. Professor Chen Zongmao identified the most appropriate weather stations to represent tea locations in China.

Over the years, I have worked with a number of people on tea-related projects. These include, in no particular order: Caleb Othieno, Bruno N'dunguru, Firmin Mizambwa, Mainul Huq, Anna Nyanga, John Kilgour, Paul Burgess, Cassian Haulle, Charles Rao, James Tucker, William Stephens, Emmanuel Simbua, Tim Harding, Kingsley Bungard, Chris Warn, Malcolm Keeley, Simon Hill, Janakie Balasuriya, Roger Smith, Andrew Bennett, Robin Mathews, Keith Weatherhead, Dunstan and Imelda N'damugoba, Ernest and Alexandra Kimambo, H. I. Kipangula, J. R. Myinga, Michael Mhosole, David Nixon, Galus Myinga, Julio Lugusi, Boniface Miho, C. P. Chelesi, R. M. Masha, David Mgwassa, C. Chambaka, R. Madege, Ernesto Ng'umbi, Badan Sanga, Rashidi Vangisda, Francis Shirima, Wilson Ng'etich, Hastings Nyirenda, Mike Green, John Tolhurst, David Debenham, Sarah Taylor, Gail Smith, Mark Dale, Ernest Hainsworth, S. W. Msomba, Ken Willson, Ray Fordham, Geoff Squire, Julius Kigalu, Peter Martin, Matt Dagg, John Templer, Adrian Whittle, Jimmy Winter, Noel Lindsay-Smith, Norman

Kelly, Rick Ghaui, Bimb Theobold, Bertie Amritanand, George Kyejo, G. M. Mitawa, Nuhu Hatibu, Ray Wijewardene, J. M. Sikira, and many, many others. It's been fun.

I thank Dr V. S. Sharma, previously director of UPASI in South India and now a private consultant. He has also been editor-in-chief of the *International Journal of Tea Science*. I am grateful to him for writing the foreword to this book. There is not much on tea that Dr Sharma has not written about during a lifetime studying the crop. He has recently synthesised this experience into an excellent small book entitled *A Manual of Tea Cultivation*, published by the International Society of Tea Science (2011). Strongly recommended. I only wish I had met Dr Sharma earlier in my career.

Last, but most important, my wife Dr Susan Carr used her considerable editorial skills and patience to improve my written English, and to ensure that the book was completed. For this sacrifice, I will always be very grateful. But all mistakes are mine alone.

Enjoy!

1 Karibuni!

Welcome to You All

Te, té, thee, thé, thea, chai, cha, ocha, chà, 茶, meng, miang, char, herbata

This is a book about research findings that can inform the practice of managing tea. But, unlike many practical tea management handbooks, it focuses on questions rather than answers. Among the basic questions that tea growers need to answer are where and how to plant tea, which varieties to plant and at what density, how to bring the plants into production, how and when to harvest, how best to sustain yields over time, and when to replant. These questions seem to be simple enough to answer, but in these examples and others, managers operating in the field or factory need to be creative thinkers, always questioning accepted practices.

The tea industry, by employing its own scientists and paying for its own research, led the way in looking to research to find solutions to technical problems in the field and in the factory (Hazarika and Muraleedharan, 2011). Yet, paradoxically, individual tea companies have often expected their managers to follow 'standing orders' prepared by a director working in a head office, usually a long way from the scene of the action. At one time, when managers were inexperienced, it was perhaps the best approach. A major downside however, was that this style of management inhibited change or experimentation, and tea yields remained low for many years.

Rates of Change

This culture of standing orders still exists in some organisations, including centrally controlled smallholder schemes. Unlike the case with modern fruit orchards, in tea plantations there has been so little visible change that a nineteenth-century tea planter would still feel very much at home, except that in some areas during the intervening years the shade trees may have disappeared and couch grass is no longer a problem.

Tea is a perennial crop. It is also long-lived; for example, a tea tree reputed to be 2700 years old is still alive in China (see Chen, 2012 for a picture). As a tea plantation can remain productive for at least 100 years, it is important to recognise that soil conditions as well as climate can change over time. The soil is a dynamic, living system in which nutrients are lost and imperfectly replaced. At the same time, the capacity of the plants to absorb nutrients from the soil solution alters as the soil, for example, becomes

Figure 1.1 A tea estate in Kenya, west of the Great Rift Valley. The best estates in Kenya still lead the way in terms of productivity (MKVC). *A black and white version of this figure will appear in some formats. For the colour version, please refer to the plate section.*

more acid and loses its structure, or becomes drier or excessively wet, or becomes compacted. With soil, the time scale for change is from weeks (dryness) to years (acidity). By contrast, the weather is constantly changing, sometimes from minute to minute, certainly from hour to hour, and the tea plant responds at a physiological level to these rapid changes in its physical environment.

We have come a long way in research as well as in our technical know-how in the last hundred years. For example, during the first quarter of the twentieth century, with limited research outputs, mean annual yields of tea in Ceylon only rose from 400 to 550 kg ha^{-1}, and in 1931 Eden felt that tea cultivation had '*now reached the stage of small yield increment necessitating accurate experimentation*'. This was at a time when statistics were being used in agricultural research in the UK for the first time to allow for experimental error. Eden was a pioneer in the application of statistics to tea research, since '*the literature of tea experiments in Java, India and Ceylon contained no reference to experimental error*'. Thus, no one knew what the size of an experimental plot should be, or the number of replications needed to allow for variability between bushes and in soil properties in order to be able to state with some degree of certainty that the results obtained were due to the treatments and not just to chance. We now take the use of statistics in agricultural research for granted, as in medical research.

Variability

The immediate effects of variability in responses due to weather conditions or soil properties often go unnoticed, or at least unreported, unless they are in response to a sudden event, like hail. This is partly because tea researchers have a tradition of establishing statistically designed, replicated field trials the results of which, usually yield, are reported once a calendar year in a standardised way in the traditional, often very tedious, annual report (not always produced on time!). Research findings from experiments of this type can appear to vary considerably from one site to another and from year to year, without explanation, making it difficult for tea scientists to make the site-specific recommendations that tea growers rightly demand.

A deeper understanding of the reasons why responses to experimental treatments differ from site to site or from year to year is needed if we are to make progress. To improve the relevance of their findings, researchers need to be able to:

- provide answers to practical problems of immediate concern to the tea industry,
- understand the mechanisms responsible so that the results can be applied with confidence to other locations, and
- communicate the results to the industry and to the wider scientific community in the most appropriate ways.

Factors Limiting Tea Yields

Why does the productivity of tea vary so much? *Average* annual commercial yields across the world are low at around 1150 kg of processed tea per hectare (Majumder *et al.*, 2012) (normal range 300–5000 kg ha^{-1}), but a commercial yield 10 times the global average has been obtained in Kenya. Why, for example, should we continue to grow tea on such an extensive scale when we are losing yield in the field? Would it not be better to intensify production methods and, by so doing, release land for other purposes?

Explaining the processes of yield development in terms of biology (e.g. growth rates), physics (e.g. responses to the weather) and chemistry (e.g. nutrient uptake) is fundamental if we (tea researchers and producers) are to continue to make progress.

First, it is necessary to be able to identify and to quantify the limiting factors to yield and to seek cost-effective solutions. Let us consider what changes have occurred over the last 50–60 years (a period which some of us have lived through), the limiting factors that they addressed, and what changes we can anticipate over the next 20 years.

Changes

Even though a tea planter from the nineteenth century might still feel at home in today's plantations, change does happen. For example, some of the changes in cultural practices that have taken place in Africa since 1960 are listed in Box 1.1. These have had an

---

Box 1.1 Changes in Tea Management Practices in Africa (since 1960): Some Examples

- Shade trees removed (nearly everywhere)
- Sulphate of ammonia replaced by NPK
- Weeding with hoes replaced by herbicides
- Seedling plants replaced by clones (or improved seed)
- Seed at stake replaced by plants in polythene sleeves
- Infilling is common
- *Pruning cycles* extended[1]
- Systematic harvest intervals based on temperature integration
- Mechanical harvesting introduced
- *Scheme plucking* (some)
- Shoot size increased from 2+ bud to 3+ bud (not everyone)
- Irrigation (where it can be justified)

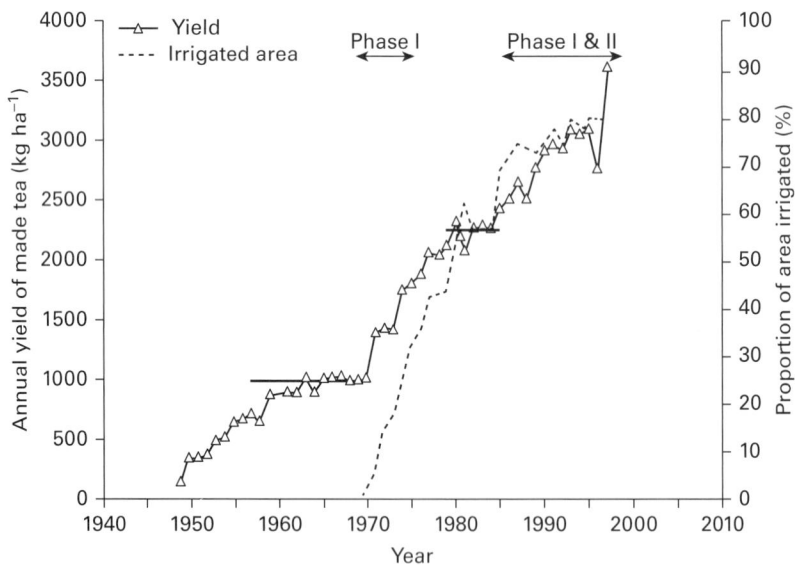

Figure 1.2 Incremental increases in annual crop yield from the same area of tea planted in the 1930s in Mufindi, Tanzania. The dotted line illustrates the introduction of irrigation. Horizontal lines indicate periods when yields appeared to stabilise.

incremental effect on crop productivity, as illustrated in the graph (Figure 1.2) showing changes since the 1950s in yield on one estate in the Mufindi District of Tanzania.

In the 1950s, yields were only about 400 kg made tea per hectare. Couch grass was a problem; it had to be manually removed by digging, which disturbed the tea bush especially the roots. The introduction of paraquat in the 1960s was a major breakthrough, enabling the couch grass to be eliminated with a herbicide. Roots were no longer disturbed at regular intervals, allowing them to proliferate in the topsoil and to extend into the surface mulch layer of organic material. This mulch is created from the prunings obtained when the tea bush is cut back at periodic intervals (2–5 years). At high altitudes in Africa, where it is cool, mulch only breaks down slowly. Water-soluble phosphate is quickly fixed in acid soils, making it unavailable to plants, but the pH in the surface layers is higher than elsewhere in the bulk soil profile. Circumstantial evidence and keen observation suggested that phosphorus remains available in the mulch layer, which stimulates its uptake (Hainsworth, 1969, 1976; Willson, 1972). In this way, a chemical developed to control weeds had contributed to the improved nutrition of tea, allowing the canopy to expand to cover the soil and, by so doing, protecting the soil from erosion, a win–win–win outcome. The forest floor had been recreated under a tea bush.

Where Have All the Shade Trees Gone?

Other concurrent changes included the removal of shade trees throughout most of eastern and southern Africa. Planters who came from India and Sri Lanka to start the embryonic tea industry had originally introduced shade into Africa. They knew nothing else at the time. But following the work of Hadfield (1968) in Assam, who recognised that shade was only likely to be beneficial in areas where daily maximum air temperatures regularly exceeded 30 °C, it was soon appreciated that in the cool, high-altitude areas of Africa shade was not necessary.

Clones were introduced during the 1960s and slowly replaced Indian hybrid seed for infilling and new plantings. The need to prune every three years, or sometimes more frequently than that, began to be questioned, and pruning rounds were extended. If a tea

Figure 1.3 Where did all the shade trees go? During the 1970s it was recognised that in the cool high-altitude areas of Africa shade was no longer necessary (MKVC). *A black and white version of this figure will appear in some formats. For the colour version, please refer to the plate section.*

plant was harvested properly, the surface of the crop canopy (the 'table') would not rise as fast, so pruning could be delayed.

The realisation that the tea canopy was increasing in height too quickly (by pruning, management mistakes were covered over!) led to other ways of managing the tea harvesting (plucking) process being considered. Scheme plucking became popular in some places. In this system, an individual plucker (or his/her family) was responsible for harvesting a fixed area of tea, for example, six 0.1 ha plots. In this way, by giving a sense of ownership to the plucker, it was expected standards would improve. However, scheme plucking was not universally adopted, since there were difficulties in managing this system.

Machine-Aided Harvesting

Shortages of labour in some places meant that mechanical harvesting became essential. Many systems, described elsewhere but often unproven in Africa and India, have been tried. These range from a simple hand-held blade, through modified hedge-cutting shears and a variety of hand-held motorised cutter bars (Figure 1.4), up to large, mobile self-propelled harvesters. Tea table height control and machine reliability became issues of concern, but the more enterprising companies have persisted with mechanisation in those countries where mechanical harvesting is not banned.

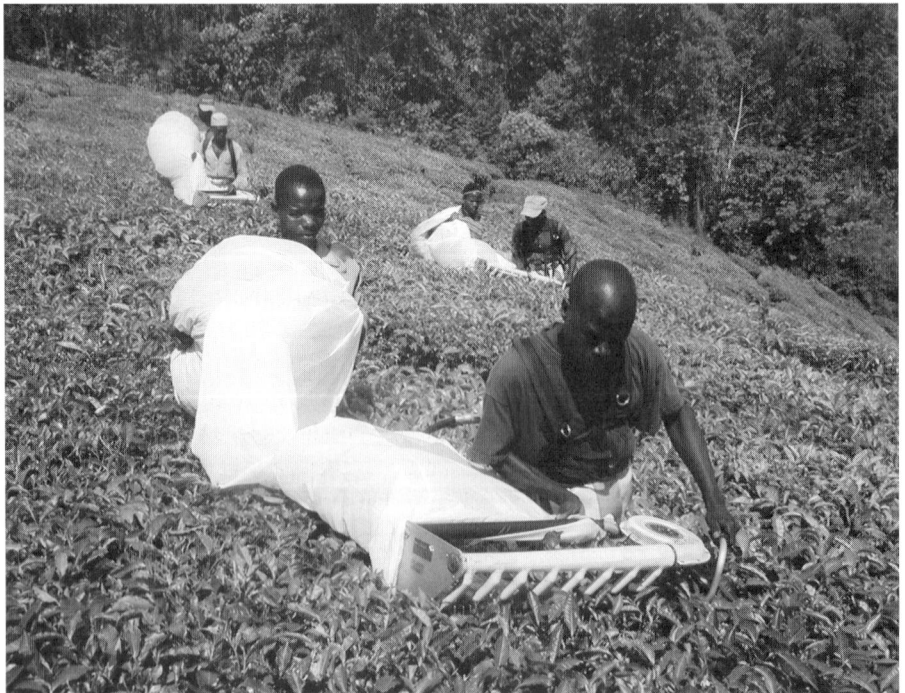

Figure 1.4 Shortages of labour in some places have meant that mechanical harvesting is essential. Many systems are now available and being tested (Uganda, MKVC). *A black and white version of this figure will appear in some formats. For the colour version, please refer to the plate section.*

Figure 1.5 A clonal field trial can occupy a large area of ground, as shown here in Kericho, Kenya. The plot size and number of replications influence the precision of trials (the ability to detect significant differences between treatments). It has been found that two replicates of 16-bush plots is a good compromise between precision and the area of trials. New modern estates are planted with the best available clones, irrigated, and harvested mechanically. The future has arrived (RHVC).

Understanding the relationships between air temperature and shoot development and shoot extension led to a method for predicting the optimum harvest interval in a rational way. This in turn reduced the need for selectivity by machines, being particularly useful at the beginning and end of a growing season when temperatures and rates of shoot development are changing (Burgess and Carr, 1998).

At the same time, the benefits of irrigation were being recognised in southern Africa. This was the 'icing on the cake', to be introduced when other limiting factors had been overcome. This allowed new estates to be developed in low-rainfall grassland areas (no need to cut down valued rainforest). These were planted with the best available clones, fully irrigated (drip, sprinkler and centre-pivot) and harvested mechanically. The future had arrived (Figure 1.5)!

The Future

What further changes can we expect in the future? Some that we might anticipate, because they are already happening in the most progressive plantations, are listed in Box 1.2. One

> **Box 1.2** Future Changes in Tea Management Practices
>
> - Unskilled labour will become skilled
> - New high-quality cultivars will be bred and released with target yields of $10 \, t \, ha^{-1}$
> - Mechanical harvesting will become common practice
> - Mechanics will become highly skilled
> - Plant populations will increase
> - Replanting will become essential
> - Production will become more intensive for some
> - Precision agriculture will become the norm
> - For others, low-input low-output production systems will continue
> - Smallholder numbers will continue to expand
> - Fuelwood will need to be managed as well as the tea
> - Domestic facilities on estates will improve in response to international audits
> - Management will become less hierarchical
> - Managers will cease to be merely messengers for 'standing orders'
> - Research will have to become more problem focused, supported by good science
> - The impact of climate change will have to be managed

can expect the divide to increase between the high-input high-output intensive producers and the low-input low-output extensive growers. For the former, intensification will be the name of the game, with the emphasis on precision agriculture. The best clones, bred for quality and yield with target yields of $10 \, t \, ha^{-1}$, will be planted at high densities ($20,000+$ plants ha^{-1}) in order to get an early return on the capital invested. Replanting after 20 years will become routine. GIS (Geographic Information Systems) will become a recognised management tool. Tea will be mechanically harvested and the transport system taking the harvested leaf from the field to the factory will be integrated through climate control with processing in the factory. As energy costs increase so will the need to manage fuel trees as efficiently as the tea. All staff will be highly trained, and housing and other staff facilities will reflect the standards of a high-quality and profitable business. The threat that climate change presents in low-altitude, already warm, areas will have to be carefully managed. In high-altitude areas a rise in temperature may be beneficial. Researchers will need to address this issue through modelling of the growth processes in tea, and predicting the impact of climate change ahead of any adverse effects. The CUPPA-TEA model provides a useful starting point (Matthews and Stephens, 1998), and the first attempts to advise on the likely repercussions of climate change on the industry have already been made (Burgess and Stephens, 2010).

Tea Is a Business

For commercial companies, the tea business objectives are to maximise profit in the short and long term, while for smallholders the primary aim is to support the family. This

---

Box 1.3 Commentary: the Tea Cycle

It is said that tea enterprises do not compete with each other, as all are supplying the same market, and individually they are tiny. This is only partly true. When times are tight, it is a case of the weakest to the wall.

Tea prices tend to fluctuate quite widely. When demand outstrips supply, prices rise quickly. Tea producers are awash with cash. Some of this is paid to shareholders, but much is retained in the business, where it is used to increase fertiliser levels and develop more plantations. Both measures increase production, the first quite quickly, the second over a longer period. Eventually these measures result in over-supply, and prices crash. Weaker enterprises then cut back on fertilisers, or cut them out alto-gether. This gradually reduces supply, but by then the new areas planted in times of plenty are coming into bearing, so that the agony continues. Eventually, though, gradually increasing demand again outstrips supply, and the cycle is repeated.

Advances in technology result in higher yields and better-quality teas, which can be summed up as higher productivity. This results in higher profits in good times, and an improved ability to survive periods of low prices. It is this ability to make profits and survive at lower price levels that drives investment, and paradoxically drives down prices.

In a free market, improved productivity always has this effect. Consider the effect of increasing productivity through innovation on the cost to the consumer of 1 Mb of computing power. We tend to take the truly amazing cost reduction as a matter of course, but it is a truism that the modern mobile phone contains more computing power than massive IBM machines of a generation ago, and at a minute fraction of the cost.

T. C. E. Congdon

is in the context of declining world market prices in real terms, the need always to reduce costs of production, and increasing competition from soft drinks. But the good news for the tea industry is that the UK, with a population of 65 million, still consumes in excess of 150 million cups of tea each day of the year. As 20–25% of the population is 15 years old or younger, this equates to about three cups per adult per day. Consumption of 'a nice cup of tea' is increasing elsewhere. Indeed, both India and China, with their expanding and increasingly middle-class populations, are consuming more of their own production, meaning that there is less tea available for export. In addition, there are the claims, with increasing amounts of supporting evidence, for the health properties of tea, which should encourage consumption in the richer countries.

In order to meet the business objectives, both the large-scale and smallholder sectors, which in any case often work together for mutual benefit, must make the best use of the available resources, such as land and the local climate, employees, and capital. Alongside this objective an additional aim should be to ensure the long-term sustain-ability of the industry, in terms of wealth creation, employment and environmental protection. For smallholders, other key issues are the need for regular, reliable and fair payments for green leaf, access to inputs and transport (including well-maintained rural road networks) when needed, good advice on tea agronomy, and the availability of high-quality planting material from local nurseries. To be successful, tea businesses will have to be modern in every sense. Managers will have to be as knowledgeable about business as their counterparts in commerce or manufacturing. In other words, the old days of 'learning on the job' are gone. New tea managers will have to be equipped with all the skills and latest tools necessary to address the increasingly sophisticated pressures that will be applied by both consumers and competitors.

This book aims to help managers acquire such skills by identifying, and where possible evaluating, the evidence from research, to allow them to question current commercial tea crop management practices (or recommendations) and reflect on if and how they might be improved.

Note

1. Technical terms (in *italic*) are explained in the Glossary.

2   The World of Tea

A Geriatric Problem

The fascinating history of tea has been described many times before in great detail, and the reader is referred to any one of numerous texts, some with intriguing titles such as *The True History of Tea* by Mair and Hoh (2009) (who would write 'the untrue history of tea'?), *Tea Classified* by Pettigrew and Richardson (2008), and *The Drink That Changed the World* by Griffiths (2007). After water, tea is believed to be the most widely consumed beverage in the world, equal to the combined total of all other manufactured drinks, including coffee, chocolate, soft drinks and alcohol. At US$1.50 per kilogram of made tea, its total annual value at the factory gate is US$7 billion. At UK supermarket checkout prices it is worth US$70 billion. Tea is big business.

The genus *Camellia* is indigenous to the forests of South-East Asia, where the plants grow into trees 9–12 m tall, or multi-stemmed shrubs. The tea plant (a species of *Camellia*) is believed to have originated within the fan-shaped area extending from the Assam–Myanmar (Burma) border in the west to Yunnan and Sichuan provinces of China in the east (*c*. 26°N), and south from this line through Myanmar and Thailand to Vietnam (*c*. 14–20°N) (Kingdom-Ward, 1950; Mair and Hoh, 2009). This is an area of monsoon climates with warm, wet summers and cool, dry (or less wet) winters. Tea from this region has been grown and consumed in China and in countries nearby for at least 4000 years.

An alternative view is that tea originated somewhere on the Mongolian Plateau of central China. According to that view, from that primary source it dispersed towards the south, reaching the secondary centre near the source of the Irrawaddy River. Further dispersal took place in three different directions along the course of three great rivers, the Yangtze, the Mekong and the Brahmaputra (Weatherstone, 1992; Sharma, 2012).

According to legend, a beverage with medicinal properties has been made from dried tea leaves in China since at least 2737 BC. There is written evidence from the period of the Tang dynasty in AD 650 that tea cultivation was widespread in most of the provinces of China by that time and the essential steps in the preparation of black and green tea had already been established (Ellis, 1997). By AD 780, a famous book entirely on the subject of tea entitled *Cha Jing* had been written. Tea is intricately associated with Chinese culture.

When the early European navigators reached the coast of China in the sixteenth century, tea was one of the first commodities that they came across. Early in the seventeenth century a regular trade in tea between China and countries in Europe was established. Tea was also re-exported to the British colonies in North America, and a dispute in 1773 over

British tax on consignments of tea was one of the events that led to the American Revolution. Annual consumption of tea in Britain increased rapidly from 2 million kg at that time to 9 million kg in 1801, with a further 6 million kg going to mainland Europe. Tea had become big business. In the 1830s there was a dispute between the principal tea trading company (the East India Company, which had a virtual monopoly) and the Chinese government. To ensure supplies of tea in the future, the British decided to attempt to grow tea in India, which at that time was part of the British Empire.

Since the mid-nineteenth century, tea has been introduced to many other areas of the world beyond China and neighbouring countries (Figures 2.1, 2.2). It is now cultivated in conditions that range from Mediterranean-type climates with hot, dry summers and wet winters, to the hot, humid tropics; from Georgia in the north (42°N) to South Africa (32°S) and New Zealand (37°S) in the south. It is now grown commercially in over 30 countries, at altitudes ranging from sea level up to 2700 m close to the equator. It is even grown in south-west England (50°N), but mainly as a small-scale novelty crop. Because of the large labour requirement, tea is still mainly a crop grown in developing countries. However, the development of mechanisation means that it is now grown in countries such as Argentina and Australia.

In 2013, the estimated total area of tea in the world was 4.14 million ha, with an annual production of processed tea of just under 5.0 million tonnes. Of these totals, 2.47 million ha (about 60% of the global total) was planted in China, producing 1.92 million tonnes (about 40%). China is followed by India (564,000 ha; 1,200,000 t), Kenya (198,000 ha; 432,000 t), Sri Lanka (187,000 ha; 340,000 t) and Turkey (77,000 ha; 235,000 t). The planted area and production data cited in this chapter have mainly been abstracted from Tables A and B in the International Tea Committee annual tea statistics report for 2013 (ITC, 2013) (Table 2.1). Elsewhere in the book, FAO data are generally used.

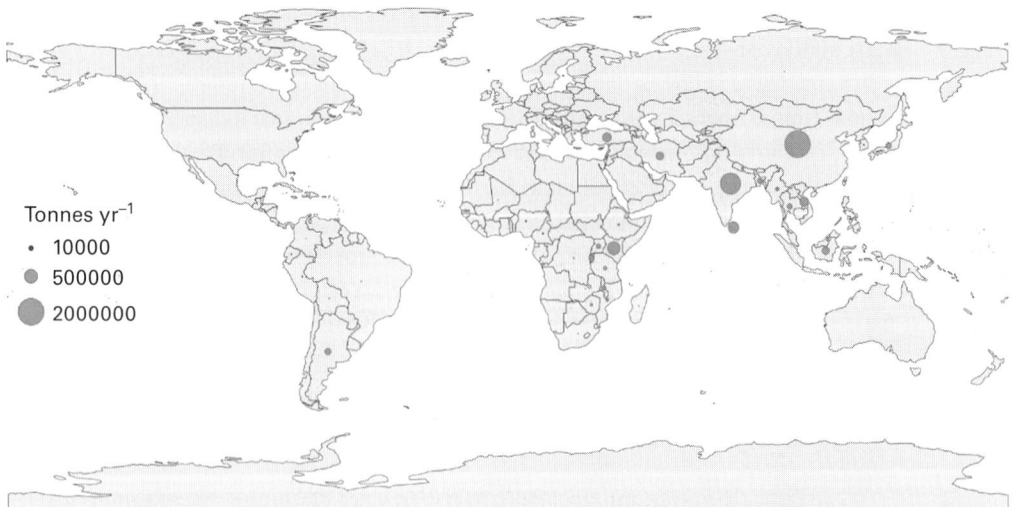

Tonnes yr^{-1}
- 10000
- 500000
- 2000000

Figure 2.1 The world of tea. An illustration showing the principal tea-growing countries in the world: the diameter of each circle is proportional to the annual production of tea in each region. The areas are based on FAOSTAT (2013) data for 2011.

Table 2.1 Area and annual production of tea in 2013 (ITC, 2013)

Country	Area of tea (ha)	Production (t)
China	2,470,000	1,900,000
India	564,000	1,200,000
Kenya	198,000	432,000
Sri Lanka	187,000	340,000
Vietnam	124,000	210,000
Indonesia	122,000	134,000
Myanmar	79,000	20,000
Turkey	77,000	235,000
Bangladesh	54,000	66,000
Japan	43,000	83,000
Argentina	42,000	85,000
Georgia	35,000	5,000
Uganda	32,000	61,000
Tanzania	23,000	32,000
Malawi	19,000	46,000
Nepal	17,000	19,000
Iran	16,000	14,000
Rwanda	13,000	25,000
Burundi	9,000	9,000
Zimbabwe	6,000	13,000
Mozambique	4,000	6,000
Malaysia	3,000	?
World totals, 2013	4.14 million ha	4.94 million tonnes

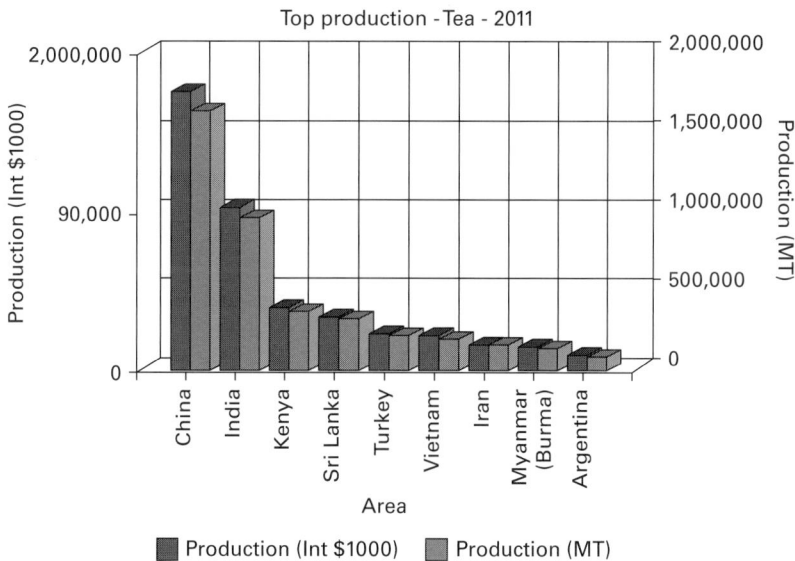

Figure 2.2 The annual tea production in the major tea-producing countries of the world, as recorded by FAOSTAT (2013) for 2011.

Box 2.1 Tea Area and Production Data: a Warning

The statistics on tea are not to be trusted. There are two principal sources, the Food and Agriculture Organization (FAO) and the International Tea Committee (ITC). They both collect data on the area of tea planted (not too difficult, one would have thought, although plant spacing and abandoned tea add complexity to the issue) and the annual production of processed tea (again easier than most crops to monitor because of the way it is marketed). Notwithstanding these apparently straightforward measures, the two sets of data rarely agree. There are also inconsistencies from year to year in the recorded statistics, while yields expressed as kilograms of made tea per hectare can sometimes be totally unrealistic. One wonders when and where the quality control measures are implemented. Rarely, I suggest. For most managers, it is just another form to complete, so it is put to one side until the deadline approaches, then some numbers are invented, or better still last year's numbers are reinstated. So the 'health warning' is: don't believe all the crop statistics that are included in this book. Why, you may be asking, include them at all if they are so unreliable? Answer: it's all we have got and they do at least give an order of magnitude to the size of the tea industry in any given country.

In this book, the FAO data are generally used in the text (unless there is a specific reference from which the data have been extracted), but Table 2.1 and most of the figures in this chapter are based on the ITC data (see Chapter 20). Spot the anomalies!

In this chapter, a brief description of the history and current status of the tea industry in individual countries within each of seven regions, namely East Asia, the Indian subcontinent, South-East Asia, the Caucasus, Africa, the Americas and Australasia, is presented. In the next chapter, the climatic conditions experienced in selected tea countries have been collated, with the data interpreted and any limiting factors identified.

The countries within each region are listed in alphabetical order.

Region: East Asia

This includes China, Japan, South Korea and Taiwan. This is the region where the cultivation of tea began and where the drinking of tea became a ritual symbolised by the classical tea ceremony.

Tea in China

According to ancient Chinese books, tea became a beverage in the Shennong times (about 2737 BC). Shennong was a legendary Chinese figure who was said to be not only the originator of Chinese agriculture but also the inventor of Chinese medicine. Tea cultivation began in south-west China in 1066 BC. Shennong tea was discovered and used during the development of Chinese primitive medicine. The history of tea has been recorded for at least 4000 years (Yu, 2012).

Figure 2.3 The principal tea-growing areas in China. Reproduced with permission from Nemec-Bochm *et al.* (2014).

China still grows and produces more tea than any other country, with an average annual yield that equates to 700–800 kg ha^{-1}. China produces and consumes seven types of tea: green tea (65% of the total in 2006), black tea (3%), oolong tea (10%), yellow tea, white tea, dark tea and reprocessed tea, including scented tea and compressed tea (10%) (Chen, 2012).

According to Forster and Etherington (1991), the enormous increase in tea production in China that occurred from the early 1960s to the mid-1980s was due largely to the tremendous expansion in the planted area of tea that occurred over this period, particularly during the Cultural Revolution. During 1971 and again in 1972, 100,000 ha of tea were planted.

An even faster rate of expansion has occurred in recent years. The area of planted tea in mainland China has increased by over 1 million ha since 2006 (from 1.43 to 2.47 million ha). To put this into perspective, it is equivalent to the total area of tea in India, Kenya and Sri Lanka planted in less than 10 years. Such figures are difficult for outsiders to comprehend, although they have been helpfully summarised (Anon., 2003; Chen, 2012; Chen Zongmao, personal communication, 2014). Actual yields in China remain low by international standards. With domestic consumption expected to increase, tea yields will need to rise substantially if China is to retain its place as a leading exporter of tea (Etherington and Forster, 1989). Premium teas (spring harvest based on one leaf and a bud, or one bud only) represent 43% of the total production but 80% of its value (in 2011). This explains, at least in part, the low yields. There is growth too in 'ready-to-drink' teas (Chen, 2012).

Tea in Japan

It is believed that tea was imported from China into Japan in the eighth century AD, although the custom of drinking tea was limited to the select upper classes until the sixteenth century, when the tea ceremony was introduced. Tea drinking then became popular amongst ordinary people, and by the mid-nineteenth century it was a daily pastime (Kodomari, 1988). The total area of tea in Japan is now about 43,000 ha. It is planted mainly along the coast of the Pacific Ocean (particularly in Shizuoka Prefecture, and Kagoshima Prefecture on Kyushu island).

Around 80,000–90,000 t of processed (green) tea are produced annually (up to 2002). Sencha is the most popular cultivar. Tea is harvested two or three times a year, the first crop (from late April to mid-May) being the most valuable. In early spring, the new shoots are susceptible to radiation frost damage. Fans are used to transfer warmer air down to the surface of the bush (Figure 2.4), or the bushes are covered.

Most of the tea in Japan is now harvested mechanically, using systems ranging from hand-held shears to large mobile machines. Virtually all the tea produced in Japan is consumed internally – none is exported (Iwasa, 1991). The development of new beverages based mainly on green tea has made tea a soft-drink leader in Japan, outranking carbonated beverages and juice.

Green, black and oolong tea are all imported. There is active research into pharmaceutical and industrial uses of tea (Miura, 2003).

Figure 2.4 Tea in Japan. Note the curved crop surface and the fans in the background to the left. These are intended to provide frost protection for the crop by drawing warm air down from above (MKVC).

Figure 2.5 Green tea is very popular in Japan, Taiwan and China (MKVC). *A black and white version of this figure will appear in some formats. For the colour version, please refer to the plate section.*

There were concerns about radioactive contamination of tea following the Fukushima Daiichi disaster.

Tea in South Korea

There is also a long history of tea production in Korea, dating back over 1300 years. New initiatives were taken in 1962 to encourage both black and green tea production. The planted area then increased in the south of the peninsula, at latitudes below *c.* 36°N, with some interruptions, from 30 ha in 1940 to about 400–500 ha in 1986. Tea-processing companies owned much of this tea. Traditional and introduced cultivars (from Japan and Taiwan) were planted (Lee *et al.*, 1988). According to the FAO the total estimated planted area in South Korea had increased by 2013 to 3800 ha producing 4600 t of processed tea.

Tea in Taiwan

Tea has been grown in Taiwan for at least 200 years. The total annual production is in the range 15,000–18,000 t, made up of black tea, green tea and several semi-fermented teas, such as pouchong and oolong, and a range of processed tea products. About 40% of the total production is now exported (previously it was as much as 80%). The total harvested area has also declined, from 45,000 ha in 1938 (Ming, 1999) to 12,000 ha in 2013, but output has been maintained. The domestic market has increased rapidly, with a corresponding decrease in exports. Taiwan is the second major exporter to make this transition from exporter to domestic consumer: Japan was the first and India is likely to be the next (Etherington and Forster, 1991).

Tea is grown on steep slopes, the principal production areas being Nantou Hsien, Taipei Hsien and Hsinchu Hsien. Annual yields of processed tea average about 1200 kg ha^{-1}. Crop distribution is reasonably even during the year, except for the period from December to March when yields are very low. The major tea cultivars grown in Taiwan are Chin-Hsin-Oolong, Chin-Hsin-Dahpan and Huand-Kan (collectively these occupy about 80% of the harvested area). By comparison, Assam-type clones occupy only 3% of the area, and cultivars newly released by the Tea Research and Extension Station about 4%.

Region: Indian Subcontinent

For the purposes of this book, the Indian subcontinent extends from Nepal in the north down to Sri Lanka in the south. In the northern area, which includes Assam, Bangladesh and West Bengal (Darjeeling), between latitudes 32°N and 22°N, the climate is characterised by a wet and hot summer and a cold, dry winter, when little crop is harvested. In contrast, South India (7–8°N) enjoys a tropical climate at low altitude extending into a subtropical/temperate climate as the altitude increases to 2400 m (Tamil Nadu and Kerala). Harvesting here continues throughout the year (Hazarika and Muraleedharan, 2011).

Tea in Bangladesh

Until partition in 1947, Bangladesh was part of India. It then became East Pakistan until 1972, when, after the liberation war in 1971, the independent state of Bangladesh was created. This meant that Bangladesh no longer had a safe internal market within Pakistan. Commercial cultivation of tea began in what is now Bangladesh in 1854. It now produces about 65,000 t of tea annually from an area of *c.* 54,000 ha (this figure is questionable, because land can be registered as tea land even if there is a very high proportion of vacancies, or if the land is being rehabilitated before being replanted) (Carr, 1988a). The soils are very variable but suitably acidic and predominantly loamy in texture. Over 100,000 people depend directly on the tea industry for their livelihoods, half of whom are women. Together with dependants, this suggests that at least half a million people benefit from employment opportunities in the industry.

Production has remained relatively static over time, while internal consumption has increased so that the quantity of tea available for export is declining. Bangladesh could soon become a net tea importer, unless yields increase. Identifying the limiting factors must be a research priority.

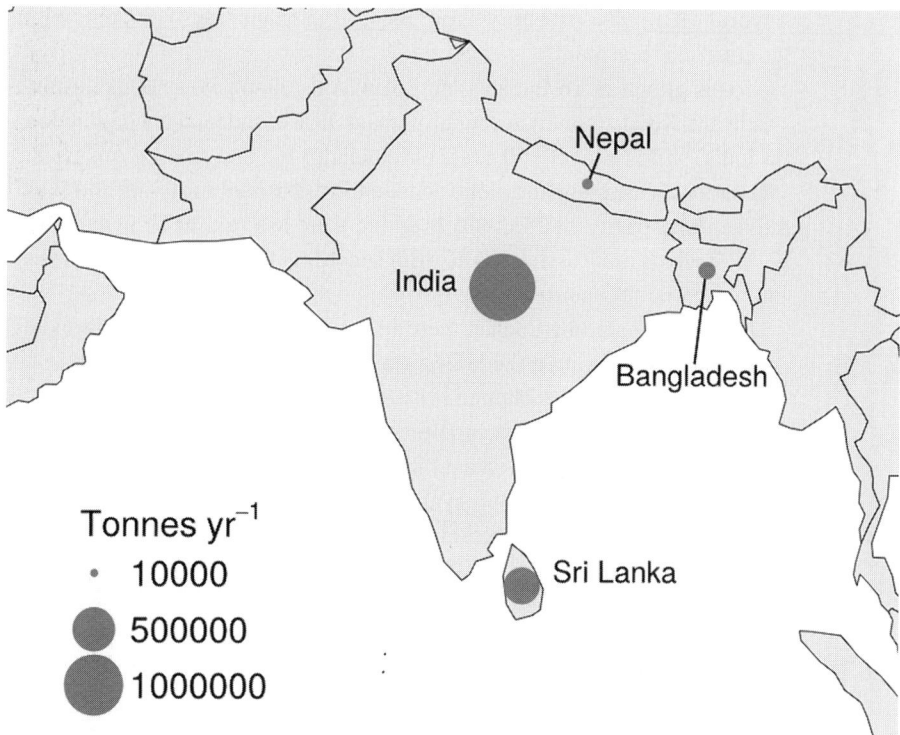

Figure 2.6 Principal tea-growing countries in the Indian subcontinent. The tea industry in India is divided into two distinct parts: South India (which produces about 25% of the national crop), and North-East India.

Tea in India

When the procurement of tea for the UK market from China became increasingly difficult, it was necessary to find an alternative source, tea drinking having become a popular English habit that had to be satisfied. India was soon identified as the obvious place to grow tea. The first trial plantings were made during the 1830s. This was soon after indigenous tea plants (var. *assamica*) had been 'discovered' growing wild in the hills close to Rangpur (Sivasagar), then the capital of Assam. Assam is located in North-East India close to the border with Bangladesh, which was then part of India. The first tea manufactured from *assamica* tea plants was successfully auctioned in London in 1839. The tea industry in Assam had begun. In the rush to expand, tea seed orchards (*bari*) were established with whatever seed was available. Some were pure China-type, some were pure Assam-type and some were deliberately interplanted with both types. The cross-pollination that followed led to the creation of Indian Hybrid Tea plants, which exhibited tremendous plant-to-plant variability within a population. This diversity was sustained by natural cross-pollination. Hybrid seed from this source became the foundation for many tea industries throughout the world. According to Ellis (1997), the creation of Indian

Hybrid Tea was probably 'the most important event in the evolution of the commercial tea plant'.

Tea is grown in 16 Indian states, of which Assam, West Bengal (north-east), Tamil Nadu and Kerala (south) account for about 95% of the total tea production. The industry in India, which is over 170 years old, includes big and small growers, and government plantations. One company alone operates 54 tea estates in Assam and West Bengal, with an annual capacity of 75,000 t. In India, there are said to be about 130,000 holdings, 92% of which are less than 10 ha. (But according to Hazarika and Muraleedharan, 2011, there are 200,000 holdings.)

In Assam, including Cachar, there are about 312,000 ha of tea, producing 500,000 t of processed tea, making it the largest tea-growing region in the world.

The first tea trial was planted in Darjeeling at Beechwood (alt. 2134 m) with seeds of China stock in 1841, but commercial production did not begin until 1870 (Hajra, 2001). Tea is now cultivated at altitudes between 800 and 2000 m, and Darjeeling has the reputation (similar to a vineyard producing vintage wine) of producing some of the most aromatic teas in the world. West Bengal produces about 276,000 t of tea from 115,000 ha.

In northern India there are about 60,000 holdings in total, with 43,000 in Assam, 9000 in West Bengal and 9000 elsewhere. Average yields in northern India are about 2100 kg ha^{-1}.

Tea was first planted in South India in 1853, and in 2013 there were about 107,000 ha of tea producing around 242,000 t of tea annually. This is equivalent to about 25% of India's total production. Yields average around 2200 kg ha^{-1}. The south-west monsoon

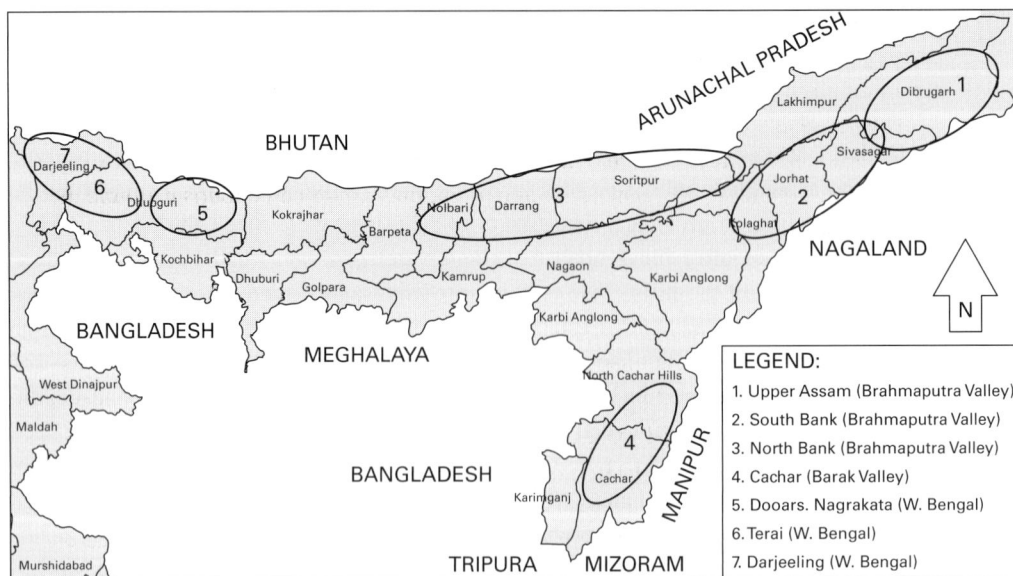

Figure 2.7 The principal tea areas of North-East India (reproduced from Jain, 1991, with permission).

Figure 2.8 A tea estate in South India. On undulating land, South India produces about one-quarter of the tea produced in the whole of India (MKVC).

dominates the weather pattern, with a dry or less wet season, which can last up to five months, from November to March.

A big change in recent years is that 77% of the total tea production is now consumed within India, leaving only 200,000 t for export (Jain, 2007). According to Hazarika and Muraleedharan (2011), the tea industry in India is facing a 'geriatric problem'. There is stagnation in productivity and no new areas are available for large-scale planting. An ambitious programme of replanting and rejuvenation pruning has begun.

In response to the changes that are occurring in the traditional tea sector, the major tea companies in India are also developing their business model by diversifying their range of products, expanding the retail sector and promoting nature-based tourism around the tea gardens.

Tea in Nepal

The tea industry began to develop in Nepal during the late 1970s and 1980s. This was after several false starts, beginning in 1863. The original pure China-type plants established at that time still exist in Ilam (alt. 1060 m). Ilam is adjacent to the tea-growing areas in Darjeeling (India). Previously a government monopoly, the tea industry has since been liberalised. High-grown tea, produced in the eastern hills on estates as well as by outgrowers, attracts premium prices. The average area of tea on a smallholding is about 0.67 ha. The land is steep. There are also plantations in the Terai flatlands. These extend from India up to the foothills of the Himalayas. A total of 19,000 t is now produced annually from a planted area of about 17,000 ha. Of this about 4000–5000 t are exported. A shortage of available land (and money) has restricted development.

Tea in Sri Lanka

The first tea was planted in Ceylon (Sri Lanka) in 1867 following the collapse of the coffee industry as a result of coffee rust disease. Sri Lanka is the fourth largest producer of tea (currently about 340,000 t each year) and the fifth largest exporter, having recently been overtaken by Kenya. In the mid-1960s Sri Lanka was the largest tea exporter in the world. The industry employs over 1 million people, directly and indirectly, the majority (75–85%) of whom are women. In 1976, many of the tea estates were nationalised before being privatised again in 1995/96. Although Sri Lanka was a pioneer in small-holder tea production, attempts to establish a smallholder tea industry in the 1950s failed. The reasons included: land tenure issues; no advisory service; no leaf collection procedures in place; no control over leaf processing. But lessons were learned, and smallholders now contribute about 74% of the national production from c. 60% of the area planted to tea in c. 400,000 holdings. The average area of a family-managed tea plot is 0.5 ha, which is often intercropped with coconuts, food or export crops.

Region: South-East Asia

This includes Indonesia, Malaysia, Myanmar, Thailand and Vietnam.

Figure 2.9 Tea-growing countries in East Asia and South-East Asia (excluding China). The tea industry in Vietnam is expanding rapidly.

Tea in Indonesia

Tea seed was successfully introduced into the Dutch East Indies (Indonesia, island of Java) from India in 1878, seed from China having failed. The area of tea at the start of the Second World War exceeded 130,000 ha (estates and smallholders). The war and subsequent security disturbances during the post-liberation period resulted in many tea plantations in Indonesia being abandoned, but recovery had begun by 1965, and in 1988 the total area of tea had reached 120,000 ha. Of this, about 44,000 ha were managed by a state-owned corporation (PTP), 26,000 ha were privately owned tea estates, and 50,000 ha were farmed by smallholders. Much of this is seedling tea, which is at least 90 years old. The total annual production in 2013 was estimated to be about 134,000 t of predominantly black tea from a planted area that by then had increased to 122,000 ha (of which more than 80% is in western Java, the balance being in Sumatra). This, together with some green tea and jasmine-scented tea (predominantly for the local market), had a total value of about US$150 million (FAOSTAT, 2013, accessed June 2013). The vast majority of the black tea is exported. About 200,000 people are employed in the industry, which is ranked as the eighth largest in the world (Darmawijaya, 1988; Arifin *et al.*, 1991).

Tea in Malaysia

There are tea estates in the Cameron Highlands in Peninsular Malaysia (altitude 900–1600 m), with a few more at lower altitudes. The largest estate is named Boh, after Bohea – the name of a mountainous district of China's Fujian Province where black tea was made and exported to Europe. Boh was established in 1929 with Assam-type plants. The total planted area is about 3100 ha, producing about 2200 t. The Cameron Highlands are claimed to have the ideal climate for tea, with well-distributed rainfall.

Tea in Myanmar

Tea is indigenous to Myanmar (formerly known as Burma), where it is called *let ta phet*, signifying its long independent association with Myanmar. *Camellia irrawadiensis* (a valuable gene-bank species), which is low in caffeine but rich in theobromine, can still be found growing wild along the Irrawaddy river valley. Tea estates are located in the fertile eastern hilly region named Sham State. The total area of tea is about 79,000 ha, producing only 20,000 t of tea, giving an average yield of only 250 kg ha^{-1}. Pickled tea (fermented wet green leaf, eaten whole), known locally as *lahpet so*, accounts for 44% of total production. It is a traditional dish in Myanmar, offered to guests at ceremonies. It is always served with a cup of green tea, made according to a local recipe. Some black tea is also produced (Htay *et al.*, 2006).

Tea in Thailand

According to Katikarn and Swynnerton (1991), tea has been grown for centuries in the northern part of Thailand by the Hill Tribes, who used it to make pickled tea called *miang* (Hajra, 2001). But further development of these smallholdings has been virtually non-existent. The tea plants are either single-stemmed trees or multi-stemmed, all grown at low plant densities. Plans for rehabilitation have been proposed. In the 1960s Chinese settlers obtained tea clones from Taiwan and a small tea industry, producing mainly green and oolong teas for local consumption, has since been developed. The ITC (2013) does not provide an estimate of the planted area, nor of production. Any tea is mostly believed to be in and around Chang Mai Province (18–19°N, alt. 1000–1400 m). The FAOSTAT tea production data for Thailand are suspect.

Tea in Vietnam

Before 1975, the development of the tea industry in Vietnam was continuously interrupted by war. Afterwards, the tea industry expanded so that by 2013 the planted area across the whole country had reached 124,000 ha producing 180,000 t of dried tea (of which a large proportion was exported). The tea industry in Vietnam is claimed to provide employment for 400,000 households together with 10,000 small crop-processing households (Dasgupta, 2007).

Tea is grown at a range of altitudes from below 300 m up to 850–1500 m in the mountainous areas, which includes Son La Province (Anon., 2013).

Region: Caucasus

This includes Azerbaijan, Georgia, Iran and Turkey.

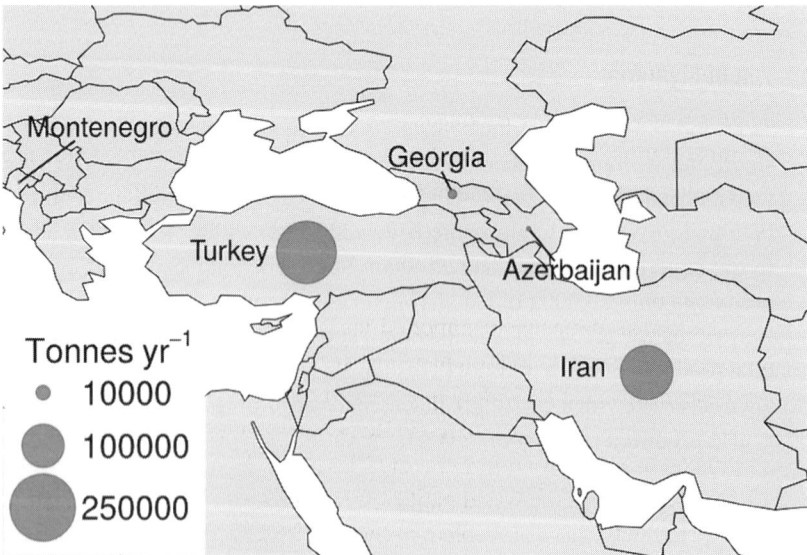

Figure 2.10 Principal tea-growing areas in the Caucasus. Turkey is the fifth largest producer of tea in the world, and the largest consumer (per head).

Tea in Azerbaijan

Following the breakup of the Soviet Union in 1991, Azerbaijan (*c*. 40°N) inherited about 13,000 ha of tea, but this has since been reduced to less than 600 ha, producing in 2010 and 2011 only about 280 t of tea annually. The FAOSTAT (2013) data need to be viewed with caution. Clearly the tea industry in Azerbaijan is still going through a bad time.

Tea in Georgia

Prince Miha Eristavi initiated the tea industry in Georgia, having first encountered tea during his travels across China in the 1830s. At the time, exporting tea seeds from China was forbidden so the prince smuggled some out of the country. These were used to create the first tea plantations. In 1864, samples of Georgian tea were exhibited at the Russian International Exhibition in St Petersburg, and by the turn of the century Georgian tea was winning prizes including a gold medal at the World Fair in Paris in 1889. In this way, the tea industry began in Georgia (Georgia About, 2013).

By the late 1920s, Georgia had become the main supplier of tea to the whole of the Soviet Union. Templer (1970) described the tea production processes on a 2000 ha (600 ha tea) state farm in Krasnodar (Russia). However, during the 1990s, following the Chernobyl nuclear accident and the breakup of the Soviet Union, many of Georgia's plantations were cut down or abandoned and the factories were allowed to deteriorate. In 2013, following privatisation, there were 19 tea-processing factories producing around 4500 t of tea, 90% of which was exported, from an estimated harvested area of 3000 ha (cf. planted area 35,000 ha). This is far below the potential production from a planted area that was once in excess of 50,000 ha (up to 67,000 ha). At its peak, 30% of the planted area was machine-harvested (Anon., 1997).

Tea in Iran

Tea was successfully established in Iran for the first time in 1900. There are numerous small farms, each with less than 1 ha of tea. The total planted area is about 16,000 ha, producing annually *c*. 14,000 t of tea (Hajra, 2001). Smuggling of tea is a problem.

Tea in Turkey

The first seedlings were imported into Turkey from Japan and China in 1888, but tea only really became important as a crop and as an industry in the 1970s after a number of false dawns. Tea in Turkey now extends from Hopa near the Russian border (41°25′N 41°27′E, alt. 10 m) and along the Black Sea as far as the Arakli river. By 1985 there were 40,000 ha of tea in Turkey, but the Chernobyl disaster meant that Turkish teas had to be withdrawn from the world market. Today 65% of all Turkish tea – primarily black tea – is produced under the auspices of the Turkish government, with the balance produced by the private sector (Cilengir, 2010). In 2013 the total production was 235,000 t from 77,000 ha (90,000 ha in 1990), making Turkey the fifth largest producer in the world, after China, India, Kenya and Sri Lanka (FAOSTAT, 2013). The yields appear to be large, averaging just over 3000 kg ha^{-1} a year, despite the short harvesting season. Tea in Turkey is predominantly a smallholder crop with more than 200,000 farmers, the majority of whom (about 80%) have a planted area of less than 0.5 ha. The tea is harvested with shears (Vanli, 1991).

Region: Africa

Tea is grown in about 12 countries in Africa, the most important being Kenya (by far), Uganda, Malawi, Tanzania and Rwanda, nearly all of which are in eastern and southern Africa. Together they produce approaching 600,000 t of processed tea each year, about 12% of the world total.

Tea in Burundi

In Burundi, tea is the second largest cash crop after coffee, and 80% comes from village plantations. It is harvested throughout the year and provides some 60,000 households with a stable and regular source of income. The total planted area is about 8800 ha, producing up to 8800 t of tea annually (a chance coincidence?). Conditions are very similar to those in adjacent Rwanda.

Tea in Cameroon

The Germans first planted tea in the Cameroon in 1914, but commercial tea production only began in 1954–1958. The annual production of black tea is currently 4500 t from a planted area of about 1600 ha. A new estate was planted in 2002 in Tole (3°45′N 11°32′E).

Tea in Ethiopia

Tea is a relatively new industry in Ethiopia, the first tea having been planted in 1928 in Gore town. In 1957 a small tea estate was started in Gumaro. The government began to fund its expansion in 1981. There are now 2700 ha of tea on two estates. Annual production of black tea now totals 6100 t. There are big plans to expand tea production in Ethiopia. A small outgrowers scheme began in 2004.

Tea in Kenya

Tea was introduced into Kenya in 1903 but it was not until the 1920s that commercial tea production began, with plantings in the Kericho and Limuru districts (and afterwards in Sotik and Nandi Hills). The Second World War interrupted the development of the industry, but over the last 50 years production has gone from strength to strength, so that Kenya is now the third largest producer of tea (nearly all black tea) in the world and the biggest exporter. There is still a strong plantation sector, with average annual

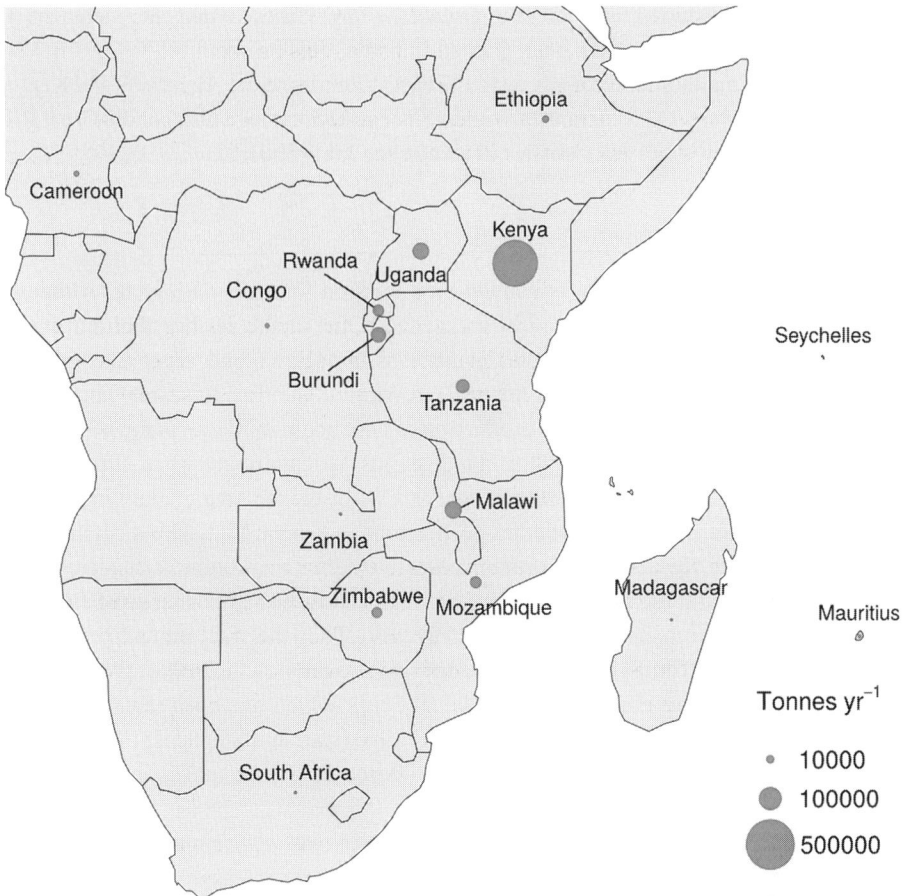

Figure 2.11 Kenya is, by far, the leading producer of tea in Africa.

yields exceeding 3000 kg ha^{-1}, but much of the growth in area has occurred following a change in the law that allowed the indigenous population to grow tea, and the establishment of the Kenya Tea Development Authority (now an agency, KTDA) in the mid-1960s.

KTDA was charged with the responsibility of developing the smallholder sector. The total area of smallholder tea exceeds 127,000 ha (serviced by over 50 factories) with in excess of 150,000 smallholders. Annual yields only average just under 2000 kg ha^{-1}, which compares with 3000 kg ha^{-1} from the estate sector. To show what is possible, an individual estate field in Kericho has yielded a record 10,000 kg ha^{-1} (Othieno, 1991; Oyamo, 1992). According to Owuor *et al.* (2007), yields from smallholdings are below what could be achieved because smallholders fail to follow the recommended cultural practices. However, KTDA teas consistently get good prices, as a result of a high standard of plucking.

In addition to the estate and smallholder sectors, there is also the Nyayo Tea Zone project. Begun in the late 1980s, the aim of the project was to surround indigenous forest with a band of tea in order to protect the forest from human encroachment. To date about 3500 ha of tea have been planted as part of this project.

Smallholder tea areas lie to the east (foothills of the Aberdares and Kenya mountains) and west (the Mau ranges, Nandi, Kisii and Kakamega hills) of the Great Rift Valley. For an analysis of tea statistics in Kenya see Jain (2001/02).

Tea in Malawi

In 1878, tea seed was planted on a mission farm near Blantyre. Although this failed, a second attempt in 1886 succeeded. One of the bushes that survived became the ancestor of the original tea plantations in Malawi, and vegetatively propagated plants from the original bush are still growing at the Tea Research Foundation of Central Africa. Smallholder tea production did not begin in Malawi until 1966, after a change in the law (as in Kenya) allowed indigenous Africans to grow tea. This sector now occupies about 2500 ha on 5000 holdings, or 13% of the tea area, mainly in Mulanje.

This is a much smaller proportion than in Kenya. Malawi is a landlocked country and, as with Uganda, transportation costs to external markets are considerable. Opportunities to expand the planted area are limited. Yields, however, increased from 940 kg ha^{-1} in 1964 up to 2200 kg ha^{-1} in 1994. By 2013, the total annual production was about 46,000 t from 18,600 ha (includes smallholders), which equates to about 2500 kg ha^{-1} overall. For comparison, individual fields, which had been replanted with clonal tea in 1972, were yielding 6500 kg ha^{-1} by 1990 (Carr and Stephens, 1992). The tea industry in Malawi employs in excess of 40,000 people, which makes it the largest employer of people in the private sector.

Figure 2.12 A tea smallholding in Malawi, where smallholder tea was a late starter (MKVC).

Tea in Mozambique

The successful establishment of tea in Malawi played a large part in stimulating a similar development in neighbouring Mozambique. Here the initial plantings began during the early 1920s in Milanje (16°05'S 35°46'E), a district bordering Mulanje (note the difference in spelling of these two similar names). The first factory was built in 1924. However, it was in the Gurue District, north-east of Milanje, that expansion occurred. Tea was also planted at Tacuane (16°22'S 36°30'E), east of Milanje. By 1967, 6000 ha had been planted, with an annual yield of 14,500 t of made tea. Soon afterwards internal strife resulted in much of the tea being abandoned, with production declining from a peak of 20,000 t in 1982 to virtually zero in 1988. In 1994, 16 of the 22 factories in the country were described as being virtually destroyed or badly damaged, and only three were operative, while many of the tea plantations had become forests (Whittle, 1999). Since then, some recovery has occurred, and 6400 t of tea was produced in 2013 from a harvested area of 3500 ha.

Tea in Nigeria

The most populated country in Africa has just one tea estate. First established on the Mambilla Plateau in the 1970s, the irrigated, mechanically harvested estate now has 850 ha of clonal tea (the cuttings were originally imported from Kenya). There is an associated outgrower scheme with about 400 ha of tea. The factory has the capacity to process 2250 t of made tea annually. The grassland plateau shares a border with Cameroon.

Tea in Rwanda

Tea growing in Rwanda began in 1952. The industry has now recovered from the civil strife (genocide) that occurred during the 1990s and produces up to 25,000 t of high-quality black tea from a planted area of about 12,500 ha. The tea is grown on hillsides at altitudes between 1900 and 2500 m, and in well-drained swamps at 1550–1800 m. There are smallholders (40% of the planted area), estates (about 30%), private growers and cooperatives. Average annual yields are 2000 kg ha^{-1}, with the best estates yielding 3500 kg ha^{-1}.

Tea in South Africa

Several attempts have been made to establish a viable and profitable tea industry in South Africa. Most of the tea estates were established in areas of high unemployment during the 1970s and 1980s in an attempt to dissuade people from migrating to urban areas. Beginning in 2003, most of the tea estates in the country went out of production, and those that remain are in a serious financial position. There are an estimated 4200 ha of tea still being harvested, producing 3800 t of tea. The main reasons for the demise of the tea industry in South Africa are the relatively high production costs and low market prices for black tea (DAFF, 2011) and, in particular, the revaluation of the rand. Previously about 10,000 t of black tea were produced annually from 9000 ha.

Tea in Tanzania

German settlers made experimental plantings of tea in Tanzania at the beginning of the twentieth century, but commercial production did not begin until 1926 (Carr, 1988b). By 1934, 1000 ha had been planted, which produced 20 t of processed tea. The bulk of production comes from estates in the south, with steadily rising yields year-on-year from existing tea bushes. New estates are being planted using clones, irrigation and mechanical harvesting. The best fields have reached 7000 kg ha^{-1}. The smallholder sector, with innovative management, reliable and timely transport of leaf to the factory, regular and transparent payments, and supported by extension staff provided and trained by the Tea Research Institute of Tanzania, with EU support, is growing rapidly, after years of neglect. In 2013, Tanzania produced 32,000 t of processed tea from 22,000 ha, up from 18,000 t in 1990. There is one organic estate.

Tea in Uganda

The first tea plants to be grown successfully in Uganda came from Kew and Edinburgh Botanic Gardens in the UK in 1900 and were planted in the Botanical Gardens in Entebbe, where they still exist as large trees. Several other introductions followed, including batches of seed from Assam (Dahootea, Dangri, Manipuri, Betjan Assam and Rajghur) and Myanmar. Of these, Betjan Assam proved to be the most successful, and the seed that grew from that batch formed the basis for several estates. After 1914, seed from these initial trials was being issued to planters in the Toro, Bunyoro and Mubende districts, but by 1924 only one estate had been planted (at Mityana). By 1939, the area of tea had increased to 1300 ha, and by 1960 it had reached 6000 ha. During the period following the coup in 1971 when Idi Amin ruled, much of the tea in Uganda was abandoned, and it was not until 1980 that it became possible to rehabilitate the tea estates. By this time, the tea had grown to 7 m tall trees (polled tea) and the factories were derelict. There was also a severe shortage of labour. Mechanical harvesting was introduced with a purpose-built machine (with a performance equivalent to 150 manual labourers) (Kilgour, 1990). Other forms of machine-aided harvesting were also tried. With the support of the Uganda Tea Development Agency, smallholder schemes ($c.$ 10,000 ha with 15,000 smallholders) contribute about 30% of the 55,000–60,000 t annual production of (black) tea.

Tea in Zambia

There is only one tea estate in Zambia, named Kawambwa. The 500 ha estate is located in Luapula Province in the north-west of the country (09°48′S 29°05′E, alt. 1250 m). It was developed initially in the 1970s, but has since been constrained by several changes in ownership, and by lack of investment. About 1000 t of tea are produced annually.

Figure 2.13 Tea plants growing into multi-stemmed trees in western Uganda, about 15 years after Idi Amin took over power and tea estates were abandoned. Tea is very forgiving and soon comes back into production when normal management practices are resumed (MKVC). *A black and white version of this figure will appear in some formats. For the colour version, please refer to the plate section.*

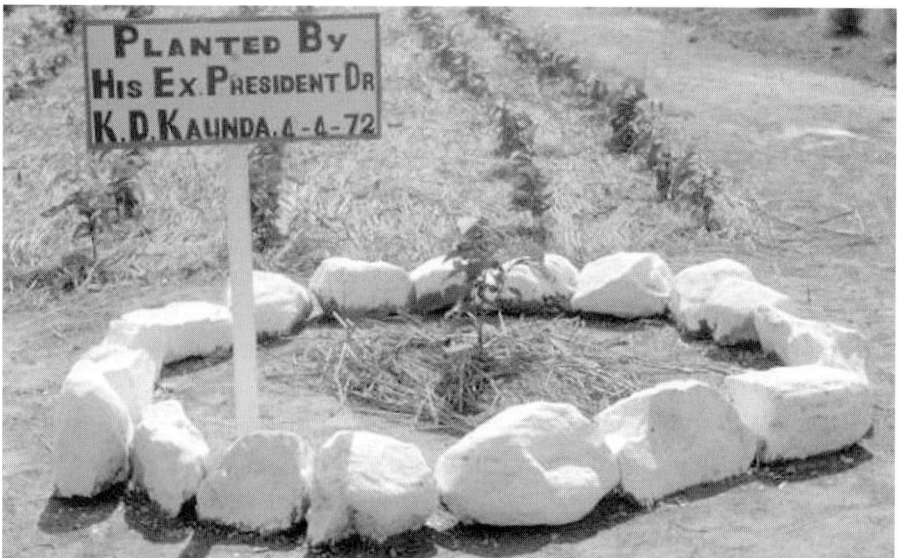

Figure 2.14 A government-protected tea plant in Zambia: a false dawn for the tea industry in that country (MKVC).

Tea in Zimbabwe

The first tea was planted in what is now Zimbabwe, formerly Southern Rhodesia, in 1900 at the Mount Selinda mission. Commercial production in the Chipinge District in the Eastern Highlands began in 1925. Slow but steady expansion in the planted area has occurred over the years, and by 2011 there was a total of about 6000 ha of tea, owned predominantly by five large-scale producers. There are also a number (30–40) of private farms (averaging 60–70 ha of tea each). In addition, there is a small but expanding smallholder sector. A total of 13,000 t of (black) tea was produced in 2013.

Region: the Americas

This includes Argentina, Brazil and the USA.

Tea in Argentina

The Argentinian government distributed tea plants from China to farmers in the northern province of Misiones, bordering Brazil, in 1924. Further inducements followed, but it was only after 1946 that development of the tea industry began. By 1958, Argentina had the largest planted area of tea in South America, a position it retains. About 85,000 t of tea are produced from a harvested area of 42,000 ha, mainly in Misiones (26°S, alt. 300–400 m) but also in Corrientes (34°S, alt. 100–150 m). Only South Africa, 32°S (where the industry is in decline), and New Zealand, 37°S (where it is embryonic), grow tea as far south as this. Mechanisation was introduced to Argentina in the 1970s, and all the tea is now harvested by machine. Most of the (black) tea is exported to the USA, where it is used for making iced tea.

Tea in Brazil

Following early initiatives by Chinese immigrants in the late nineteenth century, it was in the 1920s that Japanese immigrants processed the first green and black teas in Brazil. There are now about 4000 ha of tea, making Brazil the second largest producer in South America after Argentina, with an estimated annual production of 5000 t (published figures vary). The tea is located in the area of Minas Gerais (24°23′S 47°09′W) around Registro, south-west of São Paulo.

Tea in the United States of America

Since 1799, several attempts have been made over the years to establish a tea industry in the USA, but these have largely been short-lived, owing in part to the relatively high cost of labour and its limited availability. However, Pratt and Walcott (2012) believe that tea is poised to become a significant niche industry catering for two markets: (1) the lower-cost mass market, based on expansion of machine-harvested black tea farms successfully initiated in South Carolina (33°37′N 81°11′W) – known as the 50 ha farm model; and (2) higher-cost hand-cultivated speciality teas, by further development of the small (5 ha) tea farms that have been created in Hawaii (at altitudes up to 1200 m). There are five other places where tea is grown in the USA. They are all coastal sites: California (38°44′N 121°48′W; Yolo County); Oregon (44°45′N 123°04′W; Minto island); Washington State (48°16′N 122°40′W; Skagit Valley); Alabama (30°26′N 87°50′W; Mobile Bay), and North Carolina (7°19′N 79°03′W; Chapel Hill). Black, green and oolong teas are produced. Over 85% of the tea consumed in the USA is served as iced tea (this is based on a low-quality black tea, most of it imported). The health benefits of tea, particularly green tea, encourage its consumption. As Pratt and Walcott (2012) state, 'overall, North American conditions have proved quite challenging for the production of tea.'

Region: Australasia

This includes Australia, New Zealand and Papua New Guinea.

Tea in Australia

The first tea estate in Australia was planted in 1884 near Innisfail (17°31′S), but it was destroyed by a cyclone and the tidal wave that followed. In 1959 interest in tea was renewed and another estate was planted, also close to Innisfail. There are now about 900 ha of tea in north-east Queensland and in northern New South Wales, producing about 1700 t of green tea. The tea is mechanically harvested. Most of the tea is grown at about 15°S of the equator at an altitude of 800 m. Some tea is grown at sea level.

Tea in New Zealand

A very small (3 ha) but expanding (to 51 ha) estate is producing oolong tea Taiwan-style in the North Island (Gordonton). It represents an embryonic tea industry initiated by a Taiwanese immigrant. At 37°S, this is the southernmost commercial tea plantation.

Tea in Papua New Guinea

Large-scale planting began in Papua New Guinea in 1954. There are now about 3900 ha of tea planted in the flat Wahgi Valley (5°54′S 114°16′E, alt. 1600 m) in the Western Highlands, producing up to 6500 t of tea annually. Much of this is mechanically harvested.

Summary

These are some of the main points to emerge from this mini-review of tea production worldwide:

- Tea production has expanded away from its centre of origin and production in South-East Asia into diverse habitats since the mid-nineteenth century, a process that continues.
- With the growth of a middle class in the traditional tea regions (e.g. India and China), there has been an increase in domestic consumption, and countries that traditionally used to export now import tea. This is following a trend set by Japan.
- There is now less tea available for export from these and other countries, many of which are unable to expand their production because of a scarcity of suitable land.
- As a result, new tea-growing countries have been able to meet this international demand by slowly expanding their production (e.g. Tanzania), or in the case of Kenya rapidly and with great success.
- In many countries, much of the growth in production has come from smallholders (e.g. Kenya).
- Traditional estates/plantations now produce a declining proportion of the world's tea.
- The best estates continue to set the standards for productivity, although the best smallholders are outstanding (e.g. Kenya).
- Civil strife has adversely affected tea production in a depressingly large number of countries (e.g. Uganda, Mozambique, Bangladesh, Rwanda), with tea being abandoned in some cases for many years before being rehabilitated. The tea plant is very forgiving!
- In one tragic industrial accident at Chernobyl, radioactive contamination forced the tea industry to be abandoned in neighbouring countries (Georgia and Turkey). There are also health concerns in Japan following the Fukushima Daiichi disaster.
- Shortages of labour and/or its cost are forcing countries to consider and introduce mechanical harvesting (e.g. South India, Papua New Guinea, Tanzania, Uganda, Argentina).
- When combined with revaluation of a currency, high costs of production can force a country to abandon tea production (e.g. South Africa).
- Tea prices are still too low in real terms.
- Nevertheless, millions of people in developing countries depend on tea for their livelihoods.

3   A Changing Climate

Stay Cool!

Apart from the soil (it must be acidic: see Chapter 14) it is the climate (preferably warm and wet) and its day-to-day variability, known as 'weather', that largely determine the potential productivity of tea in a given location. In this chapter, meteorological data have been collated (see Annex) for selected tea locations within the five principal regions described in Chapter 2. Comparisons between sites have then been attempted, and limiting factors identified. The regions considered are as follows: East Asia, Indian subcontinent, South-East Asia, the Caucasus, and eastern and southern Africa. The countries mentioned in this review have been chosen because of their research record or because they demonstrate a particular issue. For example, a country might include the most northerly or southerly site, or a particularly high- or low-altitude site, or it might have been chosen on the basis of the amount/quality and availability of good weather data. Not all data sets were complete. Great care has been taken over quality control. When there was any doubt about the reliability of the data they were discarded.

Weather

Without an understanding of how the weather influences the growth and development of the tea crop, it is difficult to extrapolate with confidence from the results of research conducted at one site to somewhere else where the climate is different. With good weather data, one can seek to explain why responses to treatments vary between one year and the next, to predict potential yields, to identify any constraints to production, and to extend the results to other locations.

The two most important weather variables influencing the development and yield of tea are temperature and rainfall.[1] Temperature is largely a function of altitude, modified by proximity to a large body of water, and the season (linked mainly to latitude). Rainfall is locally variable in relation to nearby topographic features, including lakes, mountains and escarpments.

Understanding how weather influences production is important in the context of forecasting yields at factory, national, regional or world levels, and for explaining why targets have not been met, with convincing evidence! There is also the bigger question of predicting in advance the likely impact of climate change on tea productivity. Will it become too hot for tea in some low-altitude locations near the equator, and will it become possible to grow tea commercially at higher altitudes and latitudes than at present? These are questions that Laderach and Eitzinger (2011) have recently attempted to answer with respect to Kenya, where the optimum tea-producing zone is currently at altitudes between 1500 and 2100 m. By 2050, they forecast that this optimum zone could rise to 2000–2300 m. Increasing altitude compensates for increases in temperature.

As weather forecasting improves, it should become possible to plan ahead, to make rational management decisions based on the likely impact of the weather on productivity, and to minimise self-inflicted damage to the bottom line of the accounts. In view of the importance of having access to reliable weather data, it is important that the weather station is correctly sited (Figure 3.1), while the recorders need to be trained for their important responsibilities (Box 3.1).

Figure 3.1 An example of a poorly sited weather station. For example, the Stevenson screen (contains thermometers) and the evaporation pan are both far too close to the tree. Ideally, the measurements should be made over a short, well-watered, grass surface. We depend upon good weather data to judge the suitability of a site for tea (MKVC). *A black and white version of this figure will appear in some formats. For the colour version, please refer to the plate section.*

Box 3.1 A Prestigious Job

Reliable weather data can play an important role in the management of a tea farm, as well as being an essential tool to enable researchers to extrapolate results of experiments from one location to another. Weather stations need to be well sited and well maintained. Rarely is this the case.

Too often the job of recording the weather data is delegated to the most junior member of the office staff, with little or no training. Instead it should be a prestigious job, deserving of a high-status position. The data are intended to be useful, but rarely is this the case.

Region: East Asia – China and Japan

Because of their relative importance, China and Japan are considered in this section. This region also includes South Korea and Taiwan. In the FAO recorded data Taiwan is considered to be part of mainland China.

China

Tea is grown in 19 provinces in China across a range of latitudes (18–38°N) and longitudes (94–122°E), at altitudes of up to 2600 m, and currently in six climatic zones (Chapter 2, Figure 2.3). Previously, there were four designated tea-growing areas. It is useful to remind ourselves what they were:

- **South China**: this is the most southerly area, with a warm, humid tropical monsoon climate.
- **South-west China**: the oldest tea-producing area, at high altitude with undulating topography extending into the south-east corner of Tibet; subtropical, with warm, mild winters (since cold air currents from the north are blocked by mountains); *Camellia sinensis* var. *assamica* cultivars grow well here.
- **South of the Yangtze River**: this important tea production area experiences a mild spring, a hot summer, a cool autumn and a cold winter.
- **North of the Yangtze River**: this tea-growing area is situated on the northern fringe of the subtropical zone. Droughts occur during the growing season, and frost injury can occur in the winter (Yu, 2012).

Following a large expansion in the area of planted tea, these four regions have now been reduced to three, only one of which matches the previous delineation, namely, the Eastern Area (previously, South Yangtze region). The greatest expansion in planted tea has occurred in the (new) Western Area, where there are now 1.29 million ha of tea (yes, million!). In the (new) Central Area there are 750,000 ha. In the Eastern Area, where new planting has now ceased, there are 550,000 ha. This makes a grand total of 2.59 million ha (inclusive of Taiwan, 12,000 ha) which compares with 1.43 million ha in

---

Box 3.2 The North-East Monsoon in China

The north-east monsoon is important to the tea-growing regions of China. Recent studies have suggested that the start of the monsoon is getting later in the major tea-producing provinces, while the end of the monsoon is also tending to be delayed. By identifying the points where the slope of cumulative rainfall increased sharply (onset) or declined sharply (end), these two dates were estimated and the duration of the monsoon was calculated for each year over the period 1980–2011 (Nemec-Bochm *et al.*, 2014). Based on this methodology, a trend line was determined. In recent years (2001–2011) the monsoon had lasted for about 110–120 days, which is close to the minimum value recorded, whereas it had been as high as 165 days (in 1983). There was also a very slight downward trend in total rainfall (from 1500 mm) as well as monsoon rain (from 1000 mm) over the 32 years. Over the same period, the mean daily maximum temperature during the spring increased, on average, across all 15 main tea-producing regions, by about 2 °C (from 15 °C to 17 °C), while average yields also appeared to increase, from 400 to 900 kg ha^{-1}. It remains to be seen whether the conclusions drawn from this detailed analysis (suggesting climate change) can be confirmed using a different approach.

2006 (Chen *et al.*, 2012; Chen Zongmao, 2014 personal communication), an increase of 1.16 million ha in eight years.

It is impossible to represent the climate for such a vast area, but on the advice of Chen Zongmao, data from four meteorological stations have been selected for presentation, three of which are in the Western Area, namely Chengdu (Sichuan Province), Dushan (Guixhou Province) and Langzhong (Sichuan Province) (Annex 3.1). The fourth site at Yuanjiang (Hunan Province) is in the Central Area. The sites range in latitude from 23.6°N (Yuanjiang) to 31°35′N (Langzhong), and in altitude from 385 m (Langzhong) to 971 m (Dushan).

All four locations have hot or very hot summers and cold or very cold winters. For example, in Yuanjiang the mean maximum temperatures from April to August reach 34–35 °C, but 'only' 31–32 °C in Langzhong, 30–32 °C in Chengdu and 27–28 °C in Dushan. In the winter, minimum temperatures at these four sites reach on average 10 °C, 6 °C, 3 °C and 1–2 °C respectively. Winters tend to be dry from November to April. The dependence on the north-east monsoon for rain is described in Box 3.2. In the tea-growing areas, average rainfall ranges from more than 2500 mm down to 500–1000 mm (Annex 3.1).

Japan

In Japan by contrast the climate is more maritime, much of the tea being grown close to the coast in Shizuoka Prefecture, on Honshu Island, where 45% of Japan's tea is produced. Maximum temperatures rarely exceed 30 °C in the summer months and usually remain above freezing in the winter. Rain occurs throughout the year, but is greatest during the summer. The growing season lasts from April to October. Conditions are similar in Kagoshima, the second most important region for tea in Japan, but with more rain in the summer (Annex 3.2).

Region: Indian Subcontinent

This region includes North-East India (Assam and West Bengal), South India, Bangladesh and Sri Lanka (Chapter 2, Figure 2.6). Within the last decade, Jain (2007) and Hazarika and Muraleedharan (2011) have reviewed the tea industry in India.

Assam

Assam is in northern India on a high plateau that straddles the Brahmaputra River (22–27°N). Tea is concentrated in the flood plains of the Brahmaputra Valley to the north and the Barak Valley to the south (altitude *c.* 100 m). The dry or less wet season can last from October to April. The flood plains include Cachar (alt. 22 m) (Jain, 1991; Figure 3.2).

In Assam, there is a considerable temperature range during the year. In the summer, mean monthly maximum air temperatures average about 32 °C (up to 35 °C) from May

Figure 3.2 A shaded tea garden in Assam. The debate on the value, or otherwise, of shade in tea continues. But the cooling effect is probably of the greatest value to the tea in Assam and to the pluckers (RHVC).

to September, which is way in excess of the optimum temperature for tea (*c.* 27 °C) (Annex 3.3). Even in the winter months the maximum temperature never falls below about 24 °C, with daily mean values always in excess of 16 °C. Since the (assumed) base temperature for shoot development is 12.5 °C, the effective temperature is at least 3.5 °C per day (°C d).[2] Shoots will therefore continue to develop throughout the year, albeit at a slower rate in the winter.

Some people still believe that the decline in yield during the winter could be a response, at least in part, to changes in daylength (a photoperiodic effect: see Box 3.3). The alternative view is that the observed yield fluctuations are the result of a crop flow problem when, following the start of the rains, an accumulation of shoots of different ages all mature together (in the so-called 'first flush'). The common practice of skiffing in December must also influence crop distribution. There is some (limited) indication of a changing climate (Box 3.4).

Box 3.3 Seasonality of Yield in Assam and Malawi

As tea moves further from the equator the winter harvest gradually declines until, at latitudes beyond about 16°, there is almost complete winter dormancy. Changes in daylength (photoperiodism) have been used by Barua (1969) in Assam (25–27°N) and by Tanton (1982a, 1982b) in Malawi (16°S) to explain, at least in part, the depressions in yield that occur at that time. Apical bud dormancy is observed to occur before as well as during the cool season. Dormancy is initiated when the length of a winter day is less than 11.25 hours and lasts for a period of at least six weeks. However, there is very little experimental evidence to explain more than a small proportion of the reduction in yield that occurs during the winter months (Herd and Squire, 1976). Low temperatures are thought to play a bigger role, although they cannot explain the whole story, since it is cooler in equatorial tea-growing areas at high altitudes, where tea yields are relatively uniform (subject to the availability of soil water), than at lower altitudes away from the equator (latitudes > 7–8°N or S). Instead, Tanton (1992) described seasonality of yield as a crop flow problem caused by *differences* in air temperature. In Kenya, shoots take 10–14 weeks to reach a harvestable size throughout the year (when expressed as thermal time, after the accumulation of 475 °C d), low temperatures extending the length of the shoot replacement cycle. In Kenya, there is a consistent supply of shoots reaching harvestable size each month. By contrast, in Malawi shoots only take 5–6 weeks to develop in the rainy (summer) season but 10–14 weeks during the winter. The change in seasons occurs over about one month, so there is a delay while those shoots that start development at the beginning of the cool season reach a harvestable size. Nevertheless, Matthews and Stephens (1998) found it necessary, when developing a model to predict the yield distribution of tea, to include photoperiod as a variable (daylength ≥ 12.15 h no dormancy; daylength ≤ 11.25 h full dormancy). Previously, Carr (1970) had suggested that (low) soil temperatures in southern Tanzania (8°S) might play a role in controlling shoot extension rates in tea.

Box 3.4 Climate Trends in North-East India

More than 90 years of rainfall data recorded at the Tocklai Experimental Station in Assam were analysed by Baruah and Bhagat (2012) for long-term trends. There was a trend line indicating a continuous and steady decline in seasonal rainfall, particularly during the period April to June. There was also an increase in extreme events with, for example, an exceptionally dry year (2009, with 1184 mm annual rainfall) followed by a very wet year (2010, with nearly twice as much rain, 2100 mm). Similar differences were recorded in North Bengal at Terai (2563 and 4537 mm) and Nagrakata (3701 and 5130 mm). Darjeeling was the exception, with similar rainfall totals in both years (2237 and 2046 mm).

There has been a steady increase in the minimum air temperature over the last 50 years, totalling about 1 °C, while the number of days when the maximum temperature exceeded 30 °C was relatively constant at *c*. 160 days. In comparison, the number of days when 35 °C was exceeded may have increased slightly over the last 30 years, averaging *c*. 5–10 days a year but totalling about 20 days in each of three relatively recent years.

West Bengal (Darjeeling)

In West Bengal tea is grown in the alluvial plains of Dooars and Terai, and in the Darjeeling hills at an altitude from 600 m up to 2000 m (Jain, 1991). Darjeeling is situated on the southern slopes of the Himalayas where the soils are shallow and underlain by rocks. As in neighbouring Assam, there is a marked seasonal climate but, because it is at a higher altitude and in the foothills of the Himalayas (with cold air descending from the mountains), it is much cooler than Assam (Annex 3.3). During the winter, mean air temperatures fall to 12–13 °C, which is close to the base temperature for shoot extension and development. Temperatures increase to a peak in mid-summer of 22–23 °C, with maxima of 25–26 °C. There is no need for shade trees here. The winters are dry with very little rain in the five months November to March. From May to September monthly rainfall exceeds evapotranspiration.

South India

In South India (*c*. 10°N) tea is grown on steep slopes in the Nilgiri range of mountains at altitudes from 300 m up to 2500 m in the states of Kerala and Tamil Nadu (Figure 3.3). This is a much higher altitude than in Assam, and equates more closely to eastern Africa. The tea area extends along the Western Ghat mountains, from Karnataka in the north through western Tamil Nadu to Kerala in the south, a distance of nearly 500 km. North of the Palghat Gap[3] (alt. 300 m) the majority of the tea is found at an altitude of around 1500 m, up to 2500 m. To the south of the Palghat Gap the altitude averages about 1000 m. The soils vary considerably, but they are nearly all sedentary, being derived from gneissic rocks, as in neighbouring Sri Lanka. They are described as red and lateric loams

Figure 3.3 In South India, tea is grown on steep slopes in the Nilgiri range of mountains at altitudes from 300 m to 2500 m (MKVC). *A black and white version of this figure will appear in some formats. For the colour version, please refer to the plate section.*

enriched by forest humus (Barua, 1989). The tea was originally planted in areas cleared of virgin forest or, as in the case of Sri Lanka, after uprooting diseased coffee plants.

The tea areas situated on the western face of the Western Ghat mountains (e.g. the Anamala hills) get most of their rain (about 80%) from the south-west monsoon (see Gudalur, Nilgiris District, Annex 3.4). By contrast, the Nilgiri hills, facing mainly east, get more than half of their rain from the north-east monsoon (see Coonoor, Nilgiris District).

The total rainfall in tea areas that are exposed to the south-west monsoon averages more than 4000 mm, which is about twice the amount recorded in corresponding areas in Sri Lanka. In the Nilgiris, it is usually dry from June to September, but there can also be several months from November to March when rainfall is less than potential evapo-transpiration (see Coonoor, Annex 3.4).

Maximum temperatures occur mainly in the dry (or less wet) seasons and are normally in the range 25–30 °C, but can reach 35 °C in some areas (Koppa, Wayanad), with corresponding minima of 10–16 °C. Mean air temperatures are between 15 and 20 °C (Coonoor), 18 and 23 °C (Gudalur) and 18 and 22 °C (Valparal). These are all equable temperatures for tea, reflecting the similar altitude of the three weather stations chosen to represent the climate of the South Indian tea areas (*c*. 1100 m).

Bangladesh

In Bangladesh, there are two main tea zones, Sylhet (24–25°N, alt. 35 m) and Chittagong (22°22′N, alt. 29 m). Since both these locations are at similar latitudes and altitudes as the tea areas in neighbouring Assam, and there are no special geographical effects, it is not surprising that the climates are similar. The annual rainfall ranges from 1400 mm in the dry north-west (Rajshahi) to over 5000 mm in the wettest area (Sylhet). The dry season lasts from November to April.

In Sylhet, mean monthly air temperatures range from 26 to 29 °C in the summer months, and from 16 to 18 °C during the winter. These values equate to shoot replacement cycles of 30 days in the summer and 70–80 days in the winter (assuming that water is not a limiting factor), an annual total of up to nine cycles (Carr and Stephens, 1992). For comparison, the duration of a shoot replacement cycle in Kericho, Kenya, is about 110 days throughout the year, equivalent to three to four cycles in a year; in Mulanje, Malawi, 42 days in the main growing season (about four cycles) and 70–80 days in the winter (two cycles); and in Mufindi, Tanzania, from 100 to 160 days (three to four cycles).

Sri Lanka

The tea-growing areas in Sri Lanka are spread across 12 widely varying agroecological regions. For convenience, these regions are grouped into high altitude (> 1200 m), mid-altitude (600–1200 m) and low altitude (< 600 m). There are major tea-growing areas in Ratnapura (low-altitude zone), Kandy (mid-altitude zone), Talawakelle (high-altitude zone), and Passara in Uva Province in the east. As in South India, rainfall in Sri Lanka is governed both by large monsoonal systems and by local convectional storms. The south-west monsoon lasts from mid-May to September and the north-east monsoon from mid-November to February. Convectional rains can occur during March and April and again during September and October. These consist of short-duration, high-intensity thunderstorms falling in the afternoon and evening. The Uva District, like the Nilgiris, only experiences the north-east monsoon, whereas the other tea areas benefit from the south-west monsoon as well. Uva can therefore have an extended long period of dry weather, which normally lasts from June to September, but can extend from February to September (annual rainfall total averages c. 2200 mm). Although both Nilgiris and Uva are exposed to the north-east monsoon, more rain falls in the period before the start of the monsoon in Nilgiris than it does in Uva. It is at this time that teas from Uva develop a short-lived but special flavour.

Weather data from four of the Tea Research Institute of Sri Lanka's meteorological stations, representing the four principal tea-growing areas (namely, low, mid-, and high altitudes together with Passara), are shown in Annex 3.5. In the low country at Ratnapura (alt. 29 m) annual rainfall averages 3700 mm, spread fairly uniformly throughout the year. At Kandy (alt. 760 m) the total is 1900 mm, at Talawakelle (Tea Research Institute of Sri Lanka headquarters, alt. 1380 m) 1400 mm, and at Passara (alt. 1120 m) 2200 mm. January to March can be dry in Kandy and Talawakelle. In Ratnapura, it is uniformly hot throughout the year with daily maximum temperatures consistently above 32 °C, and with daily mean air temperatures averaging 27–29 °C. This corresponds to an average shoot replacement cycle duration of about 30 days. But, as the optimum temperature will have been exceeded on most days for a substantial period of time, the actual duration is likely to be greater than this. At the high-altitude site, Talawakelle, mean air temperatures are, in contrast, only 19–20 °C, and daily maxima rarely exceed 27 °C, which is close to the optimum for tea. These temperatures correspond to a shoot replacement cycle averaging about 52 days, which equates to seven cycles in a year (providing there is no limiting *soil water deficit*). In the mid-country (Kandy) there is a potential for up to 10 shoot replacement cycles. These estimates all assume that the base temperature for shoot development is 12.5 °C.

The tea industry in Sri Lanka is vulnerable to climate change. Any increase in temperature (say by 1–2 °C) at high altitudes (e.g. > 1200 m) would probably increase yields, but at lower altitudes (< 600 m), where it is already hot, a reduction in yield can be expected. Greater variability in rainfall amounts and distribution would also make the industry less competitive.

Region: South-East Asia

This region includes Indonesia, Malaysia, Myanmar, Thailand and Vietnam. Of these, Vietnam, Thailand and Myanmar have a monsoon climate, with distinct wet and dry seasons, while Malaysia and Indonesia have what can be described as generally warm and moist maritime climates. Weather data are presented for three sites, Bogor on the island of Java, Indonesia, with a long-established tea industry, Myanmar (claimed to be one home of tea) and Son La in Vietnam, a relative newcomer to the international tea scene.

Indonesia

As it is on an island, air temperatures at Bogor (6.7°S, alt. 300 m) are relatively constant throughout most of the year, with mean monthly maxima in the narrow band 23–27 °C. Minima do not fall below 17 °C, with monthly mean temperatures of 20–22 °C (Annex 3.6). Rainfall, annual total about 2069 mm, is more seasonal but monthly totals on average still exceed 100 mm. Because Bogor is relatively close to the equator, it has equable temperatures all the year round, and no regular prolonged, dry season. This means that rates of shoot development are relatively constant throughout the year, with no peaks of production (or troughs).

Myanmar

Tea is grown mostly in the Tawngpeng District of Shan State in eastern Myanmar. This is very close to Yunnan Province in China. The capital of Shan State is Taunggyi (21°N). It is at an altitude of 1400 m, and the monthly mean air temperatures are within a narrow band of 20–25 °C (Annex 3.7).

Vietnam

Son La is in the north of Vietnam (21°19′N, alt. 676 m) and experiences a seasonal climate. Mean monthly maxima reach 30 °C during the summer (April to September), which means that on individual days temperatures probably reach 35 °C, rather too hot for tea (Annex 3.8). Minimum temperatures fall to 11 °C in December/January, when mean air temperatures remain above 15–16 °C. Remember the base temperature for tea is, in general, cited as being 12.5 °C, although it varies between clones. So growth, albeit slow, continues through the winter in Son La subject to soil water availability. The winter months are quite dry, with only about 25 mm per month expected, on average, from November to February (inclusive). The total annual rainfall is about 1200 mm.

Region: Caucasus

The two principal tea-growing countries in western Asia are Turkey and Iran (which occupy positions 5 and 17 in the annual worldwide tea production league table: see Chapter 2, Table 2.1). Georgia, at 42°N, is the most northerly commercial tea area in the world, although production has now fallen below 5000 t.

Turkey

Tea in Turkey extends from Hopa near the Russian border (41°25′N, alt. 10 m) and along the Black Sea as far as the Arakli river. The climate at Hopa, being close to the modifying influence of the Black Sea, is described as relatively mild. The growing season begins in April, when the mean air temperature reaches 12.5 °C, and extends through to October. Peak temperatures occur in July and August, when daily maxima average about 28 °C. With plentiful rain throughout the year (average annual total in excess of 2000 mm) and sufficient rain, on average, in the summer months to match potential rates of evapotranspiration, growing conditions in July and August are ideal for tea (Annex 3.11).

Georgia

Most of the tea in Georgia is found south of the Caucasus Mountains towards the Black Sea. At this most northerly site (c. 42°N), temperatures in the winter can fall to –6 °C. There are usually three harvests over the summer months, the first in late May and the last in early October (Annex 3.9).

Iran

The tea industry developed in Gilan and Mazanduran provinces on the south coast of the Caspian Sea (c. 37°N) in northern Iran. The winters are cold (down to –5 °C) with frequent snow. In the summer, temperatures can reach 35 °C, when it can also be dry (Annex 3.10). Too cold and too hot for tea! Iran is a country of extremes, not an ideal place to grow tea. It is not obvious which production figures to believe.

Region: Eastern and Southern Africa

In the equatorial areas of East Africa, rainfall distribution is bimodal, with two dry, or less wet, seasons (Stephens *et al.*, 1992). Although some tea in these areas may be irrigated, drought mitigation is usually more appropriate and cost-effective. Away from the equator, there is a single rainy season (e.g. in southern Tanzania, > 8°S) and the dry season can last from May to November. The relative amounts of tea produced between and within individual African countries are indicated in Chapter 2, Figure 2.11. Kenya is by far the largest producer of tea. Across the region, the rainfall amount and distribution is locally modified considerably by local physical features such as proximity to a large body of open water (e.g. Lake Victoria) or to mountains (e.g. Mount Kenya in Kenya, the Rwenzori Mountains in western Uganda, the Usambaras in northern Tanzania and the Mulanje Massif in southern Malawi), or a 600 m escarpment (Mufindi in Tanzania). Tea is grown at altitudes ranging from 600 m (Malawi) up to 2700 m (Kenya). In the south of this region, irrigation plays an important role in sustaining the tea industry.

As a basis for comparison, weather data are presented here in the order from north to south for Kenya, Uganda, Tanzania, Malawi and Zimbabwe.

Kenya

In Kenya, tea is grown in two principal places with contrasting climates separated by the Great Rift Valley (Figure 3.4). Seasonal variability in production is greater to the east of the Rift Valley where the climate is modified by the proximity of Mount Kenya, compared to the west, which is within 100 km of Lake Victoria. East of the Rift Valley, there are two distinct dry or certainly less wet seasons, whereas to the west there is only one dry season (January to March), and that is often wet (Annex 3.12)! In the west, Kericho, at an altitude of up to 2300 m, is cool and the seasonal mean air temperatures cover a relatively narrow range (17–19 °C). To the east of the Great Rift Valley there is by contrast a distinct 'winter' season. For example, at Nyeri, mean air temperatures decline from 20 °C in February and March to 16 °C in July and August. Further east at Embu the seasonal differences are even greater (between 23 and 17 °C). Embu is 400 m lower in altitude than Nyeri (1350 m compared with 1750 m) and therefore warmer.

A detailed analysis of long-term weather data recorded at Kericho is summarised in Box 3.5. Low soil and air temperatures (altitude-dependent: see below) restrict shoot growth rates throughout most of the year, and these can dominate the effects of other treatment variables, such as fertilisers. The importance of soil temperature was demonstrated by using plastic mulches in young tea. These increased soil temperatures and shoot growth rates (Othieno and Ahn, 1980). The occasional drought (the annual total soil water deficit varied between 120 and 400 mm) can reduce yields by about 20%.

By contrast to Assam, the relatively cool conditions throughout the year in Kericho mean that the distribution of yield is comparatively uniform month on month. There is no accumulation of shoots of varying ages maturing together (except after a drought has ended).

Figure 3.4 In Kenya, tea is grown in two principal places with contrasting climates, east and west of the Great Rift Valley. This photograph shows smallholder tea west of the Rift Valley (MKVC).

Box 3.5 Climate at the Tea Research Foundation of Kenya

The headquarters of the Tea Research Foundation of Kenya (TRFK) are at Timbilil Estate in the Kericho District (0°22'S 35°21'E) of western Kenya. At an altitude of 2180 m, they are close to the top of the altitude range in Kericho District (1700–2300 m) at which tea can be grown profitably (see Box 3.7).

Twenty-one years of daily weather data (1966–1986) were analysed in terms of year-to-year variability in air and soil temperatures, the saturation deficit of the air (a measure of its dryness), solar radiation, sunshine hours, wind run, rainfall, evapotranspiration and hail (Stephens *et al.*, 1992). In addition, a number of derived variables were determined. These included thermal time (above a base temperature of 12.5 °C)[4] and the length of the shoot replacement cycle, especially for shoots initiated in January and February (the dry or less wet season). At this time of the year, in 2 years out of 10, buds will take under 85 or over 105 days to develop into shoots suitable for harvesting.

Soil temperature, at a depth of 0.30 m under short grass, was in the range 17–19 °C. Soil temperature below 20 °C can play an important role in controlling shoot extension in tea. The mid-afternoon values of the saturation deficit of the air increased from 1.2 to 1.7 kPa as the dry season progressed. Rarely was the critical

Box 3.5 Continued

value of 2.3 kPa exceeded on individual days. The total dry season saturation deficit time (analogous to thermal time) averaged 5 d kPa (range 0.3–11.3 d kPa). For comparison, in Mulanje (Malawi) the total can reach 40 d kPa, which is sufficient to restrict responses to irrigation (Carr *et al.*, 1987).

The mean daily solar radiation total was in the range 17–22 MJ m^{-2}, with total annual receipts of 67 TJ m^{-2}. Wind speeds were generally low, ranging from 120 to 160 km d^{-1}. This is equivalent to 1.4–1.9 m s^{-1} (at 1.5 m above a short grass surface). The mean annual rainfall was 2150 mm, with over 90% falling between mid-March and mid-November. Using the Penman–Monteith *ET* model (see Chapter 15, Box 15.4), potential rates of evapotranspiration were between 3 mm d^{-1} in April (the start of the rains) and 4.5 mm d^{-1} at the height of the dry season (March). The maximum potential soil water deficit totalled over the dry months was likely to exceed 350 mm 4 years in 10. In theory, at this latitude, there should be two dry seasons each year but, in 9 years out of 20, the short dry season from September to November did not occur. In 4 years out of 10, the long dry season started before 6 November or after 28 November. In 8 years out of 10, the dry season ended between 8 March and 7 April. Its average duration was 118 ± 30 days and there was a one in five chance of it exceeding 143 days. Rainfall is notoriously variable, especially at a local level where it is enhanced or reduced by nearby topographic features, including lakes, mountains and escarpments. The potential soil water deficit exceeded 350 mm 4 years out of 10. Hail is always a risk. On average there were 250 hail events each year, of which about one-quarter caused discernible damage to the bush. These resulted in an estimated 10% reduction in the total annual yield – or, expressed another way, up to 350 kg ha^{-1} a year.

Hail damage can result in an estimated 10% loss of the total annual yield in the Kericho, Sotik and Nandi Hills districts. Between the 1960s and the 1980s a major attempt was made to find an effective method of hail suppression. However, it was not possible to demonstrate enough benefit to justify the cost of routine cloud seeding (Mwakha, 1983). Frosts can also cause damage, especially in low-lying areas, where cold air drainage is restricted.

Uganda

There are predominantly two broad areas within which tea is grown in Uganda. In the east, there is tea close to the shores of Lake Victoria and to the west there is tea in the foothills of the Rwenzori Mountains, extending from Hoima in the north (1°26′N) down as far as Ankole in the south (0°34′S). There are also individual estates and smallholder schemes outside these two principal locations. A summary of the climate in Uganda is shown in Box 3.6. Temperatures are equable throughout the year, being slightly cooler in the west compared with the east. Minimum temperatures rarely fall below 11 °C in Fort

---

Box 3.6 Climates in the Tea-Growing Areas of Uganda

The tea-growing areas of Uganda are close to the equator, and the rainfall distribution is therefore bimodal, but the amounts are influenced in the east by the proximity of Lake Victoria (which also modifies the temperature regime), and in the west the Rwenzori Mountains (also known as the Mountains of the Moon, with snow on the peaks) not only influence rainfall but cold air rolling down the mountain means that it is cooler than the location and altitude would suggest. A substation of the former Tea Research Institute of East Africa, known as Rwebitaba, is sited close to Fort Portal in the Toro District of western Uganda. Reliable weather data have been collected here since 1961 despite the setbacks that occurred subsequently in Uganda.

The mean annual air temperature at Rwebitaba is 18.7 °C, which is below what would be expected at this altitude (1450 m) in Kericho (Kenya), where the equivalent temperature would occur at $c.$ 1800 m.

Box 3.7 Climate at the Ngwazi Tea Research Station, Tanzania

The headquarters of the Tea Research Institute of Tanzania (TRIT) are at Ngwazi in the Mufindi District of southern Tanzania (8°36'S 35°21'E, alt. 1900 m). Here the climate is dominated by the seasonal effects of a single rainy season lasting from November to May when the mean air temperature (T_{mean}) is 17–20 °C, a cool dry season from June to August (T_{mean} = 13–15 °C) and a warm dry season from September to the start of the rains during November (T_{mean} = 15–17 °C). In contrast to Kericho (in Kenya), the range of altitudes across the district where tea is grown is only about 150 m. Instead, the principal climate variable here is the decline in the annual rainfall with distance from the 600 m deep Uzungwa escarpment. The mean annual rainfall declines from about 1700 mm close to the escarpment edge to less than 1000 mm at the Ngwazi Tea Research Station, where the potential soil water deficit reaches 500–600 mm 6 years out of 10. The corresponding length of the dry season increases from 25 to 28 weeks (Carr and Stephens, 1992). The saturation deficit of the air at Ngwazi rarely exceeds 1.5 kPa even in October. The incoming solar radiation averages 17 MW m^{-2}, while the daytime maximum temperature at this time is below 28 °C (Annex 3.14).

Portal (west) with maxima of 26–27 °C. Rainfall is well distributed in both areas, but with two less wet seasons from December to February and in July and August, when it can be dry in some years (Annex 3.15).

Tanzania

The main centres of tea production in Tanzania are in the Southern Highlands (8–10°S), in the Mufindi, Njombe and Rungwe districts, at altitudes ranging between 1200 and 1800 m. Tea is also grown in the Usambara Mountains in the north and in Kagera. A summary description of the climate at the Ngwazi Tea Research Station is presented in Box 3.7.

Figure 3.5 Numerous small earth dams have been created in Mufindi, southern Tanzania, to store water in order to irrigate the tea during the dry season. Note also the *Hakea salicifolia* hedges planted as shelter belts across the direction of the prevailing wind. The benefits of shelter are questionable (MKVC). *A black and white version of this figure will appear in some formats. For the colour version, please refer to the plate section.*

The results of research in the 1960s and 1970s demonstrated the potential value of tea irrigation in Mufindi (Carr, 1974), and soon afterwards commercial irrigation was introduced. This has had a large impact on productivity of estates in the Mufindi (Figures 1.2, 3.5) and Njombe districts. Much of the tea in southern Tanzania is now irrigated, and indeed two of the most modern tea estates in the world can be found here in non-traditional tea areas (where the climax vegetation is miombo woodland and grassland): Kibena in Njombe District and Ngwazi in Mufindi District. The plants are all clonal, and the tea is well irrigated with water taken from storage dams, and mechanically harvested. However, smallholder tea in these two locations and further south in Rungwe District, where it is wetter and warmer, is all rain-fed.

In the northern tea areas of Tanzania, namely the East and West Usambara Mountains, and at Kagera, which is west of Lake Victoria, the distribution of rain is bimodal. Since Kagera is close to the lake it has a climate similar to that over the international border in Uganda at Salama (and Jinja). These northern sites, which are all at a similar altitude (*c.* 1200 m), have mean monthly air temperatures between 18.6 and 21.7 °C (Annex 3.14).

Rainfall is very variable in the Usambara Mountains, as indicated by the detailed analysis of the data at the Marikitanda (Amani) substation of the Tea Research Institute of Tanzania described in Box 3.8.

Box 3.8 Rainfall Variability in Northern Tanzania

The Usambara Mountains are close to the Indian Ocean. They attract the moist air coming from the ocean. The water vapour condenses as the moisture-laden air rises over the mountains and cools. The annual rainfall at Marikitanda Tea Research Station (05°08′S 38°35′E, alt. 970 m) averages 1778 mm (range 1112–2601 mm, over the 30-year period from 1967–1996), and follows a bimodal distribution pattern. There is considerable inter-seasonal variation, with an extended dry season when the short rains (due in October and November) fail, and they are then followed by the 'normal' dry season (from December to February inclusive). An analysis of 30 years' data failed to show a significant long-term trend either in the annual rainfall or in seasonal totals. That is, the slope of the trend line was not significantly different from zero ($p < 0.05$) (Hess and Kimambo, 1998). The hydrological years 1995/96 (September to August, total rain = 1112 mm) and 1996/97 (= 1085 mm) were both exceptionally dry years, with a probability of occurrence of less than 1 in 100 years. This was followed by an exceptionally wet short rains, with a probability of occurrence of only 1 in 300 years. The influence of El Niño and La Niña is suspected. The trend appears to be towards a more variable inter-annual rainfall pattern.

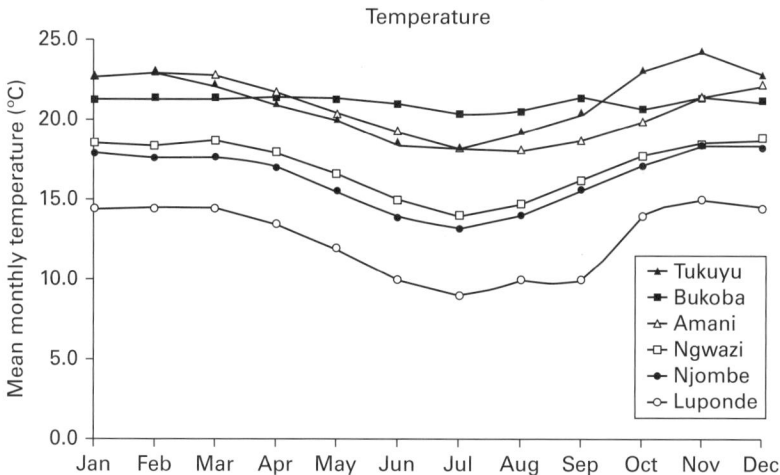

Figure 3.6 A comparison of mean air temperatures in the tea-growing regions within one country in Africa – Tanzania. These differences have to be taken into account when extending the results of research to different regions, when selecting suitable sites for growing tea, or when predicting the potential yield. For example, growing tea in Luponde will always be a challenge, because low temperatures will always dominate.

The tea areas in Tanzania cover a large geographic area and embrace a diversity of ecological habitats (Figure 3.6). At an altitude of 2200 m, the coolest site is at Luponde, where the monthly mean air temperatures fall within a narrow band between 10 °C and 15 °C. The mean monthly air temperature is only 13 °C. The number of effective 'day degrees', that is when the mean air temperature exceeds 12.5 °C, equates to a total of only 2–3 shoot replacement cycles (see Chapter 10) in a year. By contrast, other things being equal, there is the potential for 5–6 cycles at Tukuyu, also in the Southern Highlands, but at an altitude of 1200 m.

Malawi

In the tea-growing regions of Malawi (Mulanje and Thyolo districts in the south and Mzuzu in the north), there are three main seasons: the hot wet rainy season (the summer) from mid-November to April; the cool moist winter, May to August; and the hot dry months of September to mid-November (Annex 3.13). Over 80% of the tea yield is produced during the five summer months. This uneven distribution of crop contributes to the difficulty of managing tea as a business, when the logistical requirements in the field and in the factory, for example, vary so much from month to month. Several attempts have been made to explain why yields begin to decline in April and May, and remain low until September, before increasing in September and October (if there is rain or irrigation). Possible reasons for the decline in yields during the winter, including photoperiodism (Box 3.3), have long been subjects of speculation in Malawi, and also in places north of the equator such as Assam (Barua, 1969).

The local climate in the south of Malawi (16°05′S, alt. 650 m) is dominated by the proximity of the 3000 m high Mulanje Mountain. Rainfall declines with distance from the mountain, averaging about 2300 mm per year close to its base and declining to 1500 mm per year 3–5 km away. As a result, tea is only grown in a narrow band less than 10 km wide around the base of the mountain.

The different responses obtained from two similar irrigation experiments carried out at sites in Malawi and Tanzania suggest that the variation in the saturation deficit of the air during the hot dry season has a significant impact on yields (Carr et al., 1987). In Malawi, dry air (mid-afternoon saturation deficit > 2.3 kPa) restricted the responses to irrigation in some years, but in Tanzania, where the air was more humid, there were large yield increases in each year. These observations have important commercial implications in the region.

Zimbabwe

Tea is grown in the Chipinge (alt. 1130 m) and Honde Valley districts in the Eastern Highlands of Zimbabwe. The seasonal climate is similar to that found in neighbouring Malawi, with an extended dry season from May to October (Figure 3.7). It is cooler in Chipinge (mean T_{max} = 26.0 °C) during the period September to November than it is in Mulanje, Malawi (mean T_{max} = 30.3 °C, alt. 650 m), owing to the higher altitude, which suggests that tea will respond well to irrigation at this time.

Figure 3.7 Tea is grown in the Chipinge and Honde Valley districts in the Eastern Highlands of Zimbabwe. Drought is a recurring challenge, as shown here (MKVC). *A black and white version of this figure will appear in some formats. For the colour version, please refer to the plate section.*

Summary

These examples of climatic differences between sites across the tea-growing areas of the world bring out the following points of particular interest:

- Rainfall amount and distribution are considerably modified by local physical features such as a large body of open water, mountains or an escarpment.
- These features also modify the temperature regimes of the surrounding areas, for example, because of cold air descending a mountain to the foothills, or a large body of water acting like a storage heater.
- Altitude sets limits to productivity. At heights greater than about 1900–2000 m, low soil and air temperatures dominate shoot development and growth rates, and limit the number of shoot replacement cycles.
- Responses to other inputs will be dominated by the overriding influence of low temperatures on growth rates. Researchers need to be aware of this when interpreting responses to treatments.
- There is as yet no consensus view on the reasons for seasonality of production at high latitudes (> 6–7°N or S). One view is that changes in temperature create a 'crop flow problem' during the winter months as it then takes longer for a generation of shoots to reach a harvestable size. An alternative view is that apical bud dormancy is induced by changes in daylength, meaning that it is a photoperiodic response.

- Responses to irrigation vary depending on the temperature during the dry season. If daily maximum air temperatures are in excess of about 30 °C at this time, and the associated mid-afternoon saturation deficit of the air exceeds about 2.3 kPa, yield responses during the dry season will be small (but there will be an exceptionally large peak once the rains begin).
- The commercial value of tea irrigation in southern Tanzania and points south is very dependent on the prevailing dry-season air temperatures and saturation deficits. Climate change, in this case global warming, could make this analysis even more critical.
- The analysis of long-term weather data at the Tea Research Foundation of Kenya needs to be repeated at each major tea research station to provide a database against which valid comparisons can be made.
- It was difficult to obtain full sets of reliable data to include in this book. In view of the importance of their work, seven days a week, 52 weeks a year, the status of the weather station recorders needs to be improved and the value of their work recognised.
- These are a few examples of how weather data can be presented and used, so that they can serve as a basis for comparison with other sites elsewhere in the world, rather than being just routinely tabulated in annual reports.
- In some locations there is limited evidence of climate change, but there are indications of greater variability from year to year and of more extreme events.

Notes

1. Or, strictly, the soil water deficit, which at its simplest equals evapotranspiration less 'effective rainfall'.
2. These concepts are developed in detail in Chapter 9.
3. Palghat Gap is the name given to a 30–60 km wide gap in the Western Ghat chain of mountains, which separates the Nilgiri hills in the north from the Anamala hills in the south.
4. This is a term describing the sum of the difference between the base or minimum temperature at which a growth process occurs and the actual temperature (see Chapter 9).

Annex to Chapter 3 Weather Data for Some of the Principal Tea-Growing Areas of the World

Data analysis by M. Upson and M. Perez Ortola

Contents

East Asia

The bar charts show mean monthly rainfall total. The three dots are minimum, mean and maximum monthly air temperature. The bars are error bars.

EAST ASIA

Annex 3.1: China

Chengdu

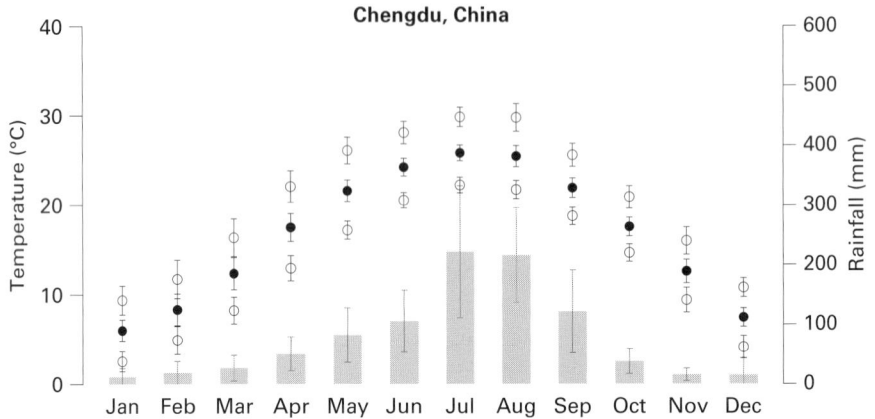

Latitude: 30.67 **Longitude:** 104.02

Altitude: 508 m

Rainfall years included: 1975 1976 1977 1978 1979 1980 1981 1982 1983 1984 1985 1986 1987 1988 1989 1990 1991 1992 1993 1994 1995 1996 1997 1998 2000 2001 2002 2003

Temperature years included: 1975 1976 1977 1978 1979 1980 1981 1982 1983 1984 1985 1986 1987 1988 1989 1990 1991 1992 1993 1994 1995 1996 1997 1998 1999 2000 2001 2002 2003 2004 2005 2006 2007 2008 2009 2010 2011 2012 2013 2014

Month	Rainfall (mm)	T_{min} (°C)	T_{av} (°C)	T_{max} (°C)
Jan	12	2.5	6.0	9.4
Feb	19	4.9	8.3	11.7
Mar	27	8.2	12.3	16.4
Apr	51	12.9	17.5	22.1
May	82	17.2	21.6	26.1
Jun	105	20.5	24.2	28.1
Jul	221	22.2	25.8	29.8
Aug	216	21.7	25.4	29.8
Sep	122	18.8	21.9	25.6
Oct	38	14.7	17.6	20.9
Nov	16	9.4	12.6	16.0
Dec	15	4.1	7.4	10.8
	924	13.1	16.7	20.6

Dushan

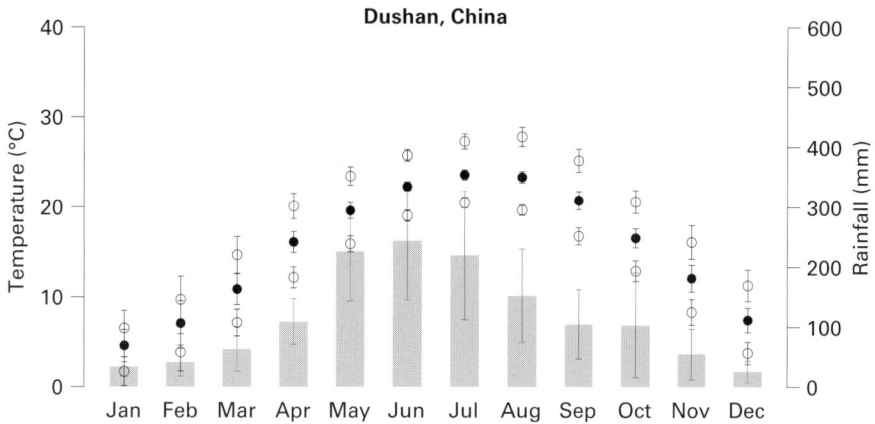

Dushan, China

Latitude: 25.83 Longitude: 107.55

Altitude: 971 m

Rainfall years included: 1975 1976 1977 1978 1979 1980 1981 1983 1984 1985 1986 1987 1988 1990 1991 1992 1993 1994 1995 1996 1997 1998 2000 2001 2002 2003 2004 2005 2006 2007 2008 2009 2010 2011 2012 2013 2014

Temperature years included: 1975 1976 1977 1978 1979 1980 1981 1983 1984 1985 1986 1987 1988 1990 1991 1992 1993 1994 1995 1996 1997 1998 1999 2000 2001 2002 2003 2004 2005 2006 2007 2008 2009 2010 2011 2012 2013 2014

Month	Rainfall (mm)	T_{min} (°C)	T_{av} (°C)	T_{max} (°C)
Jan	34	1.7	4.6	6.6
Feb	41	3.9	7.1	9.7
Mar	63	7.2	10.9	14.7
Apr	109	12.2	16.1	20.1
May	226	15.9	19.6	23.4
Jun	244	19.1	22.2	25.8
Jul	220	20.5	23.6	27.3
Aug	152	19.7	23.3	27.8
Sep	105	16.8	20.7	25.2
Oct	103	12.9	16.6	20.6
Nov	55	8.3	12.1	16.1
Dec	26	3.8	7.4	11.3
	1378	11.8	15.3	19.1

Langzhong

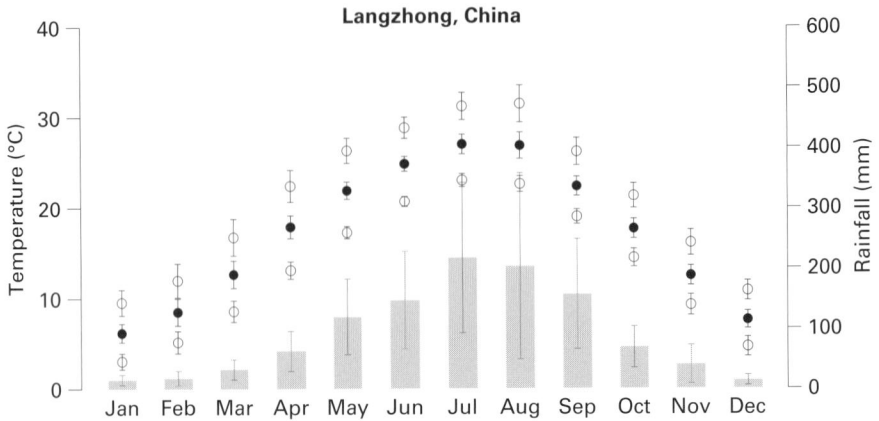

Langzhong, China

Latitude: 31.58 **Longitude:** 105.97

Altitude: 385 m

Rainfall years included: 1975 1976 1977 1978 1979 1980 1981 1982 1983 1984 1985 1986 1987 1988 1989 1990 1991 1992 1993 1994 1995 1996 1997 1998 2000 2001 2002 2003 2004 2005 2006 2007 2008 2009 2010 2011 2012 2013 2014

Temperature years included: 1975 1976 1977 1978 1979 1980 1981 1982 1983 1984 1985 1986 1987 1988 1989 1990 1991 1992 1993 1994 1995 1996 1997 1998 1999 2000 2001 2002 2003 2004 2005 2006 2007 2008 2009 2010 2011 2012 2013 2014

Month	Rainfall (mm)	T_{min} (°C)	T_{av} (°C)	T_{max} (°C)
Jan	15	3.0	6.2	9.6
Feb	18	5.2	8.5	12.0
Mar	32	8.6	12.6	16.7
Apr	62	13.1	17.9	22.4
May	119	17.3	21.9	26.3
Jun	147	20.7	24.8	28.8
Jul	217	23.0	27.0	31.2
Aug	203	22.6	26.9	31.5
Sep	157	19.0	22.4	26.2
Oct	69	14.4	17.7	21.3
Nov	40	9.3	12.5	16.1
Dec	13	4.6	7.6	10.8
	1092	13.4	17.2	21.1

Yuanjiang

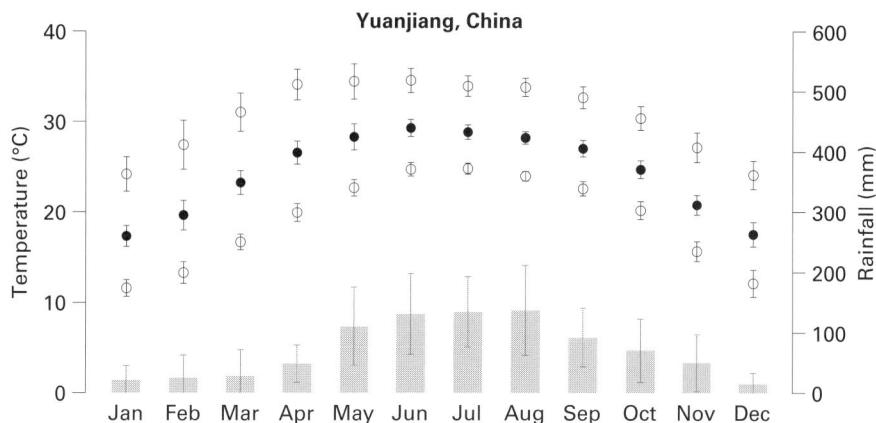

Yuanjiang, China

Latitude: 28.93 Longitude: 112.59

Altitude: 398 m

Rainfall years included: 1975 1976 1977 1978 1979 1980 1981 1982 1983 1984 1985 1986 1987 1988 1989 1990 1991 1992 1993 1994 1995 1996 1997 1998 2000 2001 2002 2003 2004 2005 2006 2007 2008 2009 2010 2011 2012 2013 2014

Temperature years included: 1975 1976 1977 1978 1979 1980 1981 1982 1983 1984 1985 1986 1987 1988 1989 1990 1991 1992 1993 1994 1995 1996 1997 1998 1999 2000 2001 2002 2003 2004 2005 2006 2007 2008 2009 2010 2011 2012 2013 2014

Month	Rainfall (mm)	T_{min} (°C)	T_{av} (°C)	T_{max} (°C)
Jan	21	11.6	17.3	24.2
Feb	25	13.3	19.6	27.4
Mar	28	16.7	23.3	31.0
Apr	49	19.9	26.6	34.1
May	111	22.7	28.3	34.4
Jun	131	24.7	29.3	34.5
Jul	135	24.8	28.8	33.9
Aug	137	24.0	28.2	33.8
Sep	92	22.6	27.0	32.6
Oct	70	20.2	24.7	30.4
Nov	50	15.7	20.8	27.2
Dec	15	12.1	17.5	24.1
	864	19.0	24.3	30.6

Annex 3.2: Japan

Kagoshima

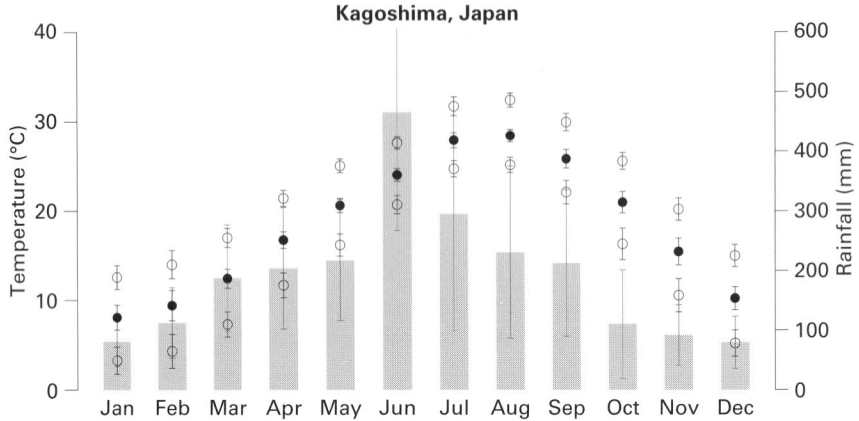

Latitude: 31.55 **Longitude:** 130.55

Altitude: 32 m

Rainfall years included: 1975 1976 1977 1978 1979 1980 1981 1982 1983 1984 1985 1986 1987 1988 1989 1990 1991 1992 1993 1994 1995 1996 1997 1998 1999 2000 2001 2002 2003 2004 2005 2006 2007 2008 2009 2010 2011 2012 2013 2014

Temperature years included: 1975 1976 1977 1978 1979 1980 1981 1982 1983 1984 1985 1986 1987 1988 1989 1990 1991 1992 1993 1994 1995 1996 1997 1998 1999 2000 2001 2002 2003 2004 2005 2006 2007 2008 2009 2010 2011 2012 2013 2014

Month	Rainfall (mm)	T_{min} (°C)	T_{av} (°C)	T_{max} (°C)
Jan	81	3.3	8.1	12.6
Feb	112	4.3	9.4	14.0
Mar	187	7.3	12.4	17.0
Apr	204	11.7	16.8	21.4
May	217	16.2	20.6	25.0
Jun	466	20.7	24.0	27.6
Jul	295	24.7	27.9	31.7
Aug	231	25.1	28.4	32.4
Sep	213	22.1	25.8	30.0
Oct	110	16.3	21.0	25.6
Nov	91	10.5	15.5	20.2
Dec	79	5.2	10.2	15.0
	2286	13.9	18.3	22.7

Shizuoka

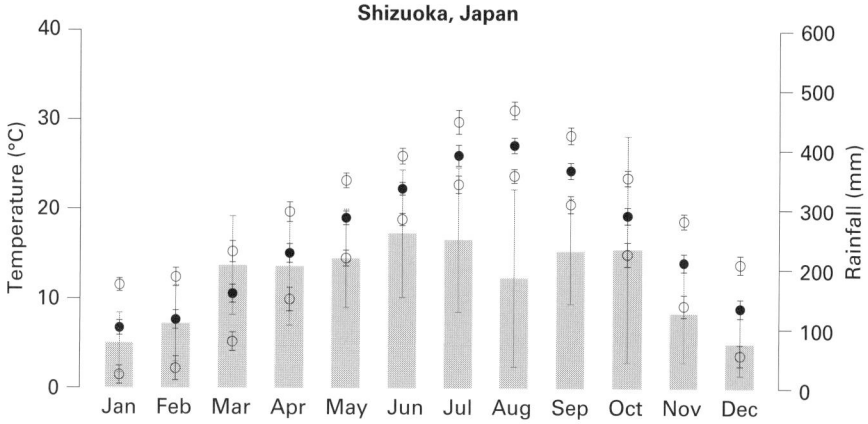

Shizuoka, Japan

Latitude: 34.98 **Longitude:** 138.4
Altitude: 16 m
Rainfall years included: 1992 1993 1994 1995 1996 1997 1998 1999 2000 2001
    2002 2003 2004 2005 2006 2007 2008 2009 2010 2011 2012 2013 2014
Temperature years included: 1992 1993 1994 1995 1996 1997 1998 1999 2000
    2001 2002 2003 2004 2005 2006 2007 2008 2009 2010 2011 2012 2013 2014

Month	Rainfall (mm)	T_{min} (°C)	T_{av} (°C)	T_{max} (°C)
Jan	76	1.4	6.7	11.5
Feb	108	2.2	7.7	12.4
Mar	206	5.2	10.6	15.3
Apr	204	9.9	15.1	19.7
May	218	14.5	19.0	23.2
Jun	260	18.9	22.3	26.0
Jul	249	22.8	26.0	29.8
Aug	185	23.8	27.2	31.1
Sep	230	20.6	24.4	28.3
Oct	233	15.0	19.3	23.6
Nov	126	9.2	14.1	18.7
Dec	75	3.7	9.0	13.9
	2170	12.3	16.8	21.1

INDIAN SUBCONTINENT

Annex 3.3: North-East India

Darjeeling

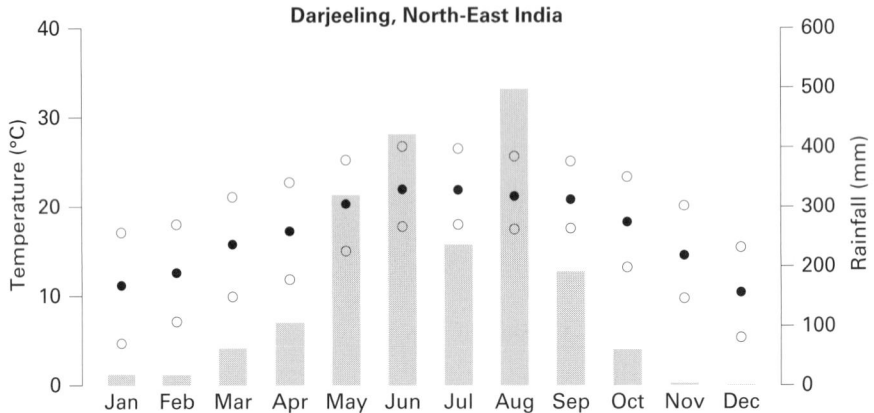

Darjeeling, North-East India

Latitude: 27.35 **Longitude:** 87.67
Altitude: 1732 m
Rainfall years included: 2015
Temperature years included: 2015

Month	Rainfall (mm)	T_{min} (°C)	T_{av} (°C)	T_{max} (°C)
Jan	19	4.7	11.2	17.1
Feb	18	7.2	12.6	18.0
Mar	62	9.9	15.8	21.1
Apr	106	11.8	17.2	22.7
May	320	15.0	20.3	25.2
Jun	422	17.8	22.0	26.7
Jul	236	18.0	21.9	26.5
Aug	498	17.5	21.2	25.7
Sep	191	17.6	20.8	25.1
Oct	60	13.2	18.3	23.3
Nov	4	9.8	14.6	20.1
Dec	1	5.4	10.4	15.5
	1937	12.3	17.2	22.2

Tezpur, Assam

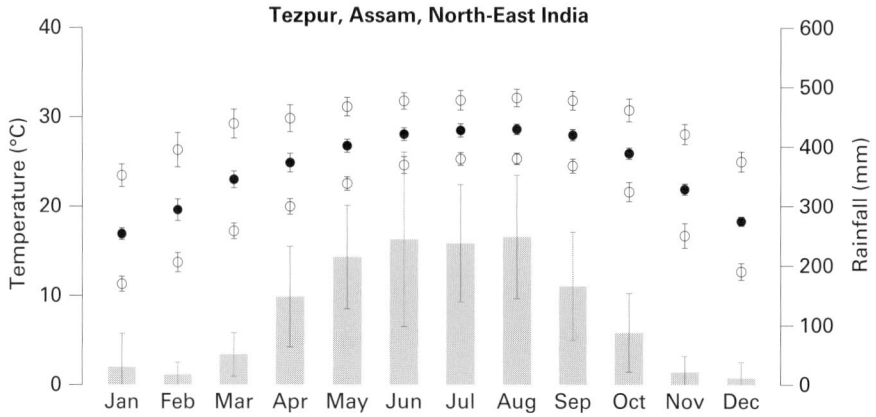

Tezpur, Assam, North-East India

Latitude: 26.62 **Longitude:** 92.78

Altitude: 79 m

Rainfall years included: 1994 1995 1996 1997 1998 1999 2000 2001 2002 2003
2004 2005 2006 2007 2008 2009 2010 2011 2012 2013 2014

Temperature years included: 1994 1995 1996 1997 1998 1999 2000 2001 2002
2003 2004 2005 2006 2007 2008 2009 2010 2011 2012 2013 2014

Month	Rainfall (mm)	T_{min} (°C)	T_{av} (°C)	T_{max} (°C)
Jan	30	11.3	16.9	23.5
Feb	17	13.7	19.6	26.3
Mar	51	17.2	23.0	29.2
Apr	148	20.0	24.9	29.8
May	215	22.5	26.8	31.1
Jun	244	24.6	28.1	31.8
Jul	238	25.3	28.5	31.8
Aug	249	25.3	28.6	32.1
Sep	165	24.5	28.0	31.8
Oct	88	21.6	25.9	30.7
Nov	21	16.7	21.9	28.1
Dec	11	12.7	18.3	25.0
	1477	19.6	24.2	29.3

Annex 3.4: South India

Coonoor

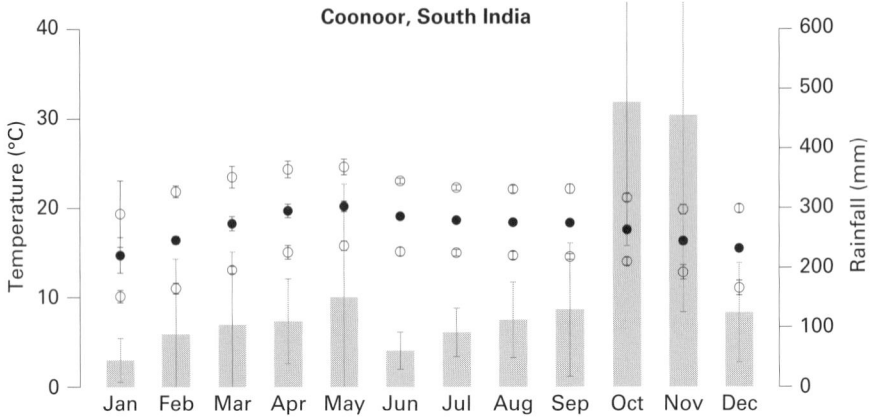

Coonoor, South India

Latitude: 11.35 **Longitude:** 76.82
Altitude: 1822 m
Rainfall years included: 2004 2005 2006 2007 2008 2009 2010 2011 2012
Temperature years included: 2004 2005 2006 2007 2008 2009 2010 2011 2012

Month	Rainfall (mm)	T_{min} (°C)	T_{av} (°C)	T_{max} (°C)
Jan	45	10.1	14.7	19.4
Feb	88	11.0	16.4	21.8
Mar	104	13.1	18.3	23.5
Apr	110	15.1	19.7	24.3
May	150	15.8	20.2	24.6
Jun	60	15.1	19.1	23.0
Jul	91	15.0	18.6	22.3
Aug	112	14.7	18.4	22.1
Sep	129	14.5	18.3	22.1
Oct	477	14.0	17.6	21.1
Nov	456	12.8	16.3	19.8
Dec	124	11.0	15.5	19.9
	1946	13.5	17.8	22.0

Gudalur

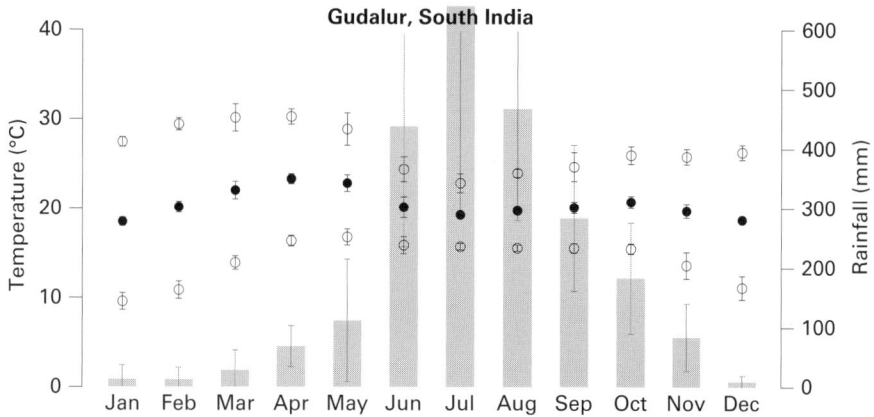

Gudalur, South India

Latitude: 11.5 **Longitude:** 76.39
Altitude: 926 m
Rainfall years included: 2004 2005 2006 2007 2008 2009 2010 2011 2012
Temperature years included: 2004 2005 2006 2007 2008 2009 2010 2011 2012

Month	Rainfall (mm)	T_{min} (°C)	T_{av} (°C)	T_{max} (°C)
Jan	13	9.6	18.5	27.4
Feb	12	10.9	20.1	29.4
Mar	28	13.9	22.0	30.1
Apr	69	16.4	23.3	30.3
May	112	16.8	22.8	28.9
Jun	437	15.9	20.1	24.4
Jul	640	15.7	19.3	22.8
Aug	467	15.6	19.8	24.0
Sep	284	15.6	20.1	24.7
Oct	183	15.5	20.7	26.0
Nov	84	13.6	19.7	25.8
Dec	9	11.1	18.7	26.3
	2338	14.2	20.4	26.7

Valparai

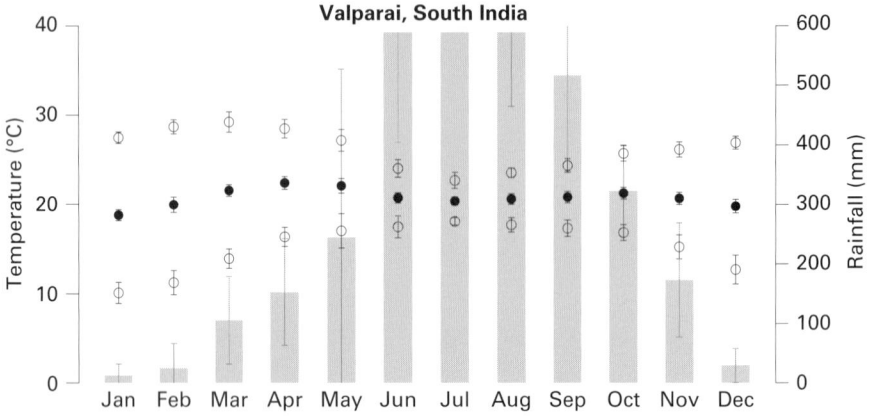

Valparai, South India

Latitude: 10.32 **Longitude:** 76.95
Altitude: 1070 m
Rainfall years included: 2004 2005 2006 2007 2008 2009 2010 2011 2012 2013
Temperature years included: 2004 2005 2006 2007 2008 2009 2010 2011 2012 2013

Month	Rainfall (mm)	T_{min} (°C)	T_{av} (°C)	T_{max} (°C)
Jan	13	10.1	18.8	27.5
Feb	25	11.3	20.0	28.7
Mar	105	13.9	21.6	29.2
Apr	153	16.4	22.4	28.5
May	245	17.0	22.1	27.2
Jun	855	17.5	20.7	24.0
Jul	1027	18.1	20.4	22.7
Aug	732	17.7	20.6	23.5
Sep	516	17.3	20.8	24.4
Oct	323	16.8	21.3	25.7
Nov	173	15.2	20.7	26.1
Dec	29	12.7	19.8	26.9
	4196	15.3	20.8	26.2

Annex 3.5: Sri Lanka

Kandy

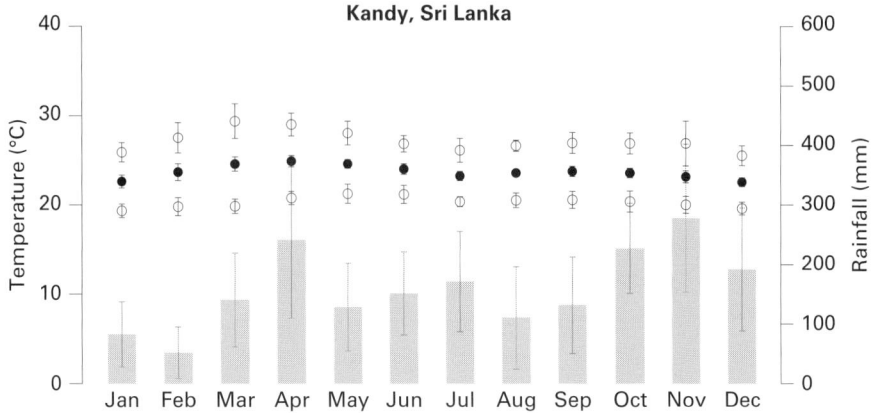

Latitude: 7.43 **Longitude:** 80.38

Altitude: 762 m

Rainfall years included: 2000 2001 2002 2003 2004 2005 2006 2007 2008 2009 2010

Temperature years included: 2000 2001 2002 2003 2004 2005 2006 2007 2008 2009 2010

Month	Rainfall (mm)	T_{min} (°C)	T_{av} (°C)	T_{max} (°C)
Jan	82	19.3	22.6	25.9
Feb	52	19.8	23.7	27.5
Mar	140	19.8	24.6	29.4
Apr	241	20.8	24.9	29.0
May	128	21.2	24.6	28.0
Jun	151	21.2	24.0	26.8
Jul	171	20.4	23.2	26.1
Aug	110	20.5	23.6	26.6
Sep	131	20.6	23.7	26.9
Oct	227	20.4	23.5	26.9
Nov	278	20.0	23.1	26.9
Dec	192	19.6	22.5	25.5
	1903	20.3	23.7	27.1

Passara

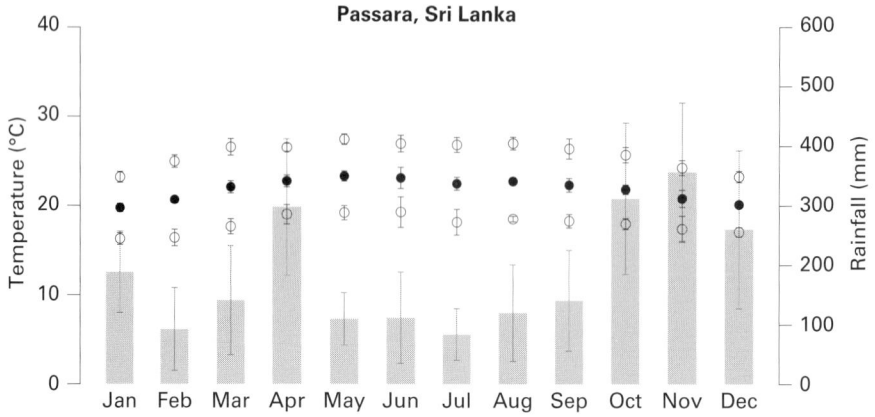

Passara, Sri Lanka

Latitude: 6.56 **Longitude:** 81.07

Altitude: 1120 m

Rainfall years included: 2000 2001 2002 2003 2004 2005 2006 2007 2008 2009 2010

Temperature years included: 2000 2001 2002 2003 2004 2005 2006 2007 2008 2009 2010

Month	Rainfall (mm)	T_{min} (°C)	T_{av} (°C)	T_{max} (°C)
Jan	189	16.3	19.8	23.2
Feb	93	16.4	20.7	25.0
Mar	141	17.7	22.1	26.6
Apr	297	19.0	22.8	26.5
May	110	19.2	23.3	27.4
Jun	112	19.3	23.1	27.0
Jul	84	18.1	22.5	26.8
Aug	120	18.5	22.8	27.0
Sep	140	18.3	22.3	26.4
Oct	312	18.0	21.8	25.7
Nov	357	17.4	20.8	24.3
Dec	260	17.0	20.1	23.3
	2215	17.9	21.8	25.8

Ratnapura

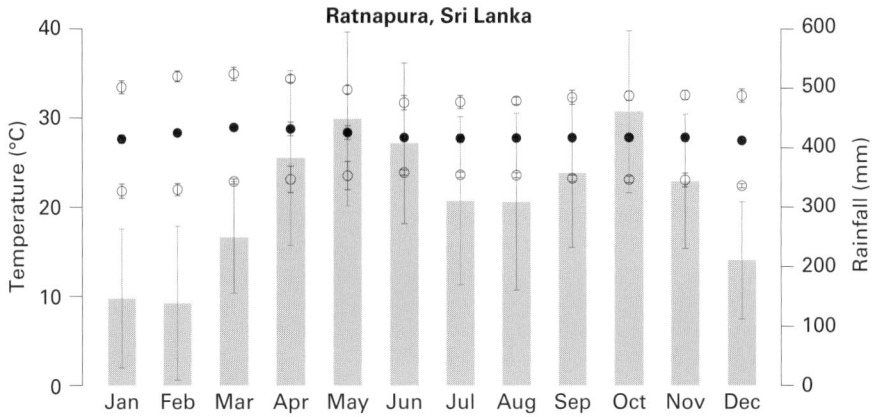

Ratnapura, Sri Lanka

Latitude: 6.41 **Longitude:** 80.4

Altitude: 29 m

Rainfall years included: 2000 2001 2002 2003 2004 2005 2006 2007 2008 2009 2010

Temperature years included: 2000 2001 2002 2003 2004 2005 2006 2007 2008 2009 2010

Month	Rainfall (mm)	T_{min} (°C)	T_{av} (°C)	T_{max} (°C)
Jan	146	21.8	27.6	33.4
Feb	139	22.0	28.3	34.7
Mar	250	22.9	28.9	34.9
Apr	383	23.1	28.8	34.4
May	449	23.6	28.4	33.2
Jun	408	23.9	27.8	31.7
Jul	311	23.6	27.7	31.8
Aug	309	23.6	27.7	31.9
Sep	358	23.2	27.8	32.3
Oct	461	23.1	27.8	32.5
Nov	343	23.1	27.8	32.6
Dec	211	22.4	27.5	32.5
	3768	23.0	28.0	33.0

Talawakelle

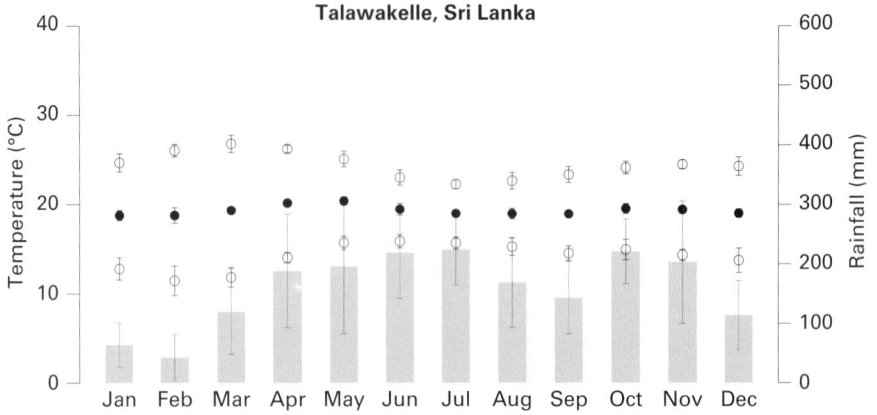

Talawakelle, Sri Lanka

Latitude: 6.54 **Longitude:** 80.42

Altitude: 1382 m

Rainfall years included: 2000 2001 2002 2003 2004 2005 2006 2007 2008 2009 2010

Temperature years included: 2000 2001 2002 2003 2004 2005 2006 2007 2008 2009 2010

Month	Rainfall (mm)	T_{min} (°C)	T_{av} (°C)	T_{max} (°C)
Jan	64	12.8	18.8	24.7
Feb	43	11.5	18.8	26.0
Mar	120	11.8	19.3	26.8
Apr	188	14.0	20.2	26.2
May	196	15.7	20.4	25.1
Jun	219	15.9	19.4	23.0
Jul	224	15.7	19.0	22.3
Aug	169	15.2	19.0	22.6
Sep	143	14.5	18.9	23.3
Oct	221	14.9	19.5	24.1
Nov	203	14.3	19.4	24.5
Dec	114	13.7	19.0	24.3
	1904	14.2	19.3	24.4

SOUTH-EAST ASIA

Annex 3.6: Indonesia

Bogor-Citeko

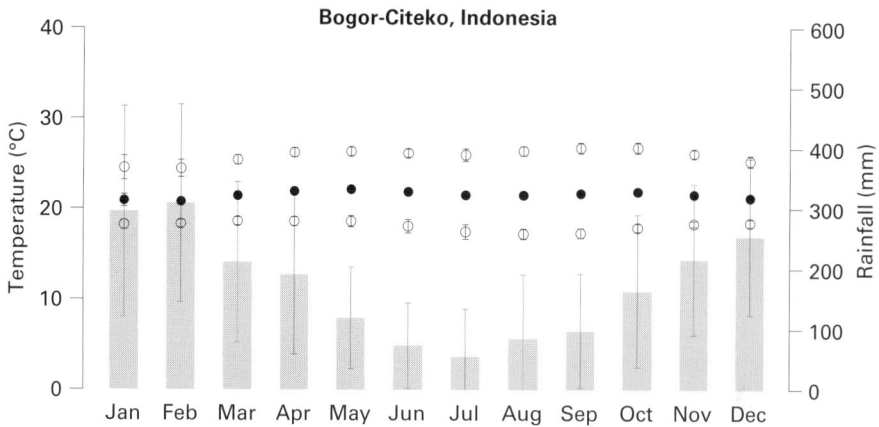

Bogor-Citeko, Indonesia

Latitude: −6.7 **Longitude:** 106.93

Altitude: 300 m

Rainfall years included: 2003 2004 2005 2006 2007 2008 2009 2010 2011 2012 2013 2014

Temperature years included: 2003 2004 2005 2006 2007 2008 2009 2010 2011 2012 2013 2014

Month	Rainfall (mm)	T_{min} (°C)	T_{av} (°C)	T_{max} (°C)
Jan	295	18.2	20.9	24.5
Feb	309	18.3	20.8	24.4
Mar	211	18.6	21.5	25.4
Apr	191	18.6	21.9	26.2
May	119	18.6	22.2	26.3
Jun	73	18.1	21.9	26.2
Jul	55	17.4	21.5	26.0
Aug	85	17.2	21.5	26.4
Sep	97	17.3	21.7	26.7
Oct	164	17.9	21.9	26.8
Nov	216	18.3	21.6	26.1
Dec	254	18.4	21.2	25.3
	2069	18.1	21.6	25.9

Annex 3.7: Myanmar

Taunggyi

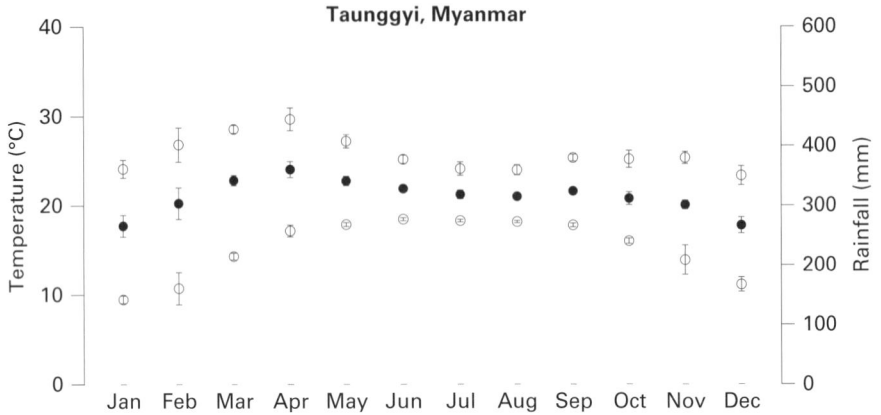

Latitude: 20.78 **Longitude:** 97.05
Altitude: 1436 m
Temperature years included: 2011 2012 2013 2014

Month	T_{min} (°C)	T_{av} (°C)	T_{max} (°C)
Jan	9.5	17.7	24.1
Feb	10.7	20.2	26.8
Mar	14.3	22.8	28.6
Apr	17.2	24.0	29.7
May	17.9	22.7	27.2
Jun	18.5	21.9	25.2
Jul	18.3	21.2	24.1
Aug	18.2	21.0	24.0
Sep	17.8	21.6	25.3
Oct	16.0	20.8	25.2
Nov	13.9	20.1	25.3
Dec	11.1	17.8	23.3
	15.3	21.0	25.7

Annex 3.8: Vietnam

Son La

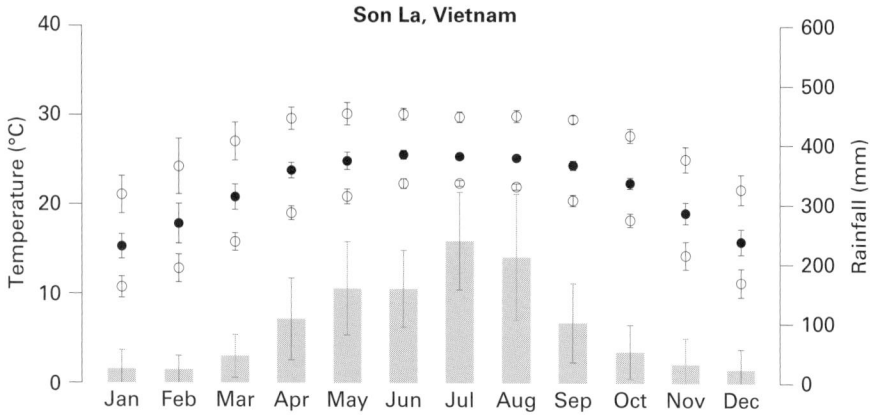

Son La, Vietnam

Latitude: 21.33 **Longitude:** 103.9
Altitude: 676 m
Rainfall years included: 1998 1999 2000 2001 2002 2003 2004 2005 2006 2007
2008 2009 2010 2011 2012 2013 2014
Temperature years included: 1998 1999 2000 2001 2002 2003 2004 2005 2006
2007 2008 2009 2010 2011 2012 2013 2014

Month	Rainfall (mm)	T_{min} (°C)	T_{av} (°C)	T_{max} (°C)
Jan	23	10.7	15.3	21.1
Feb	22	12.8	17.8	24.2
Mar	45	15.8	20.8	27.1
Apr	107	19.0	23.8	29.6
May	159	20.9	24.9	30.2
Jun	158	22.3	25.6	30.1
Jul	239	22.4	25.4	29.8
Aug	211	22.0	25.2	29.9
Sep	101	20.4	24.4	29.5
Oct	52	18.2	22.4	27.7
Nov	32	14.3	19.0	25.1
Dec	23	11.2	15.9	21.7
	1172	17.5	21.7	27.2

CAUCASUS

Annex 3.9: Georgia

Gori

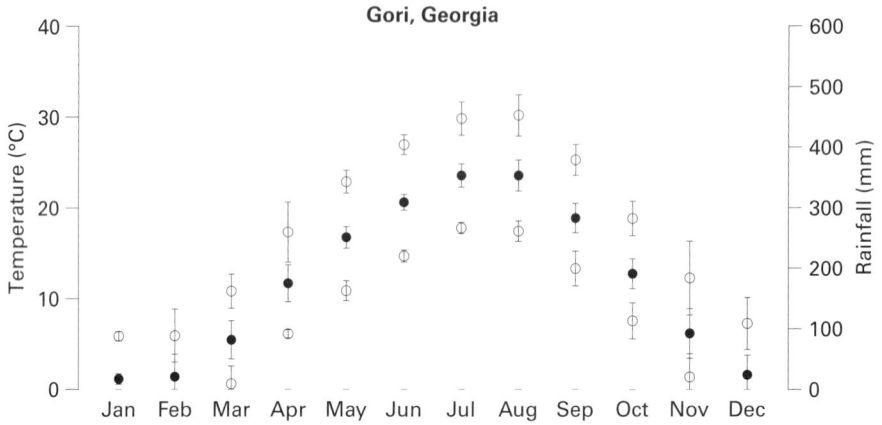

Latitude: 41.98 **Longitude:** 44.12
Altitude: 590 m
Temperature years included: 2010 2011 2012 2013 2014

Month	T_{min} (°C)	T_{av} (°C)	T_{max} (°C)
Jan	−2.7	1.2	5.9
Feb	−2.3	1.4	5.9
Mar	0.6	5.5	10.8
Apr	6.1	11.7	17.3
May	10.9	16.7	22.9
Jun	14.7	20.6	26.9
Jul	17.8	23.6	29.8
Aug	17.4	23.6	30.2
Sep	13.3	18.9	25.3
Oct	7.5	12.7	18.8
Nov	1.4	6.2	12.2
Dec	−3.0	1.6	7.3
	6.8	12.0	17.8

Tbilisi

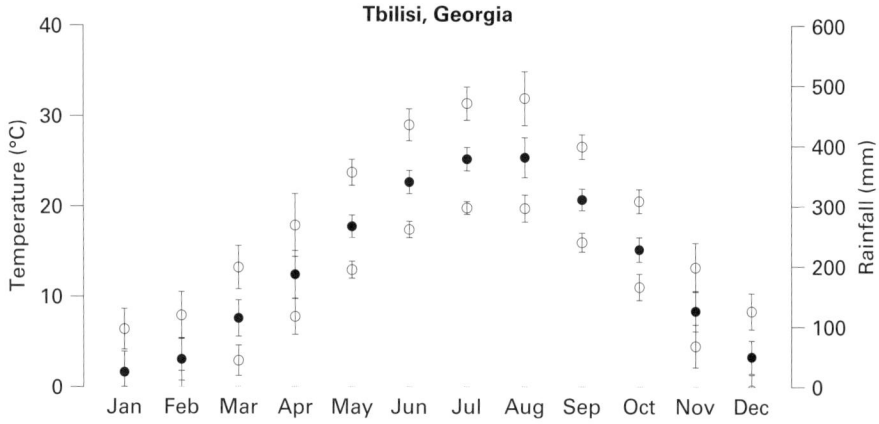

Tbilisi, Georgia

Latitude: 41.7 **Longitude:** 44.75
Altitude: 403 m
Temperature years included: 2006 2007 2008 2009 2010 2011 2012 2013

Month	T_{min} (°C)	T_{av} (°C)	T_{max} (°C)
Jan	−1.6	1.7	6.4
Feb	−0.6	3.1	7.9
Mar	2.9	7.6	13.2
Apr	7.8	12.4	17.9
May	13.0	17.8	23.7
Jun	17.4	22.7	29.0
Jul	19.8	25.2	31.3
Aug	19.7	25.3	31.9
Sep	16.0	20.7	26.5
Oct	11.0	15.2	20.5
Nov	4.5	8.4	13.2
Dec	−0.4	3.3	8.4
	9.1	13.6	19.2

Annex 3.10: Iran

Ramsar

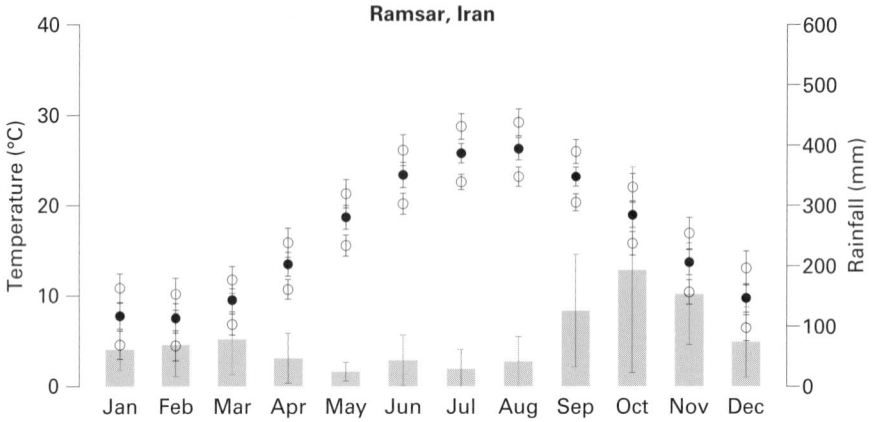

Ramsar, Iran

Latitude: 36.9 **Longitude:** 50.67
Altitude: –20 m
Rainfall years included: 1991 1992 1993 1994 1995 1996 1997 1998 1999 2000
    2001 2002 2003 2004 2005 2006 2007 2008 2009 2010 2011 2012 2013 2014
Temperature years included: 1991 1992 1993 1994 1995 1996 1997 1998 1999
    2000 2001 2002 2003 2004 2005 2006 2007 2008 2009 2010 2011 2012 2013
    2014

Month	Rainfall (mm)	T_{min} (°C)	T_{av} (°C)	T_{max} (°C)
Jan	61	4.6	7.8	10.9
Feb	69	4.5	7.5	10.2
Mar	78	6.8	9.5	11.8
Apr	46	10.7	13.5	15.9
May	24	15.6	18.7	21.3
Jun	43	20.2	23.4	26.1
Jul	29	22.6	25.8	28.7
Aug	41	23.2	26.3	29.2
Sep	126	20.3	23.2	26.0
Oct	193	15.8	18.9	22.0
Nov	153	10.4	13.7	16.9
Dec	73	6.5	9.8	13.1
	936	13.4	16.5	19.3

Annex 3.11: Turkey

Hopa

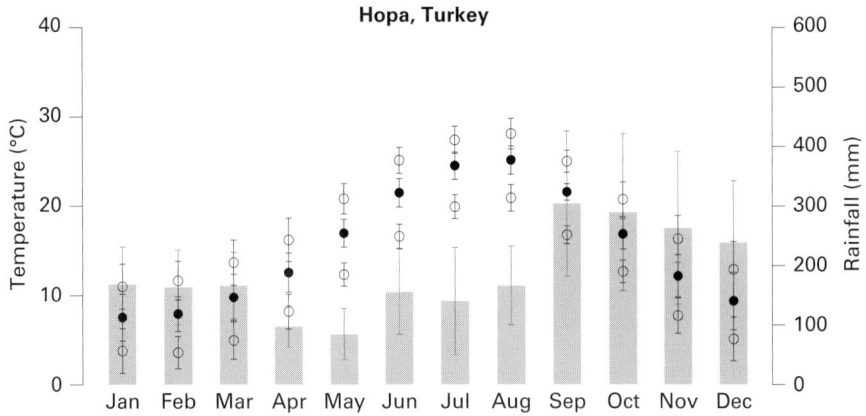

Hopa, Turkey

Latitude: 41.4 **Longitude:** 41.43

Altitude: 33 m

Rainfall years included: 1999 2000 2001 2002 2003 2004 2005 2006 2007 2008
2009 2010 2011 2012 2013 2014

Temperature years included: 1999 2000 2001 2002 2003 2004 2005 2006 2007
2008 2009 2010 2011 2012 2013 2014

Month	Rainfall (mm)	T_{min} (°C)	T_{av} (°C)	T_{max} (°C)
Jan	169	3.8	7.5	11.0
Feb	164	3.6	7.9	11.7
Mar	166	4.9	9.8	13.7
Apr	97	8.2	12.6	16.2
May	85	12.3	17.0	20.8
Jun	156	16.6	21.5	25.1
Jul	140	19.9	24.5	27.4
Aug	166	20.9	25.1	28.1
Sep	304	16.8	21.6	25.0
Oct	289	12.7	16.9	20.7
Nov	263	7.7	12.2	16.3
Dec	239	5.1	9.4	12.9
	2238	11.0	15.5	19.1

AFRICA

Annex 3.12: Kenya

Embu

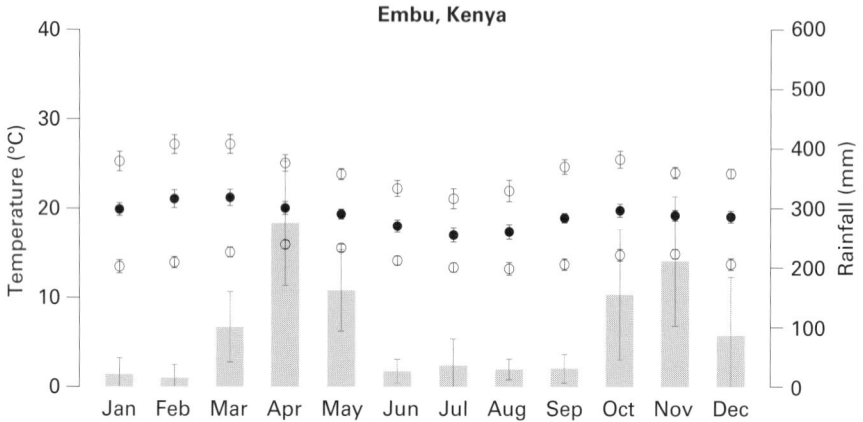

Latitude: –0.5 **Longitude:** 37.45
Altitude: 1493 m
Rainfall years included: 1981 1982 1983 1984 1985 1986 1987 1988 1989 1990
    1991 1992 1993 1994 1995 1996 1997 2001 2002 2006 2007 2011 2012 2013
    2014
Temperature years included: 1981 1982 1983 1984 1985 1986 1987 1988 1989
    1990 1991 1992 1993 1994 1995 1996 1997 2001 2002 2006 2007 2011 2012
    2013 2014

Month	Rainfall (mm)	T_{min} (°C)	T_{av} (°C)	T_{max} (°C)
Jan	21	13.5	19.9	25.3
Feb	15	13.9	21.1	27.2
Mar	100	15.1	21.2	27.2
Apr	274	15.9	20.0	25.1
May	162	15.5	19.3	23.8
Jun	26	14.1	18.0	22.2
Jul	35	13.3	17.0	21.1
Aug	29	13.2	17.3	21.9
Sep	31	13.7	18.9	24.6
Oct	155	14.7	19.8	25.5
Nov	211	14.9	19.2	24.0
Dec	87	13.7	19.1	23.9
	1146	14.3	19.2	24.3

Kericho

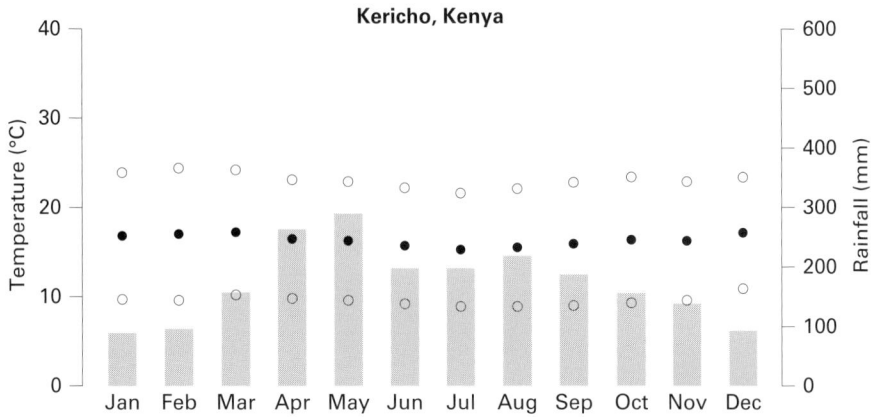

Kericho, Kenya

Latitude: −0.36 **Longitude:** 35.28

Altitude: 2178 m

Rainfall years included: 1987 1988 1989 1990 1991 1992 1993 1994 1995 1996 1997

Temperature years included: 1987 1988 1989 1990 1991 1992 1993 1994 1995 1996 1997

Month	Rainfall (mm)	T_{min} (°C)	T_{av} (°C)	T_{max} (°C)
Jan	89	9.7	16.8	23.9
Feb	96	9.6	17.0	24.4
Mar	157	10.2	17.2	24.2
Apr	263	9.8	16.4	23.1
May	290	9.6	16.2	22.9
Jun	198	9.2	15.7	22.2
Jul	198	8.9	15.2	21.6
Aug	218	8.9	15.5	22.1
Sep	187	9.0	15.9	22.8
Oct	156	9.3	16.4	23.4
Nov	138	9.6	16.2	22.9
Dec	93	10.9	17.1	23.4
	2083	9.6	16.3	23.1

Nyeri

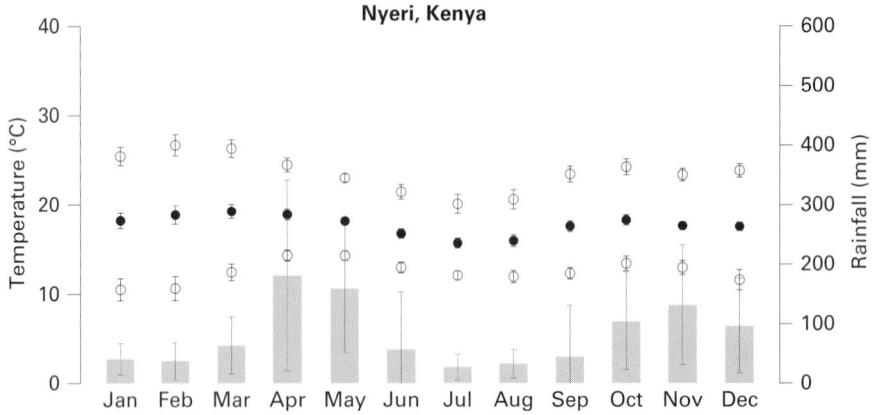

Nyeri, Kenya

Latitude: –0.43 **Longitude:** 36.97

Altitude: 1798 m

Rainfall years included: 1983 1984 1985 1986 1987 1988 1989 1990 1991 1992 1993 1994 1995 1996 1997 2001 2002 2006 2007 2008 2009 2010 2011 2012 2013 2014

Temperature years included: 1983 1984 1985 1986 1987 1988 1989 1990 1991 1992 1993 1994 1995 1996 1997 2001 2002 2006 2007 2008 2009 2010 2011 2012 2013 2014

Month	Rainfall (mm)	T_{min} (°C)	T_{av} (°C)	T_{max} (°C)
Jan	41	10.5	18.2	25.4
Feb	37	10.7	18.9	26.7
Mar	64	12.4	19.3	26.3
Apr	182	14.3	18.9	24.5
May	159	14.3	18.2	23.0
Jun	58	13.0	16.8	21.5
Jul	27	12.1	15.7	20.1
Aug	33	11.9	16.0	20.6
Sep	45	12.3	17.6	23.4
Oct	105	13.4	18.3	24.2
Nov	132	13.0	17.7	23.4
Dec	96	11.6	17.6	23.8
	979	12.5	17.8	23.6

Annex 3.13: Malawi

Mulanje

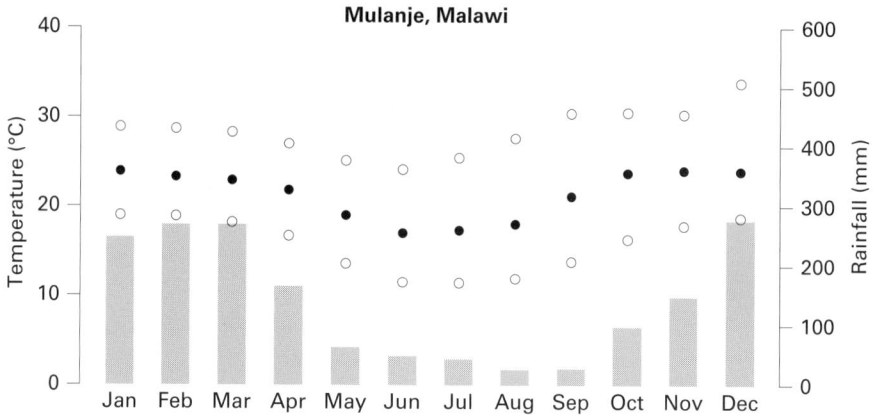

Latitude: −16.03 **Longitude:** 35.5
Altitude: 610 m
Rainfall years included: 1981 1982 1983 1984 1985 1986 1987 1988 1989 1990
Temperature years included: 1981 1982 1983 1984 1985 1986 1987 1988 1989 1990

Month	Rainfall (mm)	T_{min} (°C)	T_{av} (°C)	T_{max} (°C)
Jan	344	19.7	25.5	30.2
Feb	213	19.5	24.6	29.6
Mar	254	18.9	24.2	29.5
Apr	120	16.9	22.7	25.0
May	39	14.2	20.9	27.5
Jun	45	12.3	18.7	25.0
Jul	60	12.0	18.2	24.3
Aug	33	12.7	18.6	26.4
Sep	21	14.5	22.4	30.2
Oct	97	16.4	23.4	30.3
Nov	148	17.9	24.1	31.1
Dec	276	18.8	24.7	29.5
	1650	15.1	22.3	28.2

Annex 3.14: Tanzania

Marikitanda

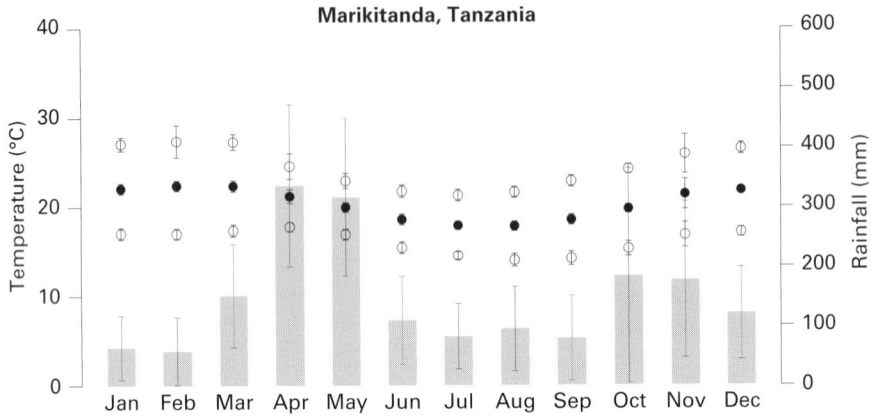

Latitude: –5.08 **Longitude:** 38.36
Altitude: 1050 m
Rainfall years included: 1967 1968 1969 1970 1971 1972 1973 1974 1977 1978
1979 1980 1981 1982 1983 1984 1985 1986 1987 1988 1989 1990 1991 1992
1993 1994 1995 1996 1997 2002 2003 2004 2005 2006 2007 2008 2009 2010
2011 2012 2013
Temperature years included: 1967 1968 1969 1970 1971 1972 1973 1974 1977
1978 1979 1980 1981 1982 1983 1984 1985 1986 1987 1988 1989 1990 1991
1992 1993 1994 1995 1996 1997 2002 2003 2004 2005 2006 2007 2008 2009
2010 2011 2012 2013

Month	Rainfall (mm)	T_{min} (°C)	T_{av} (°C)	T_{max} (°C)
Jan	63	17.0	22.0	27.1
Feb	58	17.0	22.4	27.4
Mar	151	17.4	22.3	27.3
Apr	336	17.8	21.2	24.5
May	316	16.9	19.9	22.9
Jun	109	15.4	18.6	21.7
Jul	82	14.5	17.9	21.3
Aug	95	14.0	17.8	21.6
Sep	79	14.2	18.5	22.8
Oct	184	15.3	19.8	24.2
Nov	177	16.8	21.4	25.9
Dec	120	17.1	21.8	26.5
	1770	16.1	20.3	24.4

Ngwazi

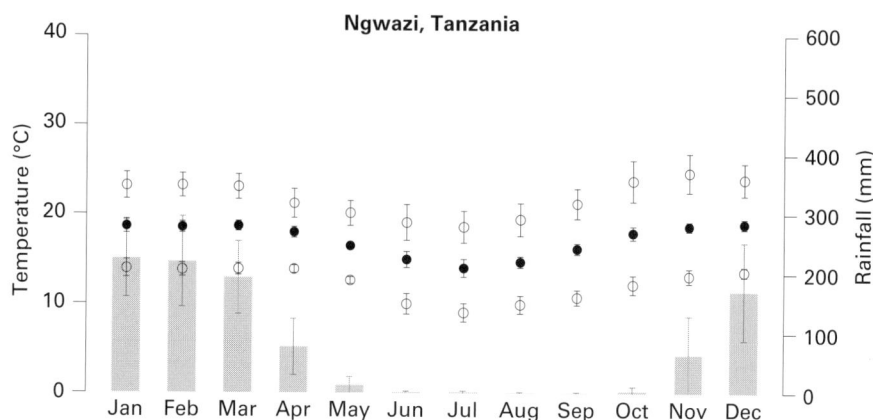

Ngwazi, Tanzania

Latitude: –8.3 **Longitude:** 35.05

Altitude: 1890 m

Rainfall years included: 1986 1987 1988 1989 1990 1991 1992 1993 1994 1995 1996 1997 1998 1999 2000 2001 2002 2003 2004 2005 2006 2007 2008 2009 2010 2011 2012 2013

Temperature years included: 1986 1987 1988 1989 1990 1991 1992 1993 1994 1995 1996 1997 1998 1999 2000 2001 2002 2003 2004 2005 2006 2007 2008 2009 2010 2011 2012 2013

Month	Rainfall (mm)	T_{min} (°C)	T_{av} (°C)	T_{max} (°C)
Jan	225	13.9	18.6	23.2
Feb	220	13.8	18.5	23.2
Mar	193	13.9	18.7	23.1
Apr	77	13.9	18.0	21.2
May	13	12.6	16.5	20.1
Jun	1	10.0	14.9	19.1
Jul	1	9.0	14.0	18.6
Aug	1	9.9	14.7	19.4
Sep	1	10.7	16.1	21.2
Oct	4	12.1	17.9	23.8
Nov	64	13.1	18.6	24.7
Dec	171	13.6	18.9	23.9
	971	12.2	17.1	21.8

Annex 3.15: Uganda

Fort Portal

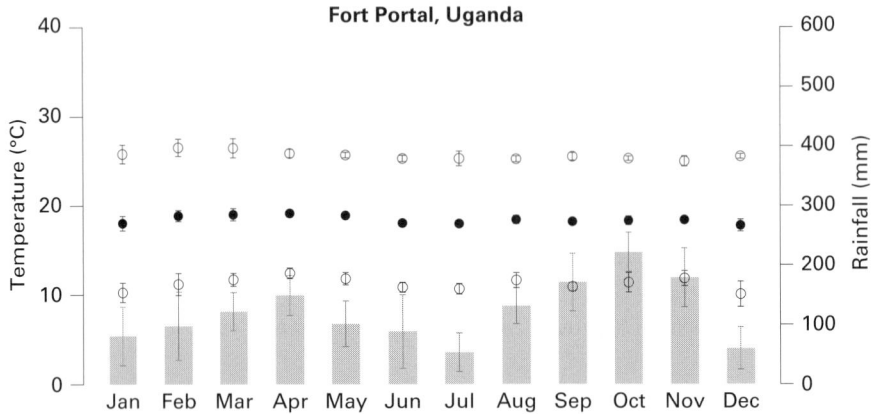

Fort Portal, Uganda

Latitude: 0.39 **Longitude:** 30.26
Altitude: 1460 m
Rainfall years included: 1972 1973 1974 1977 1978 1979
Temperature years included: 1972 1973 1974 1977 1978 1979

Month	Rainfall (mm)	T_{min} (°C)	T_{av} (°C)	T_{max} (°C)
Jan	81	10.3	18.1	25.8
Feb	98	11.2	18.9	26.5
Mar	122	11.7	19.0	26.5
Apr	149	12.4	19.2	25.9
May	102	11.8	18.9	25.7
Jun	89	10.8	18.1	25.3
Jul	54	10.7	18.0	25.3
Aug	132	11.6	18.4	25.2
Sep	171	10.9	18.2	25.5
Oct	222	11.4	18.3	25.3
Nov	179	11.8	18.4	25.0
Dec	60	10.1	17.8	25.5
	1459	11.2	18.4	25.6

Salama

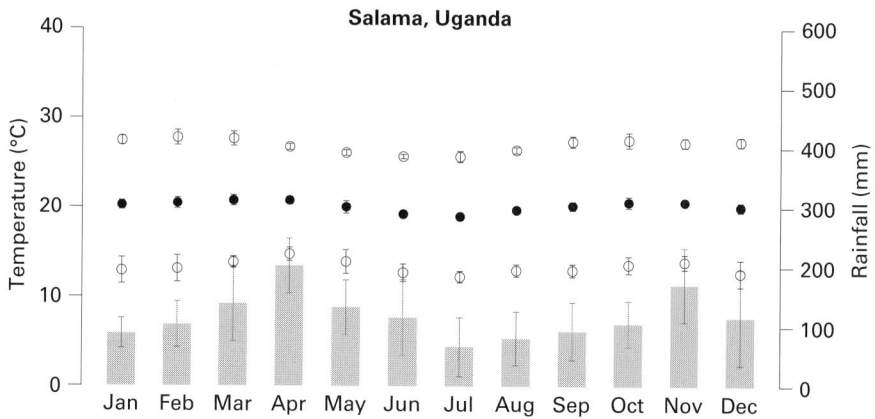

Latitude: 0.15 **Longitude:** 32.5
Altitude: 1256 m
Rainfall years included: 1968 1969 1970 1971 1972 1973 1974 1977 1978 1979
Temperature years included: 1968 1969 1970 1971 1972 1973 1974 1977 1978
   1979

Month	Rainfall (mm)	T_{min} (°C)	T_{av} (°C)	T_{max} (°C)
Jan	88	12.9	20.2	27.5
Feb	104	13.1	20.5	27.8
Mar	139	13.9	20.8	27.7
Apr	202	14.8	20.8	26.8
May	132	13.9	20.1	26.1
Jun	116	12.7	19.2	25.7
Jul	66	12.2	19.0	25.7
Aug	80	13.0	19.7	26.4
Sep	93	13.0	20.2	27.4
Oct	105	13.6	20.6	27.6
Nov	170	13.9	20.6	27.3
Dec	116	12.7	20.0	27.4
	1411	13.3	20.1	26.9

Annex 3.16: Zimbabwe

Chipinge

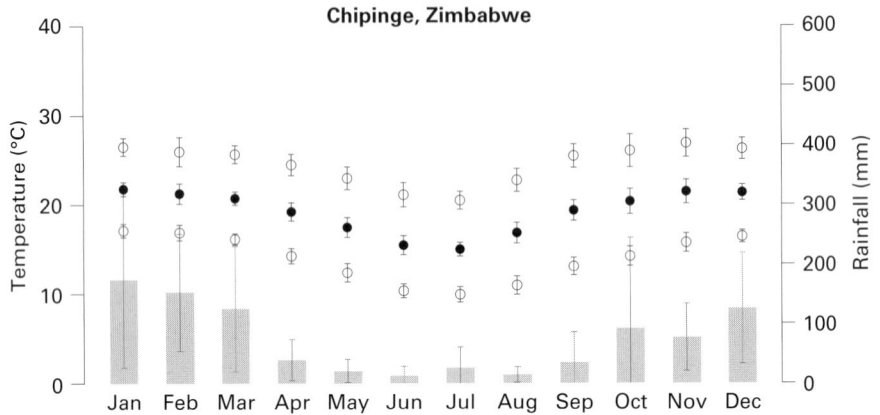

Chipinge, Zimbabwe

Latitude: −20.2 **Longitude:** 32.62
Altitude: 1132 m
Rainfall years included: 1978 1979 1980 1981 1982 1983 1984 1988 1990 1991
    1992 1993 1994 1995 1996 1997 1998 1999 2002 2012 2013
Temperature years included: 1978 1979 1980 1981 1982 1983 1984 1988 1990
    1991 1992 1993 1994 1995 1996 1997 1998 1999 2002 2012 2013

Month	Rainfall (mm)	T_{min} (°C)	T_{av} (°C)	T_{max} (°C)
Jan	174	17.1	21.8	26.5
Feb	153	16.9	21.2	25.9
Mar	125	16.1	20.7	25.6
Apr	39	14.3	19.2	24.5
May	21	12.4	17.5	22.9
Jun	13	10.3	15.5	21.1
Jul	26	9.9	15.0	20.5
Aug	14	10.9	16.8	22.7
Sep	35	13.1	19.3	25.4
Oct	91	14.2	20.3	26.0
Nov	77	15.7	21.4	26.9
Dec	126	16.4	21.3	26.2
	894	13.9	19.2	24.5

4 Taxonomic Delight

Only Plant the Best!

This chapter provides an overview of the ways in which the tea plant has been adapted to cultivation through selection from heterogeneous seedling populations, selective breeding and vegetative propagation to produce genotypes with improved yield and tea-making properties. It begins with a summary of the debate on the taxonomy of tea and ends with the question of how and why the genetic diversity should be retained when the commercial emphasis is on planting clones from the often limited number available.

Taxonomy

As mentioned in Chapter 2, the tea plant is believed to have originated in Yunnan Province in south-west China, where there is an abundance of genetic material belonging to *Camellia* L. Sect. *Thea* (L.) Dyer. (Sect. *Thea* is one of 12 sections in Sealy's (1958) *Camellia* classification system.) Although there is still a great deal of controversy/confusion with respect to the taxonomy of tea (Banerjee, 1992a), it is generally believed that existing populations of tea plants are largely derived from two original taxa. These are given varietal status within the species *Camellia sinensis* by Sealy (1958) but specific status by Wight (1962). Wight refers to these two species as *C. sinensis* (the China plant, which is a shrub, with small leaves, thought originally to have grown in the open) and *C. assamica* (the Assam plant, which is a small tree, with large leaves, thought originally to have grown in forest) (Figure 4.1). Leaf form and floral characters distinguish the two taxa (Wight, 1959a; Banerjee, 1992a). In addition, a third type, known as *Camellia assamica* ssp. *lasiocalyx* (Watt) Wight, is believed to have had an impact on the genetic makeup of cultivated tea types. This was introduced into India from South-East Asia at the beginning of the twentieth century (Singh, 2006). A small tree with a fastigiate growth habit, and broadly elliptic, yellow/light green leaves, ssp. *lasiocalyx* is similar in appearance to the Assam plant.

For convenience and simplicity, the classification proposed by Sealy is used in this book. All three types are classified under the name *C. sinensis* (L.) O. Kuntze irrespective of taxonomic variation (Paul *et al.*, 1997). As they are believed to be indigenous to China, Assam and South-East Asia, the three are referred to as China-type, Assam-type and Cambod-type, respectively (Raina *et al.*, 2012). They are largely self-sterile, so cross-breeding is normal, resulting in a cline extending in appearance from extreme

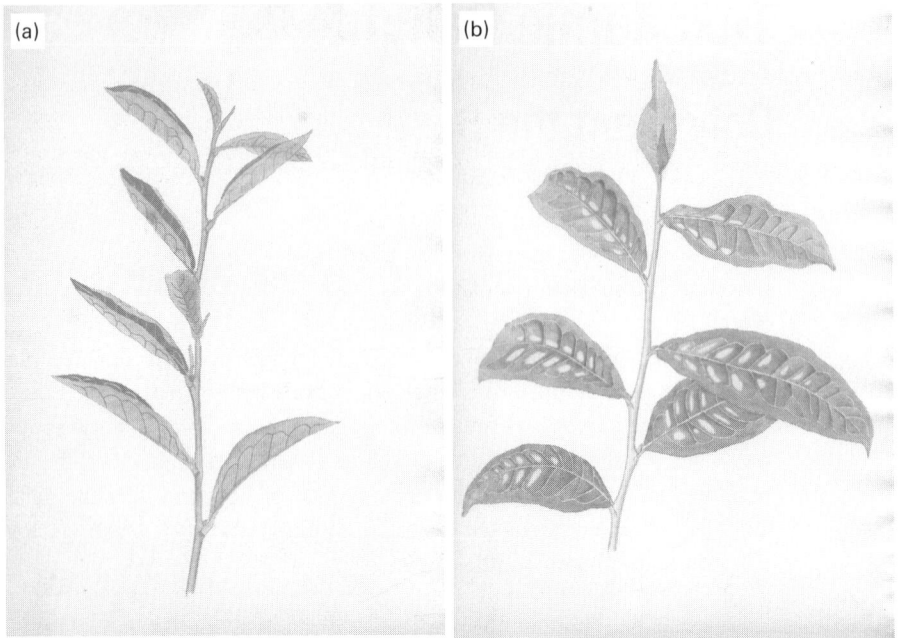

Figure 4.1 Tea shoots (*Camellia sinensis* var. *assamica*). Paintings by Mrs W. Wight, 1950. *A black and white version of this figure will appear in some formats. For the colour version, please refer to the plate section.*

'China-type' plants to those with a distinct 'Assam-type' appearance (Wight, 1959a). Individual plants of each subspecies will freely interbreed.

Diversity

Tea seed from China (*C. sinensis* var. *sinensis*) continued to be imported into India for many years after *C. sinensis* var. *assamica* had been identified in the wild in Assam. The seed from China was used to establish tea in Assam, although much of the seed was of inferior quality. When imports of tea seed from China stopped, the demand for seed was so great (as the tea industry in India and elsewhere expanded rapidly) that seed from any source was marketable. Interbreeding occurred between var. *assamica* and var. *sinensis* and later with ssp. *lasiocalyx*. Until relatively recently, most of the world's tea consisted of plants of uncertain genetic origin grown from seedlings from unknown sources.

In a very short time, hybridisation between the three types resulted in tea gardens with plants with physical characteristics ranging from pure China-type to pure Assam-type. Only in isolated locations was it possible to produce seed that was ostensibly pure *C. assamica* (Green, 1961). Tea seed of mixed origins (hybrids) was subsequently exported from Assam all over the world and formed the basis of tea industries in

---

Box 4.1 Commentary: Selection of a Mother Bush

I am not sure that the numbers stack up. If 25% of the bushes produce, say, 45% of the crop, then simply by propagating from that 25% the implication is that you would increase the yield of seedling tea by 80%: from, say, 3000 kg ha^{-1} to 5200 kg ha^{-1}. If only it were that simple. If it were the case, then selecting from the best of the 25% (assuming normal distribution) would double or treble yields of seedling tea. That may now be becoming possible, but only after selective breeding, not from selection from a wild seedling population.

T. C. E. Congdon

---

many countries, including Kenya and Sri Lanka. It also formed the basic stock from which tea clones were (and still are) selected in these countries. According to Ellis (1997), the most important event in the origin and evolution of commercial tea was the creation of Indian hybrid tea. Because of the longevity of a tea bush (>100 years) many of these bushes grown from hybrid seed of unknown parentage are still in production.

The populations of tea raised from seeds produced in a particular orchard (*bari*) had certain distinctive characteristics and became known as *jats*. This name, however, has no botanical significance, since the name given to a jat usually referred to the geographical location of the orchard/bari or to the source of the original seed (Bezbaruah and Dutta, 1977).

As a result of the interbreeding, there was great inter-bush variability in many visible physical attributes, including leaf size, posture, colour and shoot population densities. Similar diversity between bushes in the cup characteristics of the processed tea must also exist. It has been known for a long time that a small proportion of the hybrid bushes produce a large proportion of the yield. For example, Visser (1969) cited the results of studies by Wellensiek (1934) in Indonesia and by Tubbs (1938) in Ceylon. These showed that the best 25% of a seedling population yielded 44–47% of the total yield obtained from randomly selected sample areas containing 3800 and 1500 plants respectively. Similarly, the worst 25% (labelled 'passengers') yielded only 7–10% of the total. In the same way, in Kenya, the best 90 bushes selected visually for yield gave nearly four times the yields of the poorest (Green, 1971). This might be an overestimate, as the more vigorous 25% of the plants may be covering more than 25% of the ground (Box 4.1). This variability between bushes is clearly inefficient in management terms, as each bush receives the same quantity of fertiliser (in theory, anyway) and each receives the same amount of management time.

Vegetative Propagation

The first recorded attempts to propagate tea from vegetative cuttings were made in Indonesia at the Bogor Research Institute, Java, during the 1920s and 1930s.

Figure 4.2 Leaf bud cutting (MKVC). The first recorded attempts to propagate tea from vegetative cuttings were made in Indonesia during the 1920s, but it was not until 1952–1957 that clones were planted on a large scale.

A promising clone or cultivar (labelled Pasir Saronge No. 1) was soon identified and being evaluated commercially. Following a visit to Java by scientists from India, similar work was initiated at the Indian Tea Association's Tocklai Experimental Station in Assam (Tunstall, 1931). The Second World War then intervened, but research on vegetative propagation was resumed in 1946 (Wight, 1956). The first challenge for the researchers was to answer several fundamental questions. For example: What makes a successful cutting? When is the best time of the year to take cuttings? What is a suitable medium for facilitating the initiation and development of roots? How much shade is needed to protect the cutting from drying out? How and when should the shade be reduced as the plants develop?

Following a comprehensive study at Tocklai, answers to these questions were obtained and a practical methodology for propagation was developed. Single internode cuttings (Figure 4.2) were identified as being suitable for propagation, although there was plant-to-plant variability in the capacity of cuttings taken from them to develop roots (this became a criterion for the selection of a mother bush[1]); an acid subsoil proved to be a suitable rooting medium (topsoil inhibits root formation; a peat–subsoil mixture is an even better rooting medium); a woven bamboo frame with a bamboo cover (for shade) helped to protect cuttings from drying out and could be gradually raised to harden off the plants as they developed. Green (1964) described this vegetative propagation system in detail in a pamphlet intended primarily for tea growers in East Africa, while Sharma (1982) reviewed the results of research on the vegetative propagation of tea in South India.

Thus a procedure for propagation was developed, and the first clonal tea estate (Hatimara) was planted in Assam with three clones (chosen from a potential source of 8 million seedling bushes!) between 1952 and 1957.

Clone Selection

An associated challenge for researchers was to identify tea bushes (mother bushes) that had exceptional yield and tea-making qualities, and would be easy to propagate. As mentioned, there were considerable differences between individual plants grown from seed in the capacity of the leaf-bud cuttings to grow roots. In the field, promising bushes were identified using criteria such as: bush size or surface area, shoot density, pruning weight, shoot size, leaf pose, branching angle; and for processing: pubescence (preference for hairs on the abaxial surface of the leaf[2]), leaf colour (preference for light green leaves) and the fermentability of the leaf (using the chloroform test). A selection procedure recommended for use in India is shown in Box 4.2.

Similar procedures were recommended in southern Africa, with additional criteria such as bush resistance to heat and drought, ease of harvesting by hand (pluckability) or machine, and pest and disease resistance (Box 4.3).

Box 4.2 Clonal Selection Procedure: an Example from India

Field selection in India is done in the following stages, using one rejection criterion at a time (from Singh, 2006):

1. Reject any diseased bushes and those close to footpaths/roads.
2. Select for density of plucking points. Retain about 5% of bushes, i.e. 2000 out of a sample size of 40,000 bushes.
3. Select for pubescence (depends on type of tea to be produced). Retain 1000–500 bushes.
4. Prune the selected bushes, recording the weight of each bush. Reject those with below the average weight of all the pruned bushes. Retain 400–300 bushes. But at the risk of rejecting clones with a high harvest index.
5. Estimate the pruned surface area of each bush, and calculate the weight of prunings per unit area. Reject those bushes that are below the average. Retain 200–100 bushes.
6. Allow the majority of shoots to grow to tipping height and then record the weight of 'tippings'. Reject those that are slow to recover from pruning. Retain 100–50 bushes. (Can select for leaf size at same time if necessary.)
7. Take cuttings from selected bushes and conduct a replicated rooting trial in nursery beds. Select the best 40–20.
8. Bring selected mother bushes into bearing and undertake quality evaluation on at least eight occasions. Retain 20–10 bushes with above-average quality.
9. Take cuttings from selected mother bushes, observe rooting success, and establish a clonal field trial.
10. Establish long-term trials in different ecological areas. Include two or three recognised clones as controls.

It can take between 10 and 15 years to select an elite clone using this methodology.

Box 4.3 Clone Evaluation Criteria: a Case Study from Malawi

The selection procedure employed in Malawi aims at producing clones that have a high quality potential and good yields, and that are easy to manage in the nursery, field and factory (Ellis and Nyirenda, 1995). In this example, the three clones (TRFCA-PC122, TRFCA-PC123 and TRFCA-PC131) identified as suitable for release to the industry were the result of controlled cross-pollination between two superior clones. TRFCA-PC122 and TRFCA-PC123 had been selected from the progeny of TRFCA-PC1 × TRFCA-SFS204, and TRFCA-PC131 was selected from TRFCA-M9 × TRFCA-SFS204 crosses. These clones (and others) were compared with two controls (TRFFCA-PC81 and/or TRFCA-PC108) in the following ways:

1. *Oxidation* (corresponds to 'fermentation'):
   a. Original source bushes are screened using the chloroform test.
   b. At an advanced stage in the selection process the actual fermentation time of the macerated leaf obtained from a mini-processing unit (MPU) is determined in the factory (in this example, control = 70 minutes, others = 45 minutes).
2. *Cup quality*: processed leaf from the MPU is divided into two samples:
   a. One is sent to the tea brokers for organoleptic assessment and valuation, and
   b. one is analysed for theaflavin (TF) content in the research station laboratory (over three seasons).
   In addition:
   c. Shoots from source bushes are tested for TF content using the sand grinding grill method.
3. *Nursery*: rooting of cuttings in nursery must be > 85% success (actual, all > 93%).
4. *Field survival*: new selections are screened in the field for survival with and without irrigation and in different locations. If survival rate is below 90%, or below that of the control, the clone is rejected (actual, all > 96%).
5. Growth attributes:
   a. Rate and uniformity of recovery from prune.
   b. Low base temperature (as reflected by growth in the cold season).
   c. Drought tolerance (by observation).
   d. Sun-scorch damage (by observation, but it can be quantified: e.g. on a scale 0–10, clone TRFCA-PC81 is highly susceptible).
6. Ease of harvest:
   a. The internode between the third and fourth leaves should be soft, and not lignified, thereby making the shoots easier to harvest. This also increases the proportion of fine grades and reduces the fibre content in the processed tea.
   b. To differentiate shoots that are ready for harvesting from those that are not, the fourth and fifth leaves on a shoot should be nearly horizontal. This distinguishes them from the upper, younger shoots, which are more nearly upright. For example, the angle to the vertical of these leaves varied between 63° (for TRFCA-PC131) and 36.8° (TRFCA-PC81).
7. *Yield:* the average annual yield and total value after the bringing into production phase is completed.

Clones are the most efficient way of exploiting genetic variance. That is why plant breeders have successfully pursued clonal selection strategies in most of the perennial crops that can be mass-produced by conventional methods (Lockwood, 1999). The clone selection process for tea is slow. It can take a long time, typically 10–15 years, to identify a superior clone. The selection criteria for the mother bush assume that the clones of what might appear to be a vigorous bush (grown from seed) in the field will have the same attributes as the mother bush when competing with neighbouring bushes that are genetically identical. This is not always the case. Soil heterogeneity may account for yield differences of 100% between clonal bushes (Visser, 1969). Similarly, using criteria based on the growth and development of seedlings in the nursery was, at best, of limited value in the selection of clones that would be high yielding in the field (Green, 1971).

Breeding

The chances of identifying a superior clone are increased if selections are made from seedlings developed from seed produced as a result of cross-pollination between two interplanted outstanding clones (biclonal seed), or from cross-pollination between two high-performance clones from an orchard containing from three to nine clones (polyclonal seed) (Figure 4.3). Since tea is largely self-sterile, and is mainly cross-pollinated by small insects with a short flying range, a 30–50 m barrier between seed orchards or between seed orchards and tea in plucking is recommended in Malawi to protect the plants from unwanted pollen (TRFCA, 1990, Section 4.2, *Seed gardens*). In Malawi, a polyclonal seedling jat (labelled JatCJ1) out-yielded the best available seedling material at the time (Magambo), as well as the improved Assam-type jats Ramjaat and Betjan in an irrigated field trial over 11 years (up to 1978). It also out-yielded the mean of eight clones selected for quality, but yielded less than the highest-yielding clone in each year. Mean annual yields over this period were between 3100 kg ha^{-1} (Magambo) and 4300 kg ha^{-1} (best clone; s.e. ± 250). In a follow-up unirrigated trial, two biclonal jats compared favourably in yield with the two leading clones (TRFCA-SFS150 and TRFCA-SFS204) at the time (1980s), with annual yields averaging between 2000 and 2450 kg ha^{-1} (Ellis and Nyirenda, 1995).

Lockwood (1999) recommended a weighted index for the selection of clones, taking into account if possible the cost structure of the industry and the value of the processed crop in the market. In other words, there should be an economic basis to selection, and there should be transparency in the choice of selection criteria and the weighting given to individual characteristics. Cost structures will vary with management systems and over time. Breeders need to be aware of dynamic factors such as these when selecting suitable clones, especially as the tea bushes are likely to be in the ground for 50 years or more (a case for genetic modification, according to Lockwood, 1999). Based on the assumption that only 1 in 200 bushes had adequate yield and 1 in 200 had adequate quality, Wight (1956) estimated the chances of selecting an outstanding clone from field observation and testing to be 1 in 40,000 plants. According to Singh (2006), who

Figure 4.3 The chances of identifying a superior clone are increased if selections are made from seedlings developed from polyclonal seed, or better, from biclonal seed. Photograph shows a polyclonal seed orchard in western Uganda (MKVC).

claimed to have identified 110 clones tolerant of drought and five tolerant of water-logging from a survey of 7015 ha of tea, the ratio is 1 in 80,000. It is not clear how this ratio was derived. Nevertheless, the odds against successfully identifying an outstanding clone from field selection are obviously long.

To select for diverse conditions, the breeder needs to take a view on how and where the tea industry is likely to develop over the next 20 years (easier said than done), and whether to select for low-input systems (e.g. rain-fed, little fertiliser) or high-input systems (e.g. irrigated, unlimited fertiliser). Whatever the conditions, it is essential that selection experiments are managed to the highest possible standards, consistent with the prospective management regime, and that all the leaf is harvested at the optimum time intervals. Ideally, every clone should be harvested on the correct day for that clone (which depends on its base temperature for leaf emergence), but for management reasons all the clones are usually harvested on the same day, which is a compromise leading to variability in shoot-size distribution between clones. According to Lockwood (1999), all comparisons should be against recognised standard clones, with a minimum of three standards for each agroecological area and husbandry regime. Selection amongst promising but as yet unproven clones should be a priority in the short term (Box 4.4). According to Simmonds (1996), the efficiency of selection can be improved by focusing selection on superior families, although this is not normal practice.

> **Box 4.4** Suggested Procedure for Evaluating Semi-Proven Clones (after Lockwood, 1999)
>
> - Establish selection criteria and weightings to give an economic selection index.
> - Identify standard (control) clones.
> - Create an inventory of all semi-proven clones in the country/region.
> - Shortlist any candidates completing year 2 in tea breeding trials.
> - Source clones (completing year 7 or 8) from neighbouring countries/regions.
> - Agree sites (environments) and crop husbandry regimes for clone trials.
> - Establish advanced (year 8 and 9) trials for all clones in commercial use and not yet outclassed.
> - Establish year 4 and 5 trials of unproven clones that pass quality and other tests.

Genotype × Environment Interactions (G × E)

In countries where there is great diversity in soils and climate the likelihood of genotype × environment (G × E) interactions is large. By this is meant that, for example, the ranking order for a selection of clones at one site may not be the same as at another site. For example, in Sri Lanka tea growing extends from sea level to altitudes in excess of 2000 m, and from exceptionally wet areas (Talawakelle) to relatively very dry regions (Uva). Wickramaratne (1981a) recorded significant G × E interactions in annual yields from 31 clones grown in replicated trials at four sites ranging in altitude from 30 m to 1372 m. Subsequently, again in Sri Lanka, Balasuriya (1998) compared the yields of two clones grown at six different altitudes from 30 m to 1859 m. Clone TRISL-2023 was better suited to the warmer conditions at low altitudes, yielding more than TRISL-2025, but at high altitude the yields of the two clones were similar. In Kenya, where tea is also grown at a range of altitudes, Rono *et al.* (1993) and Ng'etich and Stephens (2001) in separate studies both reported significant interactions between yield from individual clones and altitude (a surrogate for temperature). In Tanzania, G × E interactions were observed between the yield of different clones and the degree of water stress imposed (Burgess and Carr, 1996a).

In a study lasting six years, Wachira *et al.* (2002) used several statistical methods to evaluate morphotypic stability in 20 tea genotypes planted in two locations east and west of the Great Rift Valley in Kenya. A standard multi-factor analysis of variance (ANOVA) test revealed that for annual yields all first-order and some second-order interactions between G and E were significant. Regression analysis (Finlay–Wilkinson model) in which yields for an individual clone were plotted against the environmental mean (represented by the mean yield for that year and site) confirmed that there was a G × E interaction. The three genotypes that performed best were AHP-S15/10, TRIEA-31/8 and BBK-35. The most stable genotypes across environments were clones BBK-35, TRFK-54/40 and TRFK-303/577 (slopes of regression were 1.11, 1.09 and 1.06 respectively). By contrast, AHP-S15/10 and TRIEA-31/8 performed relatively worse in the less favourable conditions (slopes were 1.20 and 1.25). Perhaps in the case of AHP-S15/10 it was the slow recovery from a prune that made it appear to perform relatively less well in low-yielding environments. The most stable across different environments were genotypes STC-5/3 and -11/26 (slopes of 0.80 and 0.78). The G × E analysis was complicated by the inconsistency in yield responses from year to year as the young tea plants matured. This highlights the need to test clones at multiple sites over longer periods of time if meaningful results are to be obtained. There is also the probability that there will be an interaction with the harvesting practices used. G × E interactions are only a problem when clones behave unpredictably in different environments. Otherwise, they offer opportunities that can be exploited.

Composite Plants

A further refinement to crop improvement is the composite plant in which, for example, a high-quality scion is grafted onto a vigorous, drought-resistant rootstock. There can be practical problems associated with aligning the cambium of the scion and rootstock at the time when the scion is being inserted and secured to the rootstock. This can limit the commercial uptake of this practice (Nyirenda and Mphangwe, 2000a). As in the field, mechanisation can help in the nursery. In order to increase the rate of production of composite plants, two simple mechanical tools known as the scion maker and rootstock maker have been developed by UPASI (the United Planters Association of Southern India). By using these tools the number of composite plants created in a day can be increased fourfold compared with manual production. For securing the graft partners at the graft interface, UPASI has also developed durable grafting clips.

Germplasm Collection

In addition to the tea clones released by research stations to the industry and grown commercially, tea germplasm has been collected and conserved in a number of countries, in particular in China and India.

Germplasm Collections and Clones by Region

China

Germplasm
Beginning in the 1930s, but particularly during the last two decades, tea germplasm in China has been collected and conserved (Liang *et al.*, 2012). Between 1981 and 1984, wild tea plants (the majority of which were classified as *Camellia crassicolumna* Chang (19 accessions), *C. tachangensis* F. C. Zhang (21 accessions) and *C. taliensis*

(W. W. Smith 123 accessions)), as well as new species were discovered in Yunnan. Later, new accessions were collected from the Shennongjia and Three Gorges area (before it was flooded), where 80 landraces were collected and preserved safely. Other areas investigated included Hainan, Guizhou, Guangxi, Sichuan and Shaanxi provinces, where 400 accessions were collected. Elsewhere in China, a further 396 landraces and cultivars were identified during the early 1990s, and by the end of 2010 a total of nearly 3000 accessions had been conserved in the China National Germplasm Tea Repository. These were planted in the field and also stored *in vitro*. Classifications are based on the taxonomic system proposed by Chen *et al.* (2000).[3]

A database developed by Chen *et al.* (2005) was used to describe and compare this collection of germplasm. There was considerable variability, linked in part to the region of origin of the accession. Caffeine contents of the young shoots varied, for example, between a low of 1.2% in one genotype and a high of 5.9% in another. Similarly, polyphenol contents varied from 13.6 to 47.8%, catechin contents from 82 to 263 g kg^{-1}, and amino acids from 1.1 to 6.5% (Chen *et al.*, 2012).

Clones

There are 123 national and more than 160 provincial registered tea cultivars in China, many of which have been bred by hybridisation and systematic selection. Yu (2012) has listed the cultivars that are suitable for the manufacture of particular types of tea. The most famous clones are Longjing 43 and Longjing Changye. An excellent black tea clone is Yunkang 10. There is also a temperature-sensitive, high-quality mutant in which the leaves of young shoots become white in the spring (at temperatures between 19 and 22 °C) and then turn green at temperatures of 25 °C and above (Chen *et al.*, 2012). There are still many wild tea plants in the mountain regions that have not yet been collected, and these risk extinction. It may be necessary to begin to analyse the relationships between molecular markers and important agronomic traits. Field observation is still important.

India

Germplasm

The Tocklai Tea Research Institute in Assam claims to have the largest single collection of *Camellia* germplasm in the world (Singh, 2006). There are 14 species of *Camellia* and 2532 accessions maintained at three main centres, Jorhat, Nagrakata and Darjeeling (Sharma, 2012). In addition, there are 440 accessions at the UPASI Tea Research Institute in South India (of which 63 are biclonal crosses of elite clones), 350 at the Tea Research Institute of Sri Lanka, and 320 at the Bangladesh Tea Research Institute (Sharma, 2012). Singh (2006) emphasised the need to conserve the diverse genetic base that currently exists in seedling tea fields in India and elsewhere (before they are uprooted and replanted), and to register all new cultivars (Sharma, 2012).

In India, Raina *et al.* (2012) screened 1644 accessions and clones of Indian hybrid tea, using 412 AFLP (amplified fragment length polymorphic DNA) markers amplified by seven AFLP primer pair combinations. Each genotype was individually classified as one of 15 distinct morphotypes (i.e. they were labelled, for example, as either Assam-type or China-type, or Assam hybrid or China hybrid or Cambodian hybrid, based on their physical appearance). The analysis showed clearly that accessions and clones of the same morphotype did not necessarily have the same genetic ancestry. In addition, accessions and clones from outside North-East India shared the same genetic groups (there were six in total) and clusters (16) as those from within that region. No one group was exclusive to one particular phenotype. This detailed study also demonstrated the limited genetic diversity in commercial clones compared with the great diversity existing within the tea accessions, which awaits exploitation.

Clones

In India, clones were initially selected from seedling populations in the field, and later from particular jats. Subsequent selections were made from crosses between individually selected superior clones. The shift away from field selections to family-level selection was made because it is extremely difficult to identify superior individuals in large populations, where environmental variation has an enormous influence on the performance of each bush (Smith, 1998).

The first clone to be promoted to the industry from Tocklai was 19-29-13 (stock 19, row 29, bush 13). Later renamed TV2, it was released in the 1950s. A total of 30 clones has been recommended since then by Tocklai, all labelled with the prefix TV. Of these, TV1 and TV21 are of high quality with a yield potential of 2500–2800 kg ha^{-1}, and 11 are of average quality with a high yield potential (4000+ kg ha^{-1}). The remaining clones have above-average yield potential (3000–3500 kg ha^{-1}) and quality. The clones are variously a mixture of crosses between China, Assam and Cambodian types. In addition, there are 151 clones obtained from field selections in commercial gardens. There are also 14 biclonal seed lots, labelled TS- (TRA, 2012).

In South India a plant improvement programme was initiated by UPASI in the 1960s. Superior genotypes with high yield and quality traits have been selected from the available natural germplasm and/or from controlled hybridisation. So far, 32 clones have been released to the industry. These are labelled UPASI-1 to UPASI-28 and TRF-1 to TRF-4. Among these, UPASI-3, -8, -9, -17, -25, -27, -28, and TRF-1 and -4 are high yielders, while UPASI-2, -6, -9, -20 and -26 are considered to be drought-tolerant. Most of these clones are now widely used by the industry for infilling and replanting.

Grafting of clonal plants is another technique being developed at UPASI. Drought-susceptible scions (UPASI-3, UPASI-8 and UPASI-17) are cleft-grafted onto the drought-tolerant rootstocks (UPASI-2, UPASI-6 and UPASI-9). In the field, composite plants developed through grafting are claimed to be superior to the ungrafted clones. For example, when compared with rooted cuttings, the composite plants have shown a 30% increase in productivity (further details not given). New scion clones (UPASI- 28, TRF-1 and TRF-2) and the drought-tolerant clones (TRISL-2025, UPASI-2, UPASI-9 and ATK) have been evaluated for graft compatibility.

The graft success ranged from 88% to 98% in different graft combinations. In order to register and protect the tea cultivars released by UPASI, a DUS (distinctiveness, uniformity and stability) test centre has been established on their experimental farm under the Protection of Plant Varieties and Farmer's Rights Authority (PPV and FRA), Ministry of Agriculture. Tea descriptors prepared by the Institute are being validated in the DUS centre.

Certain Sri Lankan clones (TRISL-2024 and TRISL-2025) and a few Sri Lankan estate selections (CR-6017, ATK-1 and SA-6) are also planted in southern India. Some of these clones, CR-6017, UPASI-3, TRF-2 and TRF-4, produce high-quality tea. A number of Sri Lankan biclonal seed stocks, recommended for infilling and for replanting in drought-prone areas, have also been released (namely UPASI-BSS-1 to BSS-5).

Sri Lanka

Germplasm

In Sri Lanka, over 45% of the tea area is planted with seedlings. These are highly variable, and many are now more than 100 years old. These fields are earmarked for replanting with new cultivars. There is therefore a risk of losing a valuable genetic resource. In addition, owing to repeated use of the same parents in tea breeding programmes, and the widespread use of a small number of popular clones, the genetic base of cultivated tea in Sri Lanka (and beyond) is very narrow (Gunasekara *et al.*, 2012).

Attempts are therefore being made to classify and preserve germplasm that has been introduced into Sri Lanka at different times from elsewhere in the world. For example, the ancestry of the TRI 2020 series of clones can be traced back to seeds that were collected from a single tree at the Tocklai Experimental Station in India. Other batches of seed of diverse genotype and phenotype were subsequently imported from India, Indochina (Vietnam?), Japan, Korea and Russia, some of which were used for clone selection and in controlled hybridisation programmes. This led to the release during the 1970s and 1980s of the TRI 3000 and 4000 series of clones. In addition, since the 1950s, clones have been, and continue to be, selected on estates among seedling tea plants (Gunasekara *et al.*, 2012).

In the past it has been sufficient to characterise cultivars/germplasm on the basis of their morphological characteristics. For example, Wickramaratne (1981b) used the variation in leaf characteristics to describe a clone. These included the ratio of leaf length to leaf width, the angle between leaf tip and axis, the length of the petiole and leaf colour. The principal clones grown in Sri Lanka were described in this way to aid those working in the field to recognise a clone. Subsequently, the tea descriptors developed by the International Plant Genetic Resource Institute (IPGRI, 1997) were used to define the phenotype of the Sri Lankan collection of clones. Only six of the 35 vegetative descriptors listed were found to have phenotype discriminatory value. These were serration and waviness of the leaf margin, pigmentation in the young leaf and in the leaf petiole, the size of the leaf, and the leaf angle. Even some of these descriptors can vary with the environment and with management practices.

Other types of descriptor are therefore needed in addition to the vegetative ones. In Sri Lanka, biochemical characterisation has also been used to describe accessions. Such descriptors include total polyphenols, total catechins, and amino acid and caffeine contents of the leaf. More recently, DNA marker systems have been used to measure genetic diversity or, more often, similarity between accessions. There still remains a need to relate these characteristics to agronomic traits such as disease, and pest and drought resistance, as well as yield. Cultivar-specific molecular markers are needed in order to minimise the problems of misidentification of germplasm, to enable the certification of clonal cultivars, and to facilitate plant variety protection rights. International collaboration is essential to achieve these objectives (Gunasekara *et al.*, 2012).

Clones

In the cooler high-altitude areas of Sri Lanka, China-type hybrids were planted, and in the lower, warmer areas Assam-types were used. Clonal tea was first used for replanting in the 1950s. The most popular clones selected by the Tea Research Institute of Sri Lanka came from seedlings grown from a few seeds smuggled in from Assam. In the low-altitude areas, these are TRISL-2023 and TRISL-2026 and, most recently, TRISL-2025. At high altitude, TRISL-2024 and TRISL-2025 are recommended (Sivapalan, 1991). Progress in clonal selection is indeed slow.

Africa: Kenya, Malawi, Tanzania and Uganda

Germplasm

In Kenya, AFLP markers have been successfully employed to detect diversity and genetic differentiation among Indian and Kenyan populations of tea, even those that could not be separated on the basis of morphological and phenotypic traits. Principal coordinate analysis showed that the Assam genotypes/clones selected in Kenya were indeed of Indian origin (Paul *et al.*, 1997). Similarly, based on measurements using RAPD (random amplified polymorphic DNA) markers, Kamunya *et al.* (2010a) showed that despite morphological traits suggesting great diversity between clones, they were all in fact genetically similar. The greatest disparity was between clone TRFCA-SFS150 (from Malawi) and clone TRIEA-6/8 (from Kenya). In order to ensure diversity and to minimise risk, it was recommended that farmers should plant these two clones together with clone TRFK-303/577.

Clones

In Africa, tea plantations were originally established with heterogeneous seedlings derived from seeds imported from India (although in Malawi some seeds were of Chinese origin). From the 1960s onwards, most new plantings were with vegetatively propagated clones selected for superior yield and/or tea-making qualities (Ellis and Nyirenda, 1995).

The results of trials in Kenya show that block selection (in which fewer bushes are assessed, but using a much more intensive procedure) is a more efficient way of identifying superior individuals. However, selecting from the offspring of a controlled breeding programme is an even more efficient way of identifying potentially high-performing clones. In the example from Kenya, only 9 of the 270 clones (4%) identified in a field selection out-yielded the control clone (TRIEA-31/8). With block selection, and a 100% more intensive selection procedure than that used for the field selection, the success rate was 20 out of 116 (17%), while with controlled breeding 43 out of 60 clones tested (72%) out-yielded the control (Smith, 1998).

In Kenya, as just mentioned, the genetic base is dangerously narrow. By 1996, the Tea Research Foundation of Kenya (TRFK) had identified 45 clones (over a period of 45 years), one of which was clone TRIEA-6/8; 18 had been selected from seedling tea (a limited number of sources); 24 were half-sib progenies of clone TRIEA-6/8; and three others were bred from clone TRIEA-6/8 and clone TRIEA-31/11. In total therefore 27 (or 60%) of the released clones shared a genetic pedigree with clone TRIEA-6/8. Clone TRIEA-6/8 was originally selected for the quality of its liquor; it is only a moderate yielder. None of the 47 clones combines yield with quality (Seurei, 1996).

Clones released by TRFK during the 1990s include the TRFK-303 and TRFK-347 series (Njuguna, 1993b). These were selected from the progeny of crosses made in Uganda between clone TRIEA-6/8 (maternal) and an unknown paternal parent in the 1970s. These releases were followed by selections from biclonal, full-sib progeny, namely clones TRFK-337/3, TRFK-337/138 and TRFK-338/13. To add extra quality, clones TRFK-301/4 and TRFK-301/5 include *C. sinensis* var. *assamica* ssp. *lasiocalyx* in their pedigree. The most recent releases from TRFK are clones TRFK-371/3 and TRFK-430/90. Both are considered to be resistant to drought and suited to mechanical harvesting. Clone 371/8 was released in 2012. It is suited to mechanical and hand harvesting, and can also be used for processing as green tea (Kamunya *et al.*, 2012).

In 2009, TRFK pre-released a purple tea variety (TRFK-306/1) for commercial utilisation, targeting a unique anthocyanin-rich tea market (Kamunya *et al.*, 2009a). TRFK-306/1 is thought to be an interspecific natural hybrid between a tea and a related but non-tea *Camellia* species.

In the three East African countries (Kenya, Tanzania and Uganda), the most popular clones include TRIEA-6/8, TRIEA-6/10, TRIEA-31/8, BBK-35, TN-14/3, BBT-207 (all long-established field selections) and AHPSC-12/28, selected from a cross and released in 1987. These clones were selected and evaluated by commercial companies and/or research institutes, often at a single site in Kenya. The TRFK-303 series all had clone TRIEA-6/8 as the female parent. In Kericho in Kenya, four of these clones with contrasting characteristics (AHP-S15/10, TRIEA-6/8, BBK-35, TN-14/3) were compared at a range of altitudes (1800–2200 m) and significant G × E interactions were demonstrated (Ng'etich and Stephens, 2001), emphasising the complexity of selection for the diverse geographic locations where tea is grown in eastern Africa (Wachira *et al.*, 2002).

Close to the equator in Kenya, large yields have been achieved from clonal tea under commercial conditions in Kericho (altitude 1900–2000 m). For example, clone

AHP-S15/10 is capable of yielding, at a field level, in excess of 6000 kg ha^{-1}, except in the year of prune, while annual yields of 8000–9000 kg ha^{-1} are not uncommon. It has been possible to achieve yields of 5000–6000 kg ha^{-1} in years 4 and 5 after planting (that is one year before the tea is pruned). Clone AHP-SC12/28 has also performed well, with consistent yields in the range 5500–7000 kg ha^{-1} except in the year of prune. These values are for relatively young tea. By comparison, corresponding field yields for other clones include 4000–5000 kg ha^{-1} for clone TRIEA-6/8 and clone TN-14/3, about 30% below those of clone AHP-SC12/28, and 50% below those of clone AHP-S15/10. The annual yields of 'old' seedling tea are in the range 2500–4500 kg ha^{-1}. A new clone labelled AHP-CG28U864 has also yielded well in the field, at 7000–9000 kg ha^{-1} (Carr, 1995). These figures give some idea of the yield potential of tea in these high-altitude equatorial areas.

Even more recently, Unilever Tea Kenya has announced the introduction of a new, high-yielding (14% more than AHP-S15/10 and 21% more than TRIEA-31/8) clone (labelled MRTM1). The yields were recorded for at least four years at three sites in Kericho (representing a range of altitudes). MRTM1 was also judged to be more tolerant to drought than four other clones grown commercially (Tuwei, 2008). It also performed well as a rootstock (better than EPK-TN14/3), raising the yields of scions TRIEA-31/8 and AHP-S15/10 by over 20%. Black tea quality was judged to be acceptable but not exceptional. MRTM1 was a field selection, but is now being used as a parent in a breeding programme. A clone selected from a cross between MRTM1 and AHP-S15/10 yields about 20% more than MRTM1, and is being widely planted in Kericho (R. H. V. Corley, personal communication).

In Tanzania, the most widely grown clones on commercial estates are TRIEA-6/8, BBT-207, BBK-35, TRFCA-PC81, TRIEA-31.8 and AHP-S15/10 (total area *c.* 2000 ha), while clone BBT-207 has been a particular favourite with smallholders, perhaps because they were only given a limited choice of clones (Kimambo, 2000). These clones were all first identified in the late 1950s or 1960s. As already noted for Sri Lanka, progress in plant improvement by field selection has been slow.

In Malawi, there are about 4000 ha of clonal tea (20% of the planted area) and 3000 ha of polyclonal seed. The balance is from Indian and Chinese hybrid seed. There is a small area (75 ha) of composites (see below). About 300 ha of the 2500 ha of smallholder tea is clonal. Progress in replanting has been disappointingly slow. Selection of clones from field observation, initially for drought tolerance and rate of recovery from pruning, began in Malawi in the 1950s at what later became the Tea Research Foundation of Central Africa. Clones TRFCA-SFS150 and TRFCA-SFS204 were identified in this way. Since then, the majority of the clones released to the industry have been selected from a biclonal breeding programme based at Mulanje (these clones have the prefix PC) (Whittle, 1997).

In Malawi, the most commonly planted clones (referred to locally as 'superior') are TRFCA-SFS150, TRFCA-SFS204 (both field selections, released in the 1970s), TRFCA-PC81, TRFCA-PC105, TRFCA-PC108 and TRFCA-PC110 (selections from crosses in a breeding programme, mostly released in 1981), followed by smaller areas of a large number of more recent releases, including TRFCA-PC122 and TRFCA-PC123 (released in 1994). These clones (and related composites/grafted plants) have all been selected for good field performance and high quality for planting in southern Africa (from 11°S to 32°S, alt. 600 to 1300 m) (Ellis and Nyirenda, 1995). Their properties (including observations on drought tolerance) have been described in a catalogue (TRFCA, 2000).

Apostolides *et al.* (2006) highlighted the need to develop methodologies that will enable poor-quality genotypes to be rejected early in the selection process. The need for a minimum quantity of leaf (200 g) to allow small samples of green leaf to be manufactured into black tea, and then evaluated by tea tasters, delays the time when genotypes can be eliminated (year 8 in the selection programme). The whole process is also very expensive and difficult for small independent tea research organisations to justify. New, high-throughput, low-cost but reliable screening methods are needed. These could include biochemical and molecular markers. International collaboration is advocated.

In both Malawi and Kenya, grafting of clones onto suitable rootstocks to create higher-yielding, high-quality composite plants is being promoted as a commercial practice (Kayange *et al.*, 1981; Tuwei *et al.*, 2008a) and to improve drought tolerance (Tuwei *et al.*, 2008b). Trials on grafting have also been conducted in Tanzania (Mizambwa, 2002).

Seed or Clone?

For a crop that is in the ground for so long, 'it is foolish to plant anything but the best you can find' (Singh, 2006). That must mean planting only clones – but, as Singh pointed out, 'a seed population composed of genetically distinct units is elastic and can be fitted in to a wide range of cultural and environmental conditions without much change in its overall performance.' By contrast, a clone lacks this elasticity and is therefore more sensitive to any changes in the environment, for example the climate. The advice from India is to mix clones with seed. Wight (1956) suggested planting three to five clones in one-fifth of the area of the estate. Later this was changed to a one-to-one ratio of clone to seed, but with no single clone occupying more than 10% of the area. The total displacement of seedling tea with clones is probably not desirable or ecologically good practice. But there is a contrary view, which is that estates and smallholders should not be planting seedling tea. Apple and pear have been exclusively clonal for centuries. According to this view, the area planted with a single clone should be limited, but the task of preserving genetic diversity in tea rests with the research organisations.

The number of clones needed in plantations to protect against catastrophic failure (due, for example, to pest attack), while at the same time achieving all the benefits associated with the management system, is difficult to specify. As computer simulation studies in forestry have shown, the answer depends not only on the intensity of the 'pest' attack but also on the amount of loss acceptable to management, the level of clonal resistance and the genetic basis of that resistance. The degree of risk, however, is unlikely to change significantly after the number of clones planted exceeds 30 or 40

(Bishir and Roberts, 1999). In an earlier analysis of the same topic, again in forestry, Libby (1982) came up with some unexpectedly contrary conclusions: depending on the tree management techniques employed (e.g. tree density and thinning schedule) and the risk assumptions made, monoclonal tree plantations are frequently the best strategy; mixtures of two or three clones are frequently the worst strategy, and are rarely the best; mixtures of a large number of clones are as safe as seedling populations; and finally, mixtures of modest numbers of clones (7–25) appear to provide a robust and perhaps optimum strategy. A similar analysis has not been reported for tea, but the answers to the question are clearly not straightforward.

Conclusion

Following the development of vegetative propagation of tea it became possible to select clones with above-average yields and tea-making properties, although the process of selection, and evaluation in different environments, is still slow. Some of the more successful clones were identified virtually by chance. Most of the selections have come from visual observations within populations of Indian hybrid seedlings (var. *assamica* × var. *sinensis*). These came from seed that was exported from North-East India to many of the emerging tea industries around the world, beginning in the late nineteenth and first half of the twentieth century. This is a relatively limited genetic pool. It is important that the inherent diversity within the wild populations of both Assam and China types of tea, which it is now possible to describe through molecular marker techniques, is not lost.

Summary

Looking back:

- Existing populations of tea are largely derived from two original taxa, *Camellia sinensis* var. *sinensis* and *C. sinensis* var. *assamica*.
- Hybridisation has resulted in tea gardens made up of plants with physical characteristics ranging between pure China-type and pure Assam-type.
- As a result of interbreeding, there is a great deal of variability in the yields contributed by individual bushes.
- The development of vegetative propagation led in turn to the identification and propagation of clones superior in yield and quality to hybrid seedling tea plants.
- Identifying, monitoring and evaluating new clones can take 10–15 years from conception to release to the industry.
- It is estimated that the chance of identifying an outstanding clone in mixed seedling garden populations may be as low as 1 in 80,000 bushes. So why waste time trying?
- But you never know! An observant manager in Tanzania spotted a tea bush which had no fibre in its stalks. This attribute could be useful in a breeding programme.

- The chances of identifying a superior clone are increased if the selection is made from seedlings derived from cross-pollination between two outstanding clones.
- Priority should be given to evaluating promising but unproven/semi-proven clones against standards. At the same time a breeding programme can begin.
- The diversity of soils and climates within countries/regions means that there is the possibility of genotype × environment interactions. That is, clones will not necessarily perform equally well in all environments.
- Opportunities exist to create composite plants with, for example, a high-quality scion grafted onto a vigorous drought-resistant rootstock.
- As heterogeneous seedling tea is uprooted, and the ground replanted with clones, so the genetic base of cultivated tea decreases. The resultant risk can be managed by the choice of clones.
- To reduce risk, a balance has to be maintained between (a) planting superior clones, which lack elasticity, and (b) maintaining a population of genetically diverse seedlings, which are suited to a wide range of environments, on research stations only, not commercial estates.
- In China and India, germplasm of wild tea is being collected, conserved and classified using molecular marker techniques. Physical appearance is no longer enough to differentiate clones.
- Other countries have created tea conservation orchards, where different agrotypes can be observed.

Notes

1. A tea bush from which cuttings are taken.
2. With orthodox manufacture, pubescent shoots produce silvery tips, deemed to be a desirable character in made tea.
3. Chen *et al.* (2000) proposed a new taxonomic system for *Thea*. This section was revised into five species (the three listed above together with *C. gymnogy* Chang, and *C. sinensis* (L.) O. Kuntze) and two varieties (*C. sinensis* var. *assamica* (Masters) Kitamura and *C. sinensis* var. *publimba*). This system was validated by molecular marker analysis (Chen and Yamaguchi, 2002).

5 The Well-Bred Tea Bush

Developing High-Yielding Clones

R. H. V. Corley and G. K. Tuwei

A tea grower who wishes to increase production, but who does not have access to land for expansion of planted area, has several options. First, production from existing fields may be increased by better management or more intensive harvesting (Chapters 9 and 10). Beyond that, though, replanting with material with greater yield potential should be considered. In this chapter we review past work on tea improvement, and describe our approach, with examples from the Unilever Tea Kenya breeding programme.

In areas where losses to disease or drought are significant, a regular infilling programme, usually with clonal plants, will result in a gradual upgrading of old seedling fields. Such upgrading will be very slow, though, and a progressive grower will certainly consider replanting. Replanting of tea is discussed in Chapter 6, but the essential point is that, unlike many perennial crops, tea does not have to be replanted. There is tea in India which was planted over a century ago, and is still yielding at an acceptable level. Therefore replanting can only be justified if the expected increase in yield is great enough to offset the capital cost of replanting, plus the opportunity cost of lost production before the replant reaches maturity. In Kenya in the 1990s, financial calculations indicated that yields would have to be at least doubled if replanting was to be profitable. With old seedling tea yielding an average of 3500 kg ha^{-1}, that meant that yields of more than 7000 kg ha^{-1} were the target. Few if any of the clones then being planted were capable of such yields as a long-term mature average, so if replanting was to be done, new clones would have to be developed.

To Breed or Not to Breed?

There are several ways in which improved planting material can be developed. The first decision to be made is whether to plant clones or seedlings. In tea, which is cross-breeding and highly heterogeneous, even biclonal seedling populations are very variable, and there will always be genotypes within a population which are markedly superior to the mean. Genetically mixed seedling populations will be more adaptable than clones, as Barua (1963) noted, but locally adapted clones of the best genotypes, once they have been identified, will always give higher yields than seedlings. Barua (1989) expected that total displacement of seedling tea by clones could not be ruled out, and in most tea-growing areas clones are now the sole commercial planting material.

The most widely used method for developing ortets (mother bushes) for clones is *field selection*: the identification of superior bushes in existing seedling fields. The limitations of this approach have been highlighted by several authors, who have emphasised the greater potential of breeding, by making crosses between already identified superior clones using either *biclonal seed gardens* or *controlled pollination*.

Field Selection

The first clones, described in Chapter 4, were all field selections. However, Chapter 4 also shows that there has been little further progress since those early selections. Visser (1969) estimated that only 0.01% of bushes would produce clones yielding twice the overall average yield; thus 10,000 bushes would have to be recorded to produce one useful clone. Wight (1958) estimated an even higher figure of 40,000 bushes. With 6–12 million bushes in a 1000 ha plantation (depending on planting density), many such bushes should exist, but finding them is a major problem. Selection methods have included:

- Visual selection by experts
- Asking pluckers to find high-yielding bushes
- Looking for bushes which show rapid recovery from pruning
- Looking for bushes which continue to yield during drought
- Recording yields of individual bushes in blocks of seedlings

Recording yields might appear a good approach, and significant correlations have been shown between yields of individual bushes and the clones derived from them (see *Heritability*, below). However, very large numbers must be recorded to identify even one that may have the required yield, as noted above. An important point is that the most vigorous bushes will tend to yield well at the expense of their neighbours (Cannell *et al.*, 1977).

Breeding

Clonal propagation of the best individual bushes has given significant yield increases, but even the most efficient selection cannot improve on the best that already exists in the population, so there will be a limit to progress. An alternative is to make deliberate crosses between selected clones or bushes. When two individuals are crossed, their genes are recombined in many different ways in their offspring, and some combinations may result in performance superior to that of either of the parents. This gives a breeding programme a major advantage over any method of field selection: it is possible to create new genotypes with combinations of characteristics better than anything in the starting population (this is illustrated in Box 5.1). Another practical advantage is that all recording can be concentrated in a limited area of trials, rather than extended over commercial fields.

Box 5.1 Breeding Creates New Genotypes, Which Did Not Exist in the Original Population

A number of crosses were made in 1994, from which 465 seedlings were planted out, and their yields recorded. In the graph, these are expressed relative to the standard clone MRTM1. The average yield was 81% of that of the standard clone, and more or less followed a normal distribution. Second-generation crosses were made in 2004 between several of the resulting clones, and the graph also shows the distribution of yield for 84 seedlings from the best of these crosses. The average yield for this cross was 132% of that of the standard clone, and 25% of seedlings yielded more, relative to the standard, than any in the original population.

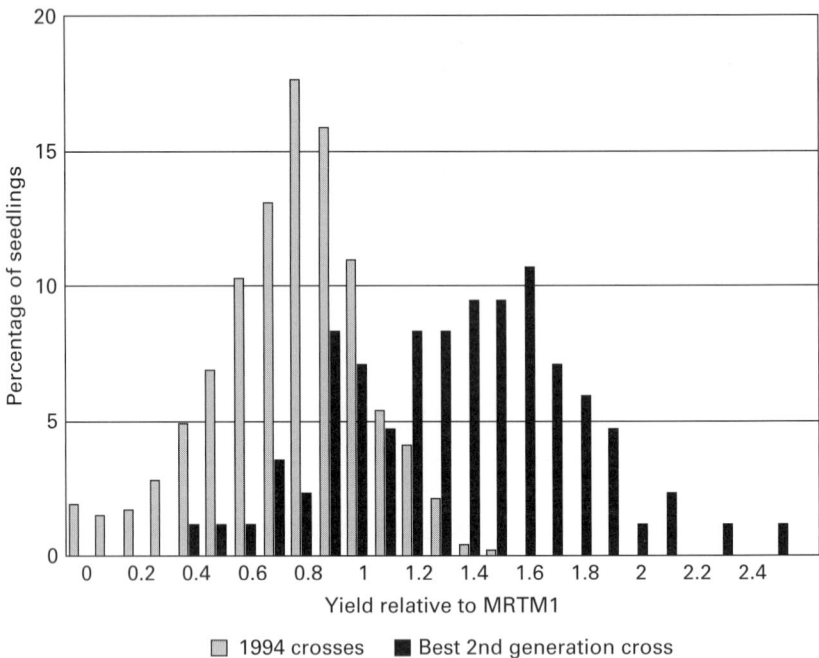

Biclonal Seed Gardens

Tea is generally self-incompatible, and self-pollination is rare (Figure 5.1). Thus if two clones are planted together and allowed to flower, any seed set is very probably a cross between the two. This is the simplest and cheapest way of making crosses, but it has the disadvantage of being very inflexible. Crosses have to be planned several years in advance, at the time the seed gardens are planted. Seed gardens also have to be separated by more than 200 m, to avoid accidental out-pollination (Muoki *et al.*, 2007). (Note that a minimum distance of 30–50 m between seed gardens was specified in Chapter 4, because pollinating insects do not fly far.) In 2011, the Tea Research Foundation of Kenya listed 11 seed gardens, including 20 parent clones (TRFK, 2011). In 1992,

Figure 5.1 A tea flower (RHVC). *A black and white version of this figure will appear in some formats. For the colour version, please refer to the plate section.*

African Highlands Plantations had 36 seed gardens, including all possible combinations of nine parents.

Once a biclonal garden has been fully exploited, a new combination can be made by cutting back the trees and grafting different clones onto the stumps, but this takes time, and the number of possible crosses remains limited.

Polyclonal seed gardens have also been used. By planting several clones together, many different crosses can be made in a limited area. Assuming that ripe seed is harvested from the trees, the female parent of individual offspring will be known but not the male parent. This is a disadvantage for inheritance studies. With molecular markers (see below) it should be possible to identify the male parents as well, but the cost for large numbers of seedlings might be prohibitive.

Controlled Pollination

This has the advantage of complete flexibility. In our programme in Kenya, over 900 different crosses have been made over a 20-year period, involving over 140 different parent clones. The disadvantage of this method is that the success rate of hand pollination is very low (Ariyarathna *et al.*, 2009), and thus the labour requirement and costs are high. Typical success rates in our programme in Kenya are given in Box 5.2. In early work, we emasculated flowers used as females (removed their stamens), to prevent accidental self-pollination (Figure 5.2). However, it is difficult to do this without damaging the flower, and fruit set was much lower than if emasculation was not done. Unemasculated flowers set some fruit through self-pollination, but we considered that

---

Box 5.2 Hand Pollination

For hand pollination, unopened flowers from the male parent are collected the evening before pollination is done. The next morning, unopened flowers of the female parent are opened, the stigmas are dusted with pollen from the chosen male parent clone, and the flower is then enclosed in a cotton bag, to prevent access by insects carrying pollen from other clones. The bags are removed after about 1 week, by which time the stigmas are no longer receptive.

  Average success rates for 140,000 pollinations over 14 years are set out below. The overall success is low, but there is wide variation between years. Some of that arises from the choice of female parents: some clones set fruit much more easily than others, as shown by the results for two clones from the 2004 programme. Once identified, clones which give poor fruit set can be limited to use as male parents in future crossing programmes.

Success rates in hand pollination: overall mean, and range of annual means, together with results for best and worst female parents in 2004 as an example

Stage	Mean (range of individual annual means)	Best clone in 2004	Worst clone in 2004
Fruit harvested, as % of pollinations	27.3 (18.4 – 38.7)	51.0	1.6
Seeds per fruit	1.80 (1.54 – 2.20)	2.20	1.0
Germination %	43.4 (20.1 – 57.3)	60.6	0
Seedlings as % of pollinations	21.3 (11.9 – 34.8)	67.9	0

---

some self-pollination was an acceptable price to pay for much improved fruit set; Visser (1969) reached the same conclusion.

  A limiting factor in any breeding programme is the time it takes to get a bush to flower (Figures 5.3 and 5.4). Yield and quality of a clone can be assessed within four years or less (see below), but before the best clones can be used for further breeding they must be allowed to grow into small trees, unplucked and undisturbed for two years or more. When a young tree starts to flower, it will produce viable pollen, so can be used as a male parent, but will often fail to set fruit when used as a female parent.

  In Box 5.3, results with different methods of clone development are compared. It is clear that the high costs of breeding work are easily justified by the high success rate compared to field selection.

Yield or Quality?

Most tea breeding programmes are aimed at increasing yield, but initial selection is often on other criteria. For example, in the programme described by Ellis and

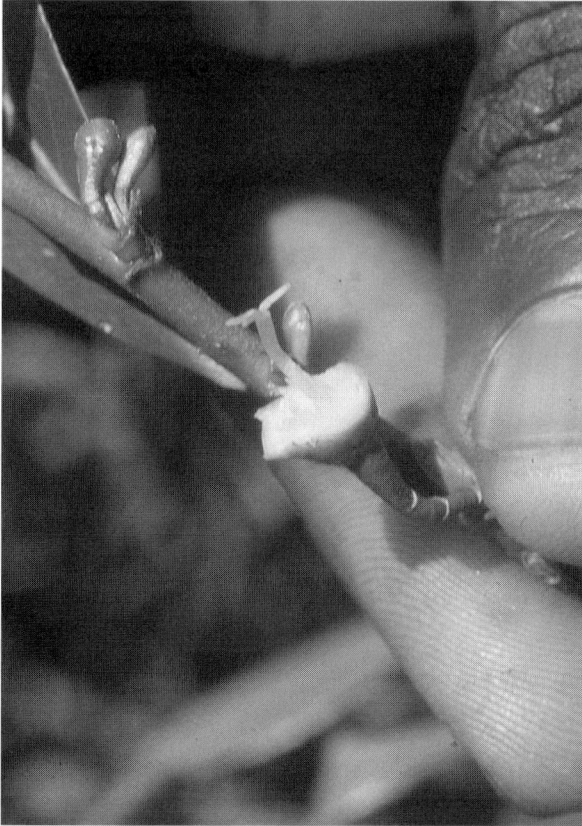

Figure 5.2 In early work in Kenya, the stamens were removed from the flowers to be used as females (emasculated). This was to prevent accidental self-pollination (RHVC).

Nyirenda (1995) emphasis was given to easy fermentation (the chloroform test: Sanderson, 1963) and good nursery growth and survival after field planting. The importance of these criteria in Malawi was explained, but there was no mention of selection for yield; evaluation of 'all traits', presumably including yield, only started 10 years after pollination of the cross.

In the scheme outlined by TRFK (1986) the chloroform test is also used, and the first clonal field trial is only maintained until the bushes have been tipped-in at the start of production. At that stage, clones with better vigour and survival than control clones are retained for the next phase of field trials, and it is suggested that the first trial can be uprooted. Thus again yield is only considered in the later stages of the programme.

As explained above, large increases in yield were the main objective of the programme in Kenya, and the emphasis was therefore on yield throughout. At its simplest, crosses were made between high-yielding clones, and the highest-yielding seedlings from those crosses selected for testing as future clones. Nyirenda and Ridpath (1984) showed that when bushes are planted at low density, to minimise competition, individual bush yield is highly correlated with bush surface area. In our programme, selection is

Figure 5.3 A mother tea bush, with flowers protected by cotton bags. The colour of the bag indicates the date of pollination. A limiting factor in any breeding programme is the time it takes to get a bush to flower. Yield and quality of a clone can be assessed within four years, but before the best clones can be used for further breeding they must be allowed to grow into small trees (Kenya, RHVC).

done after three years in the field, before competition becomes intense. To avoid selecting large bushes, which might perform less well at a higher planting density, yield per unit bush area should be considered.

Yield Components

In planning crosses, it is useful to break yield down into its components; parents can then be selected independently for these components. For example, yield is the product of shoot number and mean shoot weight (Chapter 9), so it may be worthwhile making crosses between clones with high shoot number and clones with high mean shoot weight. In our experience yield per bush is strongly correlated with shoot number, but not with shoot weight (Figure 5.5). The correlation between shoot number and mean shoot weight was low (though highly significant) in the population of Figure 5.5 ($r = -0.24$, 478 d.f., $p < 0.0001$), so it should be possible to select for the two components independently.

Dry Matter Production

Yield can also be considered as the product of total dry matter production and harvest index (the proportion of total dry matter in harvested shoots) (Chapter 7). It may be worthwhile trying to select parents independently for these components. In this respect it is worth noting that a high harvest index requires that relatively little dry matter is

Figure 5.4 (a) Close-up of tagged tea fruit, (b) with cotton bag in place and (c) after removal of cotton bag (Kenya, RHVC). *A black and white version of this figure will appear in some formats. For the colour version, please refer to the plate section.*

allocated to the bush 'frame', but such bushes may be excluded by the usual selection for vigour and rapid recovery from prune. In fact, clone AHP-S15/10, which is high-yielding and has a high harvest index (Burgess and Carr, 1996a, 1996b), shows very poor recovery from prune unless 'lung pruning' is done (see Chapter 7). Thus we question the value of the usual field selection criteria of vigour and rapid recovery from pruning. Banerjee (1992b) noted that little progress had been made towards increasing harvest index.

Drought Tolerance

Another important characteristic affecting yield is drought tolerance. It has been shown that clones differ in drought tolerance (e.g. Burgess and Carr, 1996a, 1996b; Netto *et al.*, 2010). Physiological characteristics which contribute to drought tolerance are discussed in Chapter 13, and but it has not proved easy to apply these as selection criteria. The simplest criterion is to look for bushes which continue to yield during drought.

Box 5.3 Comparison of Methods for Clone Development

- In the *field selection* work, pluckers were asked to identify high-yielding bushes; for some bushes yield was recorded for a few rounds, while for others speed of recovery from prune was evaluated.
- For *block selection*, yield was recorded for two years from plots of 480 bushes, in the same fields as the field selection work. Yields were adjusted for differences in bush area, and selections were based on yield or shoot number per unit area.
- From five different *biclonal seed gardens*, a total of 960 seedlings were recorded for one year; yields were adjusted for bush area and selections were made as for block selection.
- The seedlings from *hand pollination* were from 22 different families; cuttings were taken in the nursery, and the results shown are for all clones with at least four plants/clone. There was thus indirect selection for nursery vigour (plants giving few cuttings were excluded), but otherwise the clones were unselected.

The table shows for each method the percentage of the original seedling population tested in clone trials, and the number and percentage of clones superior to the standard clone TRIEA-31/8.

Comparison of clone development methods

Method	% selected	Clones tested	Superior to 31/8 No.	Superior to 31/8 %
Field selection	*c.* 0.05	270	9	3.3
Block selection	5	116	20	17
Biclonal seed gardens	6	60	43	72
Hand pollination	100	211	176	83

   With the same parents, hand pollination and a biclonal seed garden should give similar results. The higher success rate with hand pollination (83% superior to 31/8 with no prior selection, compared to 72% after including only the highest-yielding 6% of seedlings from the biclonal gardens) reflects the greater flexibility of the method. Hand pollinations involved 12 different parents, most of which had out-yielded 31/8 in earlier trials. In contrast, the six clones used in the five biclonal gardens gave yields similar to or less than 31/8, but these were the only biclonal gardens available at that time.

Bushes can also be scored for severity of drought symptoms (Kamunya *et al.*, 2007). In Kenya the severity of the dry season varies considerably from one year to another, with little or no drought in many years. In a year when drought does occur, yields can be compared with yields in the previous year to give an indication of drought susceptibility (e.g. Tuwei *et al.*, 2008b). These possible methods for assessing drought tolerance are discussed in Box 5.4 (Figures 5.6 and 5.7).

(a)

Mean shoot weight (g)

(b)

Shoot number per bush

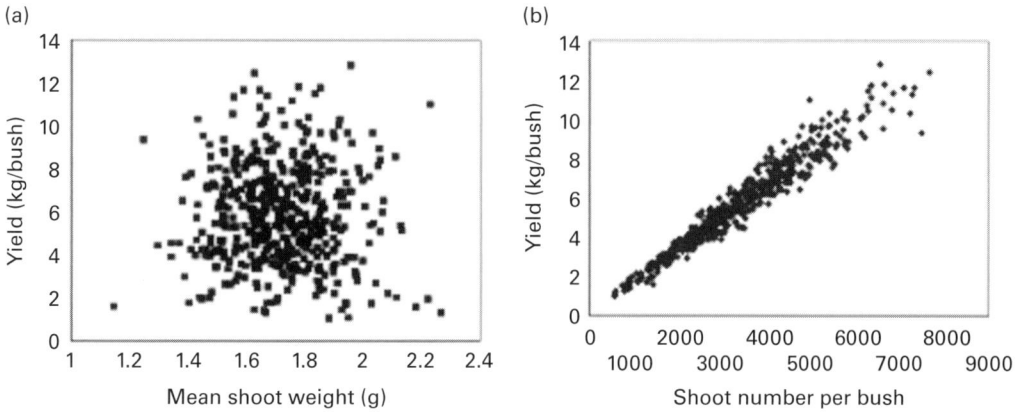

Figure 5.5 The relationships between (a) bush yield and mean shoot weight, and (b) bush yield and shoot number: green leaf yields from 480 bushes recorded for two years in Kericho, Kenya.

Box 5.4 Breeding for Drought Tolerance

In order to breed for drought tolerance, simple and reliable selection criteria are needed. In a trial at Kericho, Kenya, 34 clones and 3 graft combinations were planted in 2008, tested with and without drip irrigation. In 2012 there was a severe drought, causing an average yield depression in unirrigated plots of 37% compared to irrigated plots. For individual clones the depression ranged from 2% to 87%. Yield with irrigation and the depression caused by drought were not correlated ($r = -0.006$), indicating that they are distinct characteristics and can be selected for independently.

During the drought, the clones were scored for severity of drought symptoms, on a scale from 1 (complete defoliation) to 6 (no obvious effect). Kamunya *et al.* (2007) found that such scores were highly heritable, and the figure below shows that in our trial scores were highly correlated with the actual yield depression ($r = 0.815$, 35 d.f., $p < 0.0001$). Thus drought scores should be a reasonably reliable selection criterion.

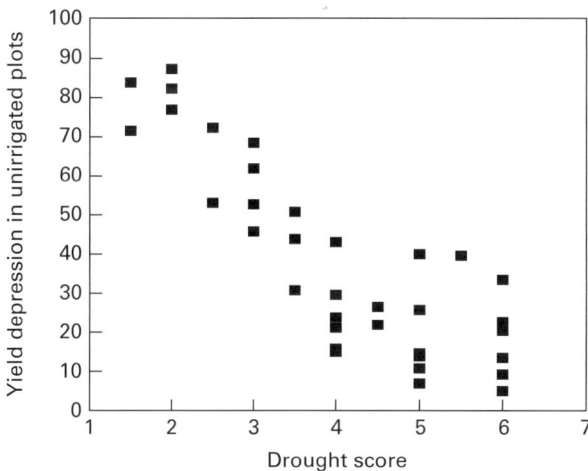

Drought score

Comparison of drought symptom scores with yield loss in unirrigated plots in 2012.

Box 5.4 Continued

In many trials we have calculated a 'drought tolerance index' (DTI) by comparing yield in a drought year with yield in the previous, undroughted year. In the same trial, the unirrigated yields in 2012 were expressed as a percentage of unirrigated yields in 2011, when the dry season was mild (average yield depression compared to irrigated plots was 9%). The figure below shows that, with one exception, the yield depression in 2012 was highly correlated with DTI (2012 yield/2011 yield). Thus DTI should also be a useful selection criterion.

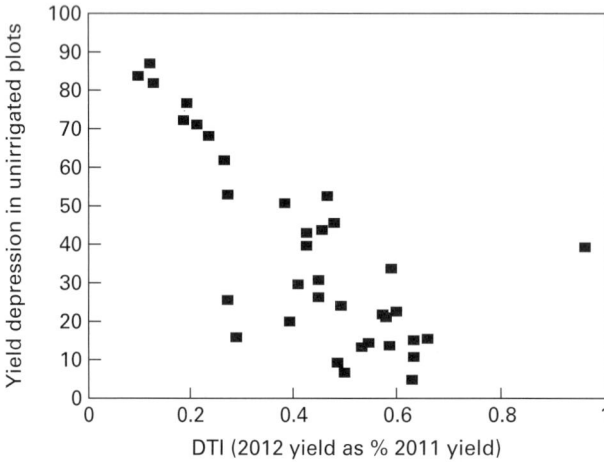

Comparison of drought tolerance index (DTI: yield in 2012 as percentage of yield in 2011) with yield loss in unirrigated plots in 2012.

Confirmation that drought tolerance, as assessed by yield depression, is a heritable characteristic comes from a comparison of breeding values (see under *Heritability*) with performance in the irrigation trial. DTIs were calculated for nearly 1000 clones, by comparing yield in 2012 with mean yield over a five-year period for each clone. The clones were selected for yield, but as yield and drought tolerance appear to be independently inherited, they may be assumed to be randomly selected for drought tolerance. Breeding values were then estimated for 22 parent clones which were also included in the irrigation trial. The correlation between breeding value from the clone trials and the 2012 yield depression in the irrigation trial was highly significant ($r = 0.720$, 19 d.f., $p = 0.00023$), as shown in the figure below (note that a high DTI corresponds to a small yield depression).

Box 5.4 Continued

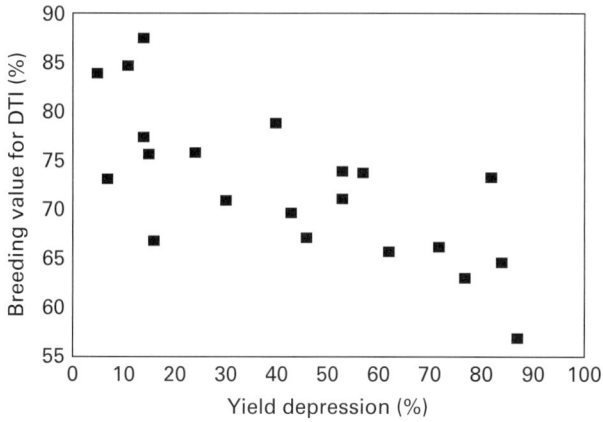

Yield depression in 2012 compared to breeding value for DTI for clones used as parents.

Figure 5.6 Clone trial highlighting the difference in drought resistance between two clones: UTK-225 drought resistant and TRIEA-31/8 (Kenya, RHVC).

Figure 5.7 Clone trials, with some clones showing severe defoliation from drought (Kenya, RHVC).

Other Stress

Tolerance to cold, frost, hail and high pH have all been considered in plant improvement programmes. Some clones, such as EPK-TN14/3, are known to tolerate high pH, and there are a few reports of cold tolerance, but little progress appears to have been made in breeding for stress tolerance.

Quality

Nyirenda and Ridpath (1984) considered that the choloroform test and theaflavin (TF) analysis had made selection for quality 'rather easy'. This may be true for Malawi teas, but Wachira (1994a) noted that Kenyan teas generally have high TF levels, so this was not a useful selection criterion in Kenya. He considered that inheritance of quality was not understood, and attempts to develop reliable criteria for selection should continue.

Quality is a subjective characteristic, ultimately reliant on tea tasters' assessments. In our programme in Kenya, yields of over 30,000 seedlings from controlled pollination, and 6000 clones derived from them, have been recorded, but it is not possible to have that number of samples individually tasted by experts. An alternative would be to use some chemical characteristic which is correlated with tasters' assessments, but despite a great deal of work by many different authors, nothing sufficiently simple to be used as a screening method has been developed. Part of the problem is that, while good tasters are consistent and reliable, they do not always agree with each other on what constitutes

Table 5.1 Studies on chemical analysis and quality

Method	Comments	Ref.
Chloroform test	Used to eliminate poor or slow fermenters	1
Natural Resources Institute (NRI) method	Pigment content of fresh leaf correlated with tasters' scores, but results not repeated outside Malawi	2
Theaflavin content	TF content correlated with tasters' estimates of value in Malawi (ref. 3), but not in Kenya (ref. 4)	3, 4
Catechin content	Catechin content of fresh leaf correlated with TF level in Malawi; too time-consuming for selection method	5
	Astringency correlated with residual catechins in made tea; too time-consuming for selection method	6
PPO/substrate ratio	Ratio of polyphenol oxidase to catechin substrate; relation with quality not clear	7
Near-infrared reflectance spectroscopy	Can be used to estimate TF levels in made tea, correlated with tasters' scores	8
Volatiles content	'Flavour index' (ratio of 2 groups of volatile compounds) correlated with tasters' scores; too time-consuming for selection method	9
Terpene index	Varies between clones, unaffected by environment, but relationship with tasters' scores not studied	10
Total polyphenol content	Positive, but not significant, correlation with tasters' scores	11
Phloem index	Presence of calcium oxalate crystals; large sampling and environmental variation	12

1, Sanderson (1963); 2, Taylor *et al.* (1992); 3, Hilton and Ellis (1972); 4, Owuor *et al.* (1986); 5, Hilton and Palmer-Jones (1973); 6, Zhang *et al.* (1992); 7, Lopez *et al.* (2005); 8, Hall *et al.* (1988); 9, Owuor (1992); 10, Owuor *et al.* (1987); 11, Obanda *et al.* (1992); 12, Wight and Gilchrist (1961).

a good clone (Corley and Chomboi, 2005). Retail teas are usually blends, with teas of different origins contributing different characteristics to the blend; thus different origins may be valued differently by tasters. Table 5.1 lists some studies on chemical analysis and quality. Leaf pubescence is said to be important for quality of orthodox tea (e.g. Banerjee, 1992b).

The objective of our programme in Kenya was defined as 'increasing yield without sacrificing quality'. To achieve this, we have concentrated on yield in the breeding and selection process, and asked tasters to assess quality only at the point when clones are selected for multi-location trials, towards the end of the process. To ensure that quality is not lost, though, the programme includes 'yield × quality' and 'quality × quality' crosses, using parent clones generally recognised as giving good quality. When selecting seedlings from such crosses, a lower yield standard is adopted than for 'yield × yield' crosses.

Pest and Disease Resistance

East African tea is fairly free of serious pests and diseases, but TRFK (2011) listed tolerance to red crevice mite (*Brevipalpus phoenicis*), scale insects (*Aspidotus* spp.) and root knot nematodes as important. *Helopeltis* is also an increasing problem. In Asia pests and diseases have much larger effects, and should certainly be considered in any plant improvement programme, but Banerjee (1992b) noted that little progress had been made in breeding for resistance.

Inheritance of Yield and Other Characteristics

Population geneticists subdivide the genetic variation in quantitative characters, such as yield, into a number of components. The most important division is between additive and non-additive variation. Additive variation is consistently transmitted from a parent to all its offspring. It results from the summation of the effects of many minor genes, and is a large component of the variation in most characters. Non-additive effects are not necessarily transmitted from parent to offspring, or not in a simple manner; such effects include maternal inheritance, dominance, 'epistatic interactions' in which one gene alters the expression of another, and various other effects. Most crossing designs allow additive and non-additive variance components to be estimated statistically. Such estimates of variance components are useful to the breeder because, whereas additive variation is easily utilised, non-additive variation may require a series of crosses before it is understood, let alone exploited.

The term *heritability* is used to describe the ratio of genetic variation to total variation (genetic plus environmental), or the degree of resemblance between relatives. It indicates the reliability of the phenotype as an indication of the genotype; possible values range between 0 (all variation environmental) and 1 (all variation genetic). Two different terms are used: *narrow-sense* and *broad-sense* heritability.

Narrow-Sense Heritability

Narrow-sense heritability (h^2) is the likelihood that offspring will inherit a particular characteristic from their parents, or the degree of resemblance between relatives. It involves additive variation only, and in a breeding programme it gives an indication of the ease with which a characteristic can be improved (see Box 5.5). Narrow-sense heritabilities have been estimated for several different characteristics of tea, as shown in Table 5.2. The heritability of yield appears to be low, but heritability estimates are specific to the population and environment studied. Kamunya *et al.* (2007) studied crosses between four clones in Kenya, while Pool (1982) used six clones in Malawi. Our estimates were from crosses between nine female and three male parents, giving 27 seedling families.

Kamunya *et al.* (2009b) found significant non-additive genetic effects for quality parameters, but variation in yield was predominantly additive. Significant maternal effects were also found. They noted that non-additive effects can be captured by clonal propagation. Lubang'a (2014) found that additive effects predominated for catechin and caffeine levels, but also found significant maternal effects.

Table 5.2 Heritability in tea

Characteristic	Narrow sense			Broad sense		
Yield (g m^{-2} yr^{-1} or kg ha^{-1} yr^{-1})	0.17	0.01	0.3–0.6	0.95	0.85	0.87
Drought tolerance (score)	—	0.25	—	0.79	—	0.49
Fermentation (chloroform test)	0.97	0.13	—	0.98	—	0.65
Theaflavin content (%)	0.28	—	—	—	0.97	—
Thearubigin content (%)	—	—	—	—	1	—
Caffeine (%)	—	—	—	—	0.86	—
Brightness (%)	—	—	—	—	0.92	—
Reference	1	2	3	2	4	5

1, Pool (1982); 2, Kamunya *et al.* (2007); 3, Unilever Tea Kenya (unpublished); 4, Paul *et al.* (1994); 5, Wachira and Kamunya (2004).

Box 5.5 Expected Selection Progress

The 'breeder's equation' can be used to estimate expected breeding progress:

$$R = h^2. \ i. \ CV$$

where R is the response to selection, as a percentage of the mean
CV is the coefficient of variation in the initial population
h^2 is narrow-sense heritability
i (taken from published tables) depends on selection intensity

From this it will be seen that selection progress will be greatest if heritability is large, variation in the initial population is large, and the selection intensity is high. Breeders commonly try to increase progress by starting with very large populations and applying a high selection intensity. However, this gives diminishing returns, because the relationship between i and selection intensity is not linear: for 5% selection, $i = 2.06$, but for 1% selection it only increases to 2.66, and for 0.1% to 3.37. Another approach is to try to increase heritability by making more accurate measurements (measurement error is included with environmental effects in the analysis). This is appropriate for perennial crops where repeated measurements are possible. In one of our trials heritability was increased from 0.3 to 0.63 by recording seedlings for two years instead of one year.

Broad-Sense Heritability

Broad-sense heritability (H^2) includes all genetic variation, non-additive as well as additive, so broad-sense heritabilities are higher than narrow-sense (Table 5.2). A clone carries all the genes of the ortet from which it was derived, so broad-sense heritability is useful when considering the selection progress achievable through clonal propagation.

If selection of seedling ortets is to be reliable, the broad-sense heritability or the correlation between ortet and clone yields must be reasonably high. Visser (1969) quoted

Table 5.3 Correlations between yields of ortets and clones derived from them (data from TRFK Annual Report, 1986)

	Trial 1		Trial 2	
Clones	Correlation	No. of clones	Correlation	No. of clones
From high-yielding ortets	0.157 ns	10	0.016 ns	7
From low-yielding ortets	0.290 ns	9	−0.554 ns	4
From all ortets	0.814 ($p = 0.002$)	19	0.863 ($p < 0.001$)	11

correlations ranging from only 0.18 to 0.31 in six groups of clones, but it is possible that this was because only clones from selected, higher-yielding ortets were included. TRFK (1986) described a study in which both high-yielding and low-yielding ortets were propagated. Table 5.3 shows that if only the high-yielding ortets were included, the correlation between ortets and clones was low. If all clones were included, however, the correlations were highly significant. In other words, recording of individual seedlings is adequate to distinguish good ortets from poor ones. This has been confirmed in Malawi: Nyirenda (1989) found a correlation of 0.53 ($p < 0.0001$) for 50 pairs of clone and ortet, and Nyirenda and Ridpath (1984) found a correlation of 0.65 ($p = 0.002$) for 20 pairs. In our programme, we have found correlations of 0.5 and above for large numbers of clones (see Box 5.9). Singh *et al.* (1993) found broad-sense heritability for yield ranging from 0.47 to 0.76 in six trials.

Breeding Value

The additive breeding value of a parent is the average value of all crosses from that parent; this is usually expressed as a difference from the overall mean for the trial or crossing programme. Once breeding values have been estimated, they may be used to predict the performance of other crosses, the expected value of a cross being the mean of the breeding values of the parents.

Inbreeding and Heterosis

When self-pollination is done, or crosses between close relatives are made, the chance that the offspring may be homozygous for harmful recessive genes is increased. Inbreeding depression is essentially the decrease in vigour which results, and is commonly observed when naturally cross-pollinated species such as tea are self-pollinated. Under natural conditions, self-incompatibility usually, though not always, prevents self-pollination (Wachira and Kamunya, 2005). The trial described by Kamunya *et al.* (2007) included four self-pollinations; on average these gave yields 11% below intercrosses between the same four clones. In our experience inbreeding depression is greater than this: backcrosses of a clone to one of its parents reduced yield by about 25%, even though a backcross involves a lower level of inbreeding than selfing.

Heterosis, or hybrid vigour, is sometimes described as the opposite of inbreeding depression, and can be defined as the superiority of the F1 cross over its better parent.

Box 5.6 Special Combining Ability

Additive breeding values for yield were calculated for all the clones used as parents in our programme. The table shows examples of average family yields relative to the yield expected from the breeding values of the parents. Clone A crossed with clone E gives the expected yield, but with clones C and D it gives average yields 14–17% above the expectation from breeding values, an example of special combining ability. Clone B, on the other hand, combines well with clones C and E, but poorly with clone D, giving an average yield below expectation.

Family mean yields, compared to expectations from breeding values of parents

	Clone C	Clone D	Clone E
Clone A	+17%	+14%	−1%
Clone B	+35%	−7%	+25%

Such figures only give part of the picture. One also needs to know the actual yield of a cross; two low-yielding parents may combine well, but still give only an average yielding progeny. The variation of yield within the progeny is also important: large variability increases the chance of finding very high-yielding individuals.

We have found that a number of crosses give average yields significantly higher than expected from the additive breeding values of the parents, and some of the best clones come from crosses exhibiting such heterosis. Such crosses are said to show 'special combining ability' (Box 5.6). Lubang'a (2014) found significant heterosis for caffeine and catechin contents.

In general, heterosis is expected to be greater the more distantly related the parents. Molecular markers (see *Biotechnology*, below) have been used to estimate genetic distance between tea clones (Balasaravanan *et al.*, 2003; Kamunya *et al.*, 2010b), but we have not found any relationship between heterosis in our programme and the genetic distances given for East African clones by Kamunya *et al.* (2010b). Nor did we find any relationship among 22 crosses where genetic distances were estimated by Lea (1998).

Maternal Inheritance

This describes the situation where a characteristic is inherited from the female parent, but not from the male. In plants it is most easily seen as a significant difference between reciprocal crosses; that is, performance depends on which parent is used as the female. Kamunya *et al.* (2009b) and Lubang'a (2014) found significant maternal effects for quality parameters, and it has been suggested that maternal inheritance might also be important for yield. In our experience, occasional differences in yield between reciprocals are seen; in one trial, out of 20 pairs of crosses made both ways, there were four pairs in which one cross yielded significantly more than its reciprocal. Two involved the same clone as female parent, but that clone was also used in four other pairs of crosses where no difference was seen. Thus we consider that, while maternal inheritance may sometimes occur, it is not predictable, and not sufficiently prevalent that it needs to be taken into account in planning a programme. Where it is feasible to make a cross in both directions, though, it makes sense to do so.

Two-Stage Selection

The selection of clones in a breeding programme is a two-stage process: first, superior seedling ortets must be selected, and then the clones derived from these are tested and the best clones selected for commercial planting. The problem of two-stage selection was discussed by Simmonds (1985). He showed that, if the ortet selection intensity is too high, most of the best clones may be eliminated at that stage. Breeders sometimes assume that higher selection intensities will lead to greater success, as is true for parent selection (see Box 5.5), but for two-stage selection Simmonds argued that the opposite is the case. Figure 5.8 illustrates the problem. The objective of two-stage selection is to identify a group of superior clones in stage 2: these are $a + b$ in the figure, but after stage 1 selection ($b + c$), a will be excluded from the stage 2 trials. If stage 1 selection is too stringent, b will be much smaller than a, and the group selected may include very few of the superior clones. This will be true unless the ortet–clone correlation is very high. With less stringent selection, though, the number of poor clones (c) tested in stage 2 may be very large, increasing trial costs. Simmonds showed how relative costs of stage 1 and stage 2 could be used to arrive at an optimal system. Box 5.7 shows an example of two-stage selection in practice.

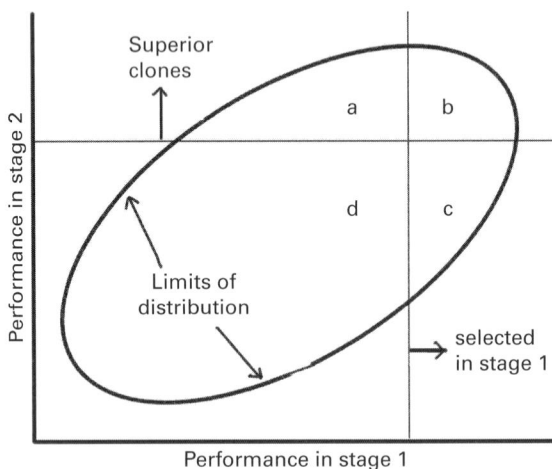

Figure 5.8 Two-stage selection (based on Simmonds, 1985). Ellipse represents two-way distribution of yield in stage 1 (seedling trial) and stage 2 (clone trial), with vertical and horizontal lines showing cut-off points for selection. For further explanation see text.

Box 5.7 Two-Stage Selection in Practice

The figure shows an example of two-stage selection. All of a population of 320 seedlings from controlled crosses were propagated as clones. The seedlings were recorded for one year, and the clones for five years. The ortet–clone correlation was 0.50 (318 d.f., $p < 0.0001$). If the target group of clones is defined as the best 2%, and if the best 5% of seedlings had been selected, none of the superior clones would have been included. To include a majority of the superior clones, it would have been necessary to test nearly 30% of the seedling population as clones. Based on this and other similar studies, we aim each year to include at least 20% of the seedlings from hand pollination in clone trials, and preferably 25%.

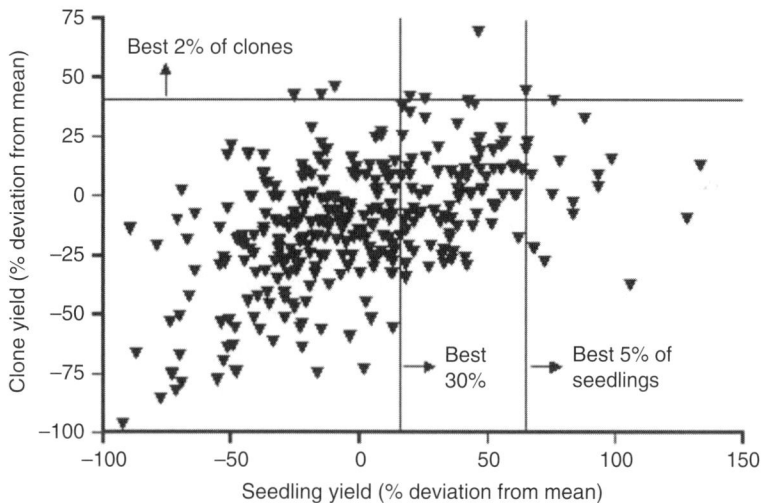

An important point to note from this figure is that the range of yields for seedlings tends to be much greater than that for the clones derived from them. While the best 2% of clones yielded at least 40% more than the mean, the best 5% of seedlings yielded at least 65% more than the mean. This is because environmental variation (mainly soil) has a greater influence on a single seedling than on the average for several plants of a clone in replicated plots.

How Long Will It Take?

Banerjee (1992b) indicated that clone development might take as little as 8–10 years, but this was from field selection. Ellis and Nyirenda (1995) set out the time scale for a tea breeding programme (Table 5.4). They noted that the question of yield differences was not given high priority; only at stage 4, starting 10 years after pollination of the cross, was there evaluation of 'all traits' (see discussion of

Table 5.4 Time schedule for a tea breeding programme (Ellis and Nyirenda, 1995)

Stage	Year	Operation
1	0	Controlled cross of two selected clones
	1	Seedlings in nursery
	2	Select on vigour, line out in the field
	3	Select on quality (chloroform test), leaf size, recovery from prune, vigour and *Helopeltis* resistance
2	4–5	Select on rooting potential and nursery growth
	5	Plant 2 × 8 plants per clone in the field
3	6–8	Select on survival, regrowth, quality (theaflavin content), pest and drought tolerance, nursery performance
4	8–9	Plant 5 × 6 plant plots, replicated five times, at several sites
	10–15	Evaluation on all traits
5		One or two clones on limited release

selection criteria above, under *Yield or Quality?*). Clones may be released after about 15 years.

In the scheme outlined by TRFK (1986), yield is also only considered in the later stages of the programme. It is not clear what period of recording is currently used by TRFK, but one of the clones released in 2006 (Kamunya and Wachira, 2006) was first planted in a clone trial in 1994, and originated from a seedling population planted in the mid-1980s. Thus the total interval between crossing and release may be up to 20 years. The time scale for our programme in Kenya is shown in Box 5.8.

For selection of ortets in the field, Visser (1969) indicated that yield over 6–10 plucking rounds was correlated with yield over a full three-year cycle. However, he concluded from the poor correlations between ortet and clone yields (0.18–0.31 for six groups of clones) that much of the variation in yield between bushes was environmental. As noted above, Visser's poor results may have been because he only included selected ortets, with a narrow range of yields; others have found much higher correlations (see *Broad-Sense Heritability*, above). However, results with mature bushes may not be relevant to recording of young seedlings in a breeding programme. Results from our programme are described in Box 5.9; we are not aware of any other information on selection of young seedlings.

A further question is how long clone trials must be recorded to give reliable results. Data from Njuguna (1989) and from TRFK Annual Reports show significant correlations between first- and second-cycle yields, ranging from 0.70 to 0.98; for 17 trials, the average correlation was 0.88. Sebastiampillai and Solomon (1976) argued that a single year would give a good indication of potential yield; their results are summarised in Table 5.5, and show that even three months' recording gave a reasonable correlation with total yield.

The duration of recording in our clone trials is considered in Box 5.9. We conclude that preliminary selection after one or two years is certainly possible. If cuttings of the best clones are taken into the nursery at that stage, a further year's data will be available

Box 5.8 Time Scale for Clone Development

In our programme we have given greater emphasis to yield than some other programmes; as noted at the beginning of this chapter, yields must be roughly doubled if replanting is to be profitable. The time scale is outlined below. The use of flush-shoot cuttings, first described by the Tea Research Foundation of Central Africa (Mphangwe and Nyirenda, 2001a, 2001b; Kamunya and Wachira, 2003) allows yield recording of seedlings to continue without a break. This, together with reduced periods of recording as discussed in Box 5.9, has allowed the release of clones as early as the 10th year after the initial cross was made.

Year	Operation
0	Controlled crossing of selected clones
1	Seedlings in nursery; micro-manufacture for preliminary quality evaluation
2	Field planting of seedlings; at least 100 per cross, if available; include standard clones
3	Yield recording of seedlings
4	Flush-shoot cuttings from best 30% of seedlings to nursery; continue yield recording
5	Highest-yielding 20% of clones to phase 1 field trial, based on two years' records from seedlings; two replicates of 16-bush plots, plus standard clones
6–9	Yield recording of phase 1 clone trial; mini-manufacture from highest yielders for quality assessment
7–8	Cuttings from clones which yield sufficiently better than standard clones
8–9	Plant phase 2 multi-location trials with best clones; two replicates of 16-bush plots at four or five sites, representing important agroecological zones
	Prune phase 1 clone trial; record trash weight and speed of recovery
	Possible release for planting in same zone as phase 1 clone trial
10–13	Yield recording of phase 2 multi-location trials
	Continue recording phase 1 trial for second cycle
10–11	Plant multiplication blocks of best clones
12–13	Release for commercial planting

We investigated the possibility of taking cuttings from nursery seedlings before field planting. This potentially shortens the time scale even further, but the number of cuttings available was small and variable, and the resulting clone trials were not very satisfactory.

A limiting factor in a breeding programme is the time it takes to bring a clone to flowering. Once good clones have been identified, cuttings must be planted out in a 'breeding block', and left to grow and flower. This may take up to four years. Grafting onto stumps of older bushes did not appear to accelerate flowering. We have tested paclobutrazol to induce flowering, but any effect was small. An alternative which does give earlier flowering is to allow established bushes in clone trials to grow unplucked; this disrupts yield recording, so can only be done when recording of the trial has finished. Also the selected clones will be widely scattered, so the bushes are best used only as male parents; it is preferable to keep all female parents in a single breeding block, where they are easily monitored, and can be irrigated to prevent fruit loss during drought.

Box 5.9 Duration of Recording in Trials

For seedlings from controlled crosses, we find that the correlation between ortet and clone is higher if the ortets are recorded for two or three years rather than one (see below). Thus while we make preliminary selections after one year, and take flush-shoot cuttings into the nursery, recording continues, so that before field planting two years' data are available (see Box 5.8). At that stage up to a third of the clones in the nursery will be discarded. Interestingly, the correlations with the third year of seedling yields only were just as high as with the first three years. This implies that correlations with mature bushes would probably also be high.

Correlations between ortets and clones in relation to duration of recording: 48 seedlings selected by stratified random sampling to cover full range of yields

Seedlings recorded for	Clones recorded for		
	2 years	3 years	4 years
1 year	0.442**	0.468***	0.439**
2 years	0.506***	0.593***	0.559***
3 years	0.552***	0.676***	0.660***
3rd year only	0.534***	0.691***	0.702***

** $p < 0.01$; *** $p < 0.001$.

The duration of recording in clone trials is also an important factor in the overall time scale. The increase in correlation with ortet yields when the clone trial was recorded for three years rather than two (above) implies that two years' recording may not be sufficient. The table below shows results of a clone trial which was recorded for a total of 12 years (three pruning cycles). The correlation between total yield and early yield increased as the number of years increased. Yield over the first pruning cycle (four years) was very highly correlated with total yield, and the two best clones for total yield were also the two best over the first cycle. We conclude that four years' data should be adequate. In practice, as shown in Box 5.8, the phase 1 clone trial may be recorded through two complete cycles, while the multi-location phase 2 trials are recorded for one cycle.

Correlations between early and later yield in a clone trial: 15 clones, recorded for a total of 12 years (three pruning cycles)

	Correlation
First year's yield and total (12 years) yield	0.524 ($p < 0.05$)
First 2 years' yield and total yield	0.764 ($p < 0.001$)
First 3 years' yield and total yield	0.876 ($p < 0.0001$)
First 4 years' yield (1st cycle) and total yield	0.936 ($p < 0.0001$)

In another study, we found that the mean correlation of first-year yield with the total yield over five years was 0.62, in 13 clone trials each with between 40 and 64 clones; for yield in the first two years the mean correlation with total yield was 0.80.

Table 5.5 Clone trial recording in Sri Lanka (Sebastiampillai and Solomon, 1976): 31 clones and one seedling population recorded for seven years at two sites

	Site 1		Site 2	
	No shade	Shade	No shade	Shade
Recording period	Correlation with total yield over seven years			
First 3 months	0.832	0.850	0.690	0.787
First 6 months	0.795	0.873	0.725	0.793
First year	0.866	0.882	0.834	0.846
1st cycle (3 or 4 years)	0.949	0.941	0.895	0.913

by the time of field planting, or by the time that a multiplication plot starts to produce cuttings for commercial planting.

The plot size and number of replications influence the precision of trials (the ability to detect significant differences between treatments). TRFK (1986) recommended that field trials should include three replicates of 60-bush plots. Ellis and Nyirenda (1995) used five replicates of 30-bush plots (Table 5.4). We have found that two replicates of 16-bush plots is a good compromise between precision and area of trials. In most such trials the coefficient of variation is below 10%, and the least significant difference between clones is less than 15% of the mean. Within each trial, the best clones are selected by a 'subset selection' procedure (Van der Laan and Verdooren, 1990), which involves identifying the group of clones which are not significantly different from the best.

A further point to be considered is the number of sites over which clone trials should be replicated. There is no simple answer to this. There are several reports of significant genotype × environment interactions in tea (Wickramaratne, 1981a; Rono et al., 1993; Ng'etich and Stephens, 2001; Wachira et al., 2002; Makola et al., 2013), so the aim should be to cover the range of environments over which the clones might be planted. Variation in altitude (temperature) and rainfall distribution should be considered, as should major differences in soil type. Wickramaratne (1981a) made the important point that it may be better to use locally adapted clones as standards in clone trials, rather than a single standard at all locations.

Is Grafting Useful?

Grafting offers another way of combining the characteristics of two different genotypes: for example, a scion with good quality might be grafted onto a high-yielding rootstock (Figure 5.9). Cleft or saddle grafting of single-node cuttings is very easily done (Kayange, 1990), with high success rates (Tuwei et al., 2008a). Kayange et al. (1981) showed that the yield of two low-yielding clones could be increased by over 40% by grafting onto vigorous rootstocks, without loss of quality. Even greater yield

Figure 5.9 A further refinement to crop improvement is the composite plant in which, for example, a high-quality scion is grafted onto a vigorous drought-resistant rootstock. This photograph shows an established cleft graft (Kenya, RHVC).

increases were reported by Satyanarayana *et al.* (1991) in South India. In Malawi, Pool and Nyirenda (1981) showed that the effect of grafting was to increase shoot numbers, and thus yield, without loss of quality. In Kenya, Tuwei *et al.* (2008a) confirmed that the effect of grafting was on shoot number, but found significant stock × scion interactions, as did Ellis and Nyirenda (1995). Not all scions gave yield responses even on the best rootstocks. Lower-yielding clones tended to give larger responses as scions, but there were some low-yielding clones which did not respond. Thus stock–scion combinations must be tested individually before commercial planting is undertaken, which extends by several years the time required to develop planting material.

Among rootstocks, Tuwei *et al.* (2008a) found that one of the best was TRFCA-PC 87, which gave an average yield increase of 18% with 20 different scion clones

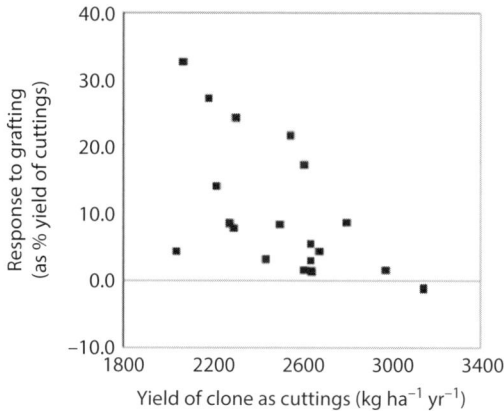

Figure 5.10 Yield increase following grafting, in relation to ungrafted yield (from Tuwei *et al.*, 2008a).

(Figure 5.10). This clone has also been recommended as a rootstock in Malawi (Ellis and Nyirenda, 1995). Tuwei *et al.* looked, without success, for characteristics which might be used to predict the performance of clones as rootstocks. Without the ability to predict performance, identification of new rootstocks is very laborious, with no guarantee of success.

In Malawi, where there is a regular and long dry season, rootstock selection has specifically concentrated on drought tolerance (Harvey, 1988). In South India Satyanarayana *et al.* (1991) obtained yield increases during dry periods from grafted plants. In Kenya, Tuwei *et al.* (2008b) showed that drought tolerance could be improved by grafting onto appropriate rootstocks, with the most drought-susceptible scions showing the greatest benefit. They found that rootstock drought tolerance was correlated with xylem water potential (Chapter 13), while scion tolerance was correlated with stomatal conductance.

Although drought tolerance has received much emphasis, it is important to recognise that some rootstocks can increase scion yield, independent of any drought tolerance effect. In Kenya, droughts do not occur every year, so the ideal rootstock will give a yield increase in non-drought years, as well as improving drought tolerance.

Non-Conventional Breeding

Interspecific Hybrids

Crosses between *C. sinensis* and other *Camellia* species have been attempted. Fertility of such crosses is usually very low, but some interspecific clones have been developed. Bezbaruah (1987) described a clone (TV24) from a cross between *C. irrawadiensis* and *C. sinensis. C. irrawadiensis* and *C. taliensis* are of interest because both species have very low caffeine content (Wachira, 1994a); tea quality is poor, but hybrids with *C. sinensis* offer the possibility of a naturally low-caffeine tea (Ogino *et al.*, 2009).

Polyploids

A normal diploid plant has two sets of chromosomes, but sometimes, through errors in cell division, the number of sets may be increased to three (triploid) or four (tetraploid). In many plant species, triploids and tetraploids tend to be more vigorous than normal diploids. They are usually sterile, but for tea, where the product comes from vegetative growth, this would not matter. The expected increase in vigour has stimulated interest in these polyploids, but results of trials are equivocal. Singh (1980) found that tetraploid plants had larger and heavier leaves than diploids, while Sarmah and Bezbaruah (1984) showed that 20 triploid clones out-yielded three diploids. Wachira (1994b) showed that on average triploids had heavier shoots than diploids, but smaller shoot numbers and lower yield, though the best triploid did out-yield most of the diploids. Wachira and Ng'etich (1999) found that 12 diploids out-yielded 11 triploids, though the best triploids out-yielded some of the diploids. A single tetraploid yielded less than all but two of the triploids.

Mutation Breeding

X-rays, gamma-rays or chemical mutagens can be used to cause mutations, thus increasing genetic variability. However, any mutations will be random, and the great majority will be harmful, or at best neutral, in their effect. The scale of screening necessary to identify useful mutations argues against this technique.

  We consider that there is still so much scope for yield and quality improvement by conventional breeding that work on unconventional methods would be an unnecessary diversion of effort.

What Can Biotechnology Contribute?

There are two aspects of modern biotechnology relevant to tea improvement. Marker-assisted selection has the potential to increase the rate of progress in a breeding pro-gramme. DNA transformation or genetic engineering (also misleadingly called genetic modification) offers the possibility of introducing completely new characters.

Marker-Assisted Selection

Variation results from differences in the genes, which carry instructions for the synthesis of enzymes (proteins). With molecular markers, variation in the DNA itself is studied. DNA markers can be used for confirmation of pedigree or legitimacy, for assessing genetic diversity, or for 'marker-assisted' selection of parents. Selection based on DNA should be more reliable than conventional selection based on phenotype, because the latter is affected by environment as well as genotype. Markers linked to (on the same chromosome as) specific genes can now be identified, and there is much overlap between marker techniques and the methods of genetic engineering.

The DNA sequences identified with markers are carried on the chromosomes, and so are inherited according to Mendel's laws. If markers linked to useful traits such as yield or disease resistance can be identified, then marker-assisted selection becomes possible. Such selection can be done for characters that are not being expressed phenotypically; for example, disease resistance could be selected for in an area where the disease did not occur. The first step is to develop a linkage map, identifying groups of markers that are on the same chromosome and tend to be inherited together. The closeness of the linkage, calculated statistically, shows the relative positions of the markers along the chromosome. Once a linkage map has been developed, a set of markers that covers all chromosomes with reasonably even distribution can be chosen, and studied for linkages with useful characters.

One of the first linkage maps for tea was described by Hackett *et al.* (2000). Kamunya *et al.* (2010b) identified several markers linked to yield quantitative trait loci (QTLs), but the clones studied were planted at two sites, and none of the QTLs was common to both sites, indicating a genotype × environment interaction. Such linkages also tend to be population-specific, so that a linkage found in one cross may not be seen in others. Kamunya *et al.* (2011) found markers linked to total polyphenol content, and to tolerance to root knot nematode and red crevice mite. Zheng *et al.* (2015) identified genes which were expressed after a cold treatment, so might be related to cold tolerance.

Mphangwe *et al.* (2013) found markers linked to quality, yield, and tolerance to drought, high and low temperature and *Phomopsis theae*. The type of marker used, RAPD, does not always give repeatable results, but the potential of marker-assisted selection is clear. In principle, it allows the screening of nursery plants or even germinated seeds for yield, thus accelerating clone development. In effect, marker-assisted selection could be the first stage of a two-stage selection scheme (see above).

Genetic Engineering

The ability to add, delete or modify individual genes by DNA transformation opens numerous possibilities. In other crops, DNA transformation has been used to change product quality, and to introduce pest, disease and herbicide tolerance. In a recent review of 147 studies, Klümper and Qaim (2014) found that, on average, adoption of genetically engineered or 'genetically modified' (GM) varieties had reduced pesticide use by 37%, increased yields by 22%, and increased farmers' profits by 68%. Yield and profit gains were higher in developing countries than in developed countries. Since 1996, over 1.5 billion hectares of GM crops have been grown worldwide, with no reliable reports of any environmental harm. Thus it appears that the 'anti-GM' stance of some countries and non-government organisations (NGOs) has no justification (Leyser, 2014).

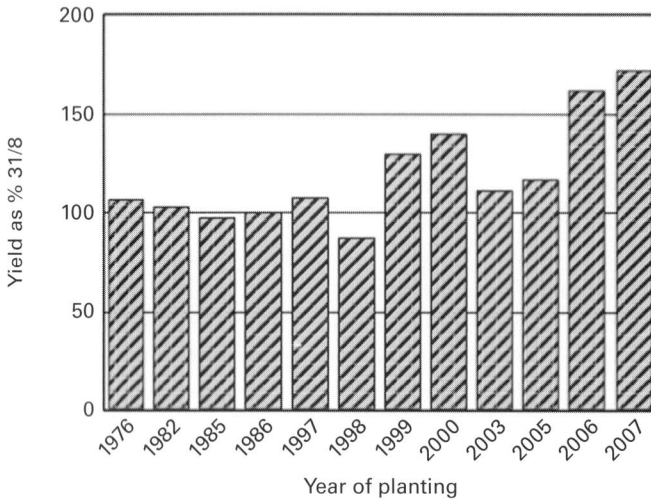

Figure 5.11 Yields of the best Tea Research Foundation of Kenya clones in trials at Kericho, Kenya, based on at least two years' records, relative to standard clone TRIEA-31/8. (From TRFK annual reports; data not available for some years.)

Mondal *et al.* (2001) demonstrated the feasibility of transforming tea, producing plants with a gene for antibiotic resistance and a marker gene. Shi *et al.* (2011) identified most of the genes involved in flavonoid, theanine and caffeine synthesis, a necessary first step if tea quality is to be modified by genetic engineering. Mohanpuria *et al.* (2011) used 'RNA interference' to suppress caffeine synthesis, and obtained plants with 44–61% lower caffeine levels. Some progress is being made, therefore, but it is likely to be many years before any GM tea varieties are commercially available.

What Can We Expect in Future?

Success in a tea breeding programme can be judged by the performance of the best clones produced. In Malawi, one of the main objectives was to improve quality (Ellis and Nyirenda, 1995). This appears to have been achieved, with one estate quoted as selling its tea for a higher price than the same grade of Indian tea. One of the clones from the breeding programme yielded 24% more than an early clone from field selection, and further progress has undoubtedly been made in the last two decades.

From annual reports of the TRFK, it appears that good progress has also been made in Kenya in recent years. Figure 5.11 shows yields of the best clones in trials in the Kericho District, by year of planting, relative to the standard clone TRIEA-31/8.

Progress in our programme over the last 20 years is summarised in Box 5.10. By 2015, clone MRTM1, a field selection, and two clones from the breeding programme had each been used for replanting over more than 500 hectares.

Box 5.10 Progress in Clone Yield in Unilever Tea Kenya Programme

When the UTK plant improvement programme started in the late 1980s, current commercial planting in UTK was predominantly of clone TRIEA-31/8, which was adopted as a trial standard. However, clone MRTM1, a field selection, was found to out-yield TRIEA-31/8 by about 33% (average of eight trials), and eventually replaced 31/8 as the trial standard. Clone 31/8 is drought-susceptible, but MRTM1 is highly drought-tolerant (Tuwei, 2008), so any clones which out-yield MRTM1 are likely also to be reasonably drought-tolerant.

In the figure below, the difference of 33% between TRIEA-31/8 and MRTM1 has been used to give the yields relative to TRIEA-31/8. It is clear that the initial target of doubling yields has been comfortably achieved.

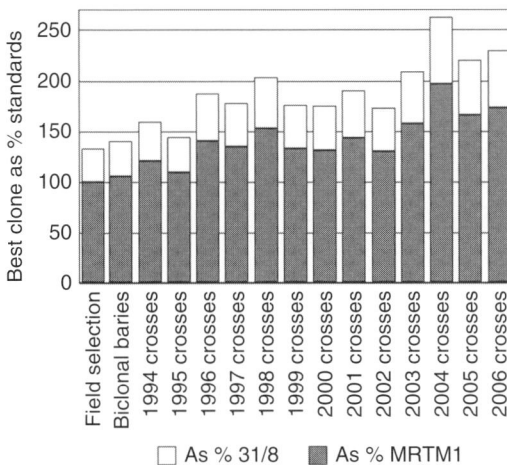

Yield of best clone from each year's crosses, relative to standards. Earlier years' data are from phase 2 clone trials, later from phase 1 (see Box 5.8).

In 2010 there was no dry season, and several clones yielded over 8000 kg made tea per hectare in the third or fourth year after planting, indicating the potential of the material.

Summary

- Many clones have been identified by field selection, but a breeding programme is much more likely to give superior material.
- Controlled pollination has high costs because of low efficiency, but gives greater flexibility than biclonal seed gardens.
- Selection for yield is straightforward, but no good method of selecting for quality has been developed.

- Developing clones from seedlings is a two-stage process. Because the correlation between seedling and clone yields is fairly low, if a high selection intensity is applied at the first stage, most of the best clones may be eliminated.
- The time from making a cross to releasing a new clone can be reduced to about 12 years.
- Composite (grafted) plants offer possibilities of increasing both yield and drought tolerance.
- Molecular markers will start to play a part in tea breeding in the near future.
- In our breeding programme in Kenya, we have doubled yields in 20 years.

Acknowledgements

We are grateful to Unilever Tea Kenya for support for the programme, and for permission to publish. F. K. K. Kaptich, J. K. arap Rono, J. Kenduyiwa, K. C. Chomboi, M. C. Langat and B. G. Smith have all contributed to the work described.

6 Planting and Replanting

Who Ever Said This Was Simple?

Crop establishment begins in the nursery, with the propagation of cuttings (or the sowing of seeds), and continues with plant management in the nursery (including judging the optimum intensity of shade), followed by land preparation and field planting. Establishment is judged to have ended when the young plants have been brought into production, the leaf canopy is covering the ground, and any infilling is complete. To support new tea developments and major replanting programmes, cheap and quick methods for propagating large numbers of clonal plants, including composites, are needed. This chapter includes a review of the experimental evidence on which to judge the value or otherwise of traditional and more recent methods of vegetative propagation. Much has been written on the propagation of tea by heterogeneous seed. This chapter will consider only vegetative propagation.

Replanting incurs a direct cost as well as a loss in revenue, and ways are needed to minimise the time taken to reach a positive return on investment. This could involve manipulating plants while they are still in the nursery to encourage branching, and it certainly means the intensification of tea crop management by, for example, reducing the amount of soil handling in the nursery, and increasing the planting density. Management issues include deciding (1) which fields to replant first, (2) the need or otherwise for tea soil rehabilitation, (3) how best to clear and replant old seedling/moribund tea, and (4) what clones to plant, and at what density. The methods available to facilitate crop development and to bring plants into production once they are in the field are reviewed in Chapter 7, and the impacts of these methods on root growth are covered in Chapter 8.

Nurseries

For any large-scale tea development, a nursery is needed where plants can be propagated prior to planting in the field. This is essentially a separate business exercise while the land preparation is under way. In addition, there is always a continuing need for a small nursery to supply plants for infilling, or for minor expansion. The requirements for large-scale uprooting and replanting of moribund, or relatively low-yielding, tea are similar to those needed for a new development. Most smallholder schemes have a central nursery that supplies and sells plants to the farmers (Figure 6.1). Depending on the conditions imposed by such schemes, smallholders may have their own nurseries as well (Figure 6.2).

The practicalities of planning or building a nursery are common across the tea world, and these are summarised in Box 6.1. Since the amount of shade needed at different times

Figure 6.1 Plants being delivered to smallholders (Tanzania, MKVC).

Figure 6.2 A smallholder-managed tea nursery in Tanzania. Women play an important role in the tea industry (MKVC).

Box 6.1 Tea Nurseries: Issues to Consider

1. Choice of site:
   - Drainage: a gentle slope is ideal
   - Access to water
   - Road access
   - Security
2. Construction of nursery:
   - Overhead/high shade (common), or low shade
   - Light intensity (25–40% of full light at midday)
   - Light distribution (even)
3. Nursery beds:
   - Width (1 m); raised (0.1 m); length (up to 30 m)
   - Direction (across slope, angle 10 degrees)
   - Drainage
4. Preparation of mother bushes:
   - Healthy bushes
   - Pruned (six months before cuttings needed)
   - Well-watered (not droughted)
5. Containers:
   - Polythene sleeves
   - Clear plastic preferred
   - Size (mini, standard, medium, jumbo)
   - Holes in base (drainage)
   - Soil mixture (subsoil cap 1/3: topsoil or mixture 2/3)
   - Firming soil
6. Propagation:
   - Single/double/multiple nodes or pluckable shoots in trays of miniature pots ('Speedling trays')
   - Fungicide
   - Keep moist
   - Use a nail to create a hole for the cutting (to minimise damage to shoot)
   - Wet beds
   - Beware mutual shading
   - Seal with polythene sheet (250–500 gauge) in cool locations, or
   - Mist with knapsack sprayer (every 30 minutes) where it is warm
7. Culture:
   - Reduce shade gradually (light meter)
   - Remove polythene gradually
   - Plants ready for planting in field after 9–24 months (temperature-dependent)

during the propagation and hardening-off processes is an important variable affecting the quality of the plants, it is a surprise that the way shade is managed in the nursery is still largely subjective. No one appears to have attempted to quantify the degree of shade needed at different times and to specify this in terms of the light intensity, which can be easily measured with a light meter. Apart from the introduction of vegetative propagation, polythene sleeves, clear plastic sheeting and, in some places, miniature pots, few advances appear to have been made over the last 50 years in the way most nurseries are managed although its complexity is acknowledged (Figure 6.3; Box 6.2).

Figure 6.3 (a) No one appears to have attempted to quantify the degree of shade needed at different times and to specify this in terms of the light intensity. (b) It is important to move the plants at regular intervals to stop them getting too big and to limit the degree that the roots escape from the polythene sleeve. (c, d) Plants have been graded in size prior to the larger ones going to the field (East Africa, MKVC). *A black and white version of this figure will appear in some formats. For the colour version, please refer to the plate section.*

Box 6.2 Commentary: Nurseries

The balance between light intensity, temperature and water is crucial. Too much shade lowers temperatures and delays rooting: too little shade can raise the temperatures too high and cook the cuttings, so ventilation is needed to allow the hot air to escape, while maintaining high light levels. But this necessitates additional irrigation. Another advantage of a high light regime is a reduced period of hardening-off. A heavier shade regime reduces temperatures, so ventilation is not needed, and less irrigation is required. For a busy manager this has its attractions, although rooting and growth will be slower, and hardening-off will take longer.

T. C. E. Congdon

Propagation

The most widely used system of vegetative propagation in use today is that based on a cutting consisting of a leaf together with its axillary bud and 30–60 mm of leafless stem below the leaf. A bi-node cutting consists of two leaves and their axillary buds plus the internodal stem. There are now standard procedures for preparing cuttings, rooting the stem in acid subsoil over topsoil in polythene sleeves, and growing-on healthy plants.

Recent developments in plant propagation, which have largely been led by scientists in Malawi, include the use of harvested shoots (known as 'pluckable shoots') as alternatives to single leaf-bud cuttings for the mass propagation of plants, the manipulation of plants in the nursery to improve lateral bud development prior to planting in the field, and the promotion of composite plants.

Pluckable Shoots

Replanting seedling tea with larger-yielding and better-quality clones is normally a slow process. This is due in part to shortages of plants. This and other related issues, including the loss in revenue after uprooting and the length of the payback period, are considered below. One of the reasons for the lack of planting material is the method used to raise plants. For example, in the seasonal climate of Malawi, relying on mature green stem cuttings limits the period for propagation to March to June. Although using brown stem cuttings and immature terminal shoots is also a successful way to propagate tea, not enough extra plants can be obtained in the limited window of opportunity for planting. These are the issues that led Nyirenda and Mphangwe (1998a) to investigate the possibility of using so-called 'pluckable shoots' for propagation in order to facilitate the large-scale production of clonal plants for replanting moribund, or just low-yielding, seedling tea. Some years earlier, in South India, Satyanarayana and Ilango (1993) had shown that shoots with up to four leaves taken from bushes under regular plucking were capable of producing roots (Figure 6.5).

In preliminary trials in Malawi three clones were tested as well as two shoot types:

* *Clones*: TRFCA-PC81, TRFCA-SFS150 and TRFCA-PC108
* *Shoot types*: three leaves with a terminal bud, and three leaves without a terminal bud.

The shoots were propagated either:

* inside a *polythene tunnel*, or
* with *misting* (applied every 30 minutes).

All the shoots tested had fully lignified tissue in that part of the stem that was placed in the rooting medium.

After five months, more shoots had developed roots in the polythene tunnel (overall mean = 64% success) than with misting (44% success). The most successful treatment combination was:

Figure 6.4 A tea nursery in Tanzania (MKVC).

Figure 6.5 Using pluckable shoots allows you to produce many more plants quickly, compared with leaf bud cuttings. (Tanzania, MKVC).

- Clone TRFCA-SFS150 in a polythene tunnel with shoots without the terminal bud (81% success).

The least successful was:

- Clone TRFCA-PC108 with misting and shoots with or without the terminal bud (both treatments 27% success only).

Using a rooting hormone (indole butyric acid, IBA) improved the success rates in all the treatment combinations tested (range 81–97% success), except with TRFCA-PC81. It is estimated that by using pluckable shoots for propagation 1 ha of tea could provide over 12 million shoots each year, which is enough to plant 1000 ha. This assumes a plant density in the field of 10,000 plants ha^{-1}, 250 shoots $bush^{-1}$, 3500 kg ha^{-1} of processed tea, 700 shoots kg^{-1} (green leaf) (Nyirinda and Mphangwe, 1998b).

Subsequent research and commercial experience in southern Africa has confirmed the value of using pluckable shoots for propagating tea. Benefits include the option to plant out earlier than is normal for conventional plants and faster establishment in the field. Plants raised from pluckable shoots were generally taller, had more branches and greater root and shoot dry weights (as measured 16 months after propagation) than those raised from conventional cuttings. The shoot:root ratios (based on the dry mass) were similar. In the field, roots of four-year-old plants reached depths of 1.2 m regardless of the method of propagation. In all cases, 70% of the total root dry mass was concentrated in the top 0.30 m of soil (Chapotoka *et al.*, 2001). Shoots harvested during the 'off' (dry) season in Malawi (May to December) developed the best roots. Pluckable shoots can be harvested throughout the year. Nyirenda and Mphangwe (2001) have described the methodology in detail.

In addition, by using trays of miniature pots,[1] rather than mini- or standard pots, the volume of soil needed is considerably reduced and space in the nursery/polythene tunnels is saved (576 cuttings m^{-2} compared with 200 or 100 m^{-2} with mini-pots or standard 90 mm diameter pots respectively). And, because there is less soil to handle, costs are reduced. There are risks associated with planting small plants (60–100 mm tall) with a small root volume into the field. A high level of post-planting care is needed to minimise plant losses. Mulching is advised, and irrigation may be necessary to ensure good field establishment (McCubbin and Steenkamp, 1998). This will add to the cost of this method of propagation, which could make the practice uneconomic. The aim of a nursery is to produce plants that establish quickly, survive well and yield early.

The pluckable shoot technique has also been evaluated in a nursery trial in Kenya. This involved comparing five different clones and four cutting types: single-node cuttings, shoots with three leaves (either leaf numbers 1, 2 and 3, or 2, 3 and 4) and shoots with two leaves (numbers 3 and 4). All three shoot types, representing pluckable shoots, were considered to be suitable alternatives to single-node leaf-bud cuttings. All five clones responded well (Nyabundi, 2009).

It is not just in Africa that the pluckable shoot technique has been found to be useful. For example, in Assam, Saikia and Sarma (2011) successfully raised and established plants grown from pluckable shoots (two leaves and a bud; clone TV1 and, especially successful, clone TV23). Benefits included marginally improved success rate in the nursery, taller plants with more roots in the top 100 mm of soil and, as in Malawi, availability of shoots at any time during the harvesting season This appears to be a technique worthy of wide uptake.

Nursery Manipulation

Nursery manipulation is a generic term used to describe the range of practices aimed at improving lateral branch development in the nursery. It includes pruning, nipping and tipping of nursery plants. Studies in Malawi highlighted issues with restricted development of both the shoot and root system as a result of such practices, leading to smaller and more compact plants. In some cases the root systems were reduced by 50%, while there was only marginal improvement in the development of lateral branches.

Mphangwe and Nyirenda (2000) have described the residual effects of treatments like these on field performance. All the plants tested had very few branches at the time of field planting, although those that had been 'nipped' in the nursery had marginally more branches than those left untouched. But this minor benefit of nipping had been lost by 4–6 months after planting out. Field survival differed between the two clones tested. Clone TRFCA-SFS150 survived well regardless of nursery manipulation (> 90% survival), whereas both pruning and nipping reduced the survival of clone TRFCA-PC108 plants (by about 10 percentage points, from c. 80% to 70%) in all three size groups (small, medium and large) compared with the 'untouched' control treatment. There were no significant or consistent yield differences between treatments in either the first or second seasons after planting. It should be noted that the apparent benefits from nursery manipulation can be more effectively achieved by removing the central growing point (de-centring) from young plants in the field or by bending and pegging (see Chapter 7). This applies especially to clones that exhibit strong apical dominance (such as TRFCA-SFS150).

For ease of management, Nyirenda and Mphangwe (1998b) emphasised the importance of planting clones that are compatible. For example, clones should have similar fermentation times, rates of recovery from pruning, and leaf pose. Another example is that in order to facilitate mechanical harvesting and to minimise the amount of cut-leaf entering the factory, all the recently released clones from the Tea Research Foundation of Central Africa have horizontal maintenance leaves.

Composite Plants

Grafting is a very ancient practice, which is applied to fruit trees in all parts of the world. Its application to tea is relatively recent. A composite tea plant consists of a scion, which will develop into the branches and leaves, grafted on to a rootstock, which will form the roots and the lower part of the frame. The primary purpose of combining two clones in this way is to combine the beneficial attributes of a scion, which may be high-quality tea, with those imparted by the rootstock, which may for example be vigour, or drought resistance. Much of the early research on grafting was undertaken in Malawi and in South India, and more recently in Kenya.

Grafting is also practised in plant breeding and vegetative propagation programmes where there is a need to obtain clonal seed or cuttings quickly. When the (seedling) rootstock of a tea tree, which has reached seed-bearing size, is grafted with a clonal scion, a large number of cuttings can be produced in less than two years, and seed in less than three years (although the time to flowering is not affected). Clonal seed was produced in this way from grafting carried out in the late 1960s/early 1970s by the Tea Research Institute of East Africa in western Uganda (Templer, 1971). The (hybrid) seed (female parent TRIEA-6/8) was transferred to Kenya for evaluation. It was not until 25 years later, in 1994, that the Tea Research Foundation of Kenya released five 303-series clones to the industry. A lot had happened in both countries, and to both organisations, in the meantime, but the seed orchard in Uganda had survived. You have to be an optimist to be a tree breeder!

For normal field planting, composite plants are created in the nursery by cleft-grafting, or chip-budding, fresh cuttings of the scion to the rootstock (see Chapter 5, Figure 5.9). Sharma (1982) perfected the cleft grafting technique in South India. The bottom of the stem of a single-node cutting to be used as the scion is tapered into a wedge. A corresponding cleft is cut into the stem of the rootstock cutting. The scion and the rootstock are then immediately united and tied with a polythene strip or, as now recommended in South India, a plastic clip (Satyanarayana and Sreedhar, 1995). Chip-budding is an alternative, less common way of combining the scion to the rootstock. It involves removing the bud from the scion along with a 200 mm strip of bark, and then inserting that end into a corresponding slot of similar size created in the internode of the rootstock. Polythene tape is again used to secure the two parts. The pluckable shoot technique has also been successfully developed for the production of composite plants (Nyirenda and Mphangwe, 2000a, 2000b).

Harvey (1989) reviewed the history of composite tea development in Malawi. In the 1960s, the original objective was to use grafting to convert old tea bushes of seedling origin into clonal tea (but on individual seedling rootstocks). This approach was not successful, nor was chip-budding onto two-year-old rootstocks. In the 1970s, the break-through came when chip-budding onto *unrooted* cuttings was tried and found to be a successful way of producing composite plants. The advantages included: the union was strong; no suckers were formed after the rootstock shoots had been cut back; and the technique was relatively easy to use (Kayange, 1988).

This collective experience in India and Africa confirmed that a rootstock clone can influence the vigour of the scion. For example, rootstock TRFCA-MFS87 increased the yield of TRFCA-PC1 and TRFCA-SFS204, although it had no effect on the yield of scion clone TRFCA-MT12. Tea quality was not affected. The yield increase was the result of an increase in the number of shoots, not the shoot size or its rate of development (Kayange *et al.*, 1981). Another promising rootstock is clone TRFCA-PC87.

In South India, Satyanarayana *et al.* (1991) found that the compatibility between a scion and a rootstock varied, as indicated by yields, drought tolerance and dry matter production (measured above the height of pruning). Promising combinations included UPASI-3 (scion) on UPASI-2 and UPASI-6 (both rootstocks); UPASI-8 on UPASI-2 and UPASI-9; and UPASI-17 on UPASI-2 and UPASI-9. A large proportion of tea in South India is now being planted with composites.

Composites are also being planted commercially in Kenya. This follows research undertaken by Unilever Tea Kenya Ltd (as mentioned in Chapter 5). The drought tolerance, as defined by a drought index, of susceptible scions/clones was improved, and yields were increased as a result of an increase in shoot number (as was also the case in Malawi). Useful rootstocks included EPKTN-14/3, China 3-1, BBT-1 and TRFCA-PC81. The most productive scion clone was TRIEA-31/8. Yield increases of 400–500 kg ha^{-1} (*c.* 10%) were obtained when this clone was grafted onto rootstock EPK-TN14/3. In addition to those clones already mentioned, clones TRFCA-MFS87, TRFCA-PC87, TRFK-7/7, TRFK-14/22, TRFCA-PC138 and TRFCA-PC141 (from Kenya and Malawi) were all considered to be potentially good rootstocks. They combined compatibility with a large range of scions, with yield benefits as well as with drought-tolerance characteristics (Bore, 2008; Tuwei *et al.*, 2008a, 2008b).

In Sri Lanka, 32 scion–rootstock combinations were compared over one four-year pruning cycle. The two highest-yielding composites, even during the dry season, were scion clone TRISL-2023 cleft-grafted onto rootstock clone CY-9, and clone TRISL-2028 onto clone DN (Pallemulla *et al.*, 1992).

Composite plants are not always a success. Previous work in Kenya had, for example, failed to show any yield advantages over a nine-year period with chip-budding of four well-known scions (clones TRIEA-6/8, AHP-SFS15/10, EPK-14/3 and TRFCA-SFS150) onto three rootstocks (clones TRIEA-7/14, TRIEA-31/8 and STC-5–3) (Bore and Njuguna, 1995). Similarly, Mizambwa (2002) was only able to demonstrate small but insignificant yield benefits with composite plants in a drought/irrigation experiment in Tanzania. The choice of scion and rootstock is critical, and there does not yet appear to be a way of identifying which clonal combinations will make a good composite plant. Long-term field evaluation trials are still needed.

The propagation method choices are summarised in Table 6.1. Full details are available in practical manuals published in most tea-growing areas, such as the *Tea Planter's Handbook* in Malawi and southern Africa, the *Tea Growers' Handbook* in East Africa, *Tea Growing* and the *Bangladesh Tea Handbook* in Bangladesh, the *Handbook on Tea* in Sri Lanka, and the equivalent publications elsewhere.

Summary: Propagation

- Vegetative propagation based on leaf-bud cuttings has become the standard way of propagating tea.
- Since the introduction of plastic, tea nurseries have changed very little in recent years.
- No one seems to think it is necessary to monitor light intensity, as a way of managing shade in the nursery.

Table 6.1 Tea plant propagation: choices to consider

Type of cutting	Advantages	Disadvantages
Single leaf-node bud	Reliable	Not enough plants for large-scale development
Bi-node	More rapid development	Fewer plants
Multiple node	As above	Still fewer
Pluckable shoots	Large scale production of plants possible	Misting may be necessary in nursery
Composites	Need to define benefits; combine two attributes	Slow, expensive
'Speedling' trays	Less soil; less space needed	Small plants; care in field during establishment

- Propagating 'pluckable shoots' offers a means of substantially increasing the number of clonal plants for large-scale development or replanting programmes.
- Miniature pots offer ways of reducing the volume of soil needed in the nursery, and save space. Extra care is needed in transplanting in the field with this approach.
- There appears to be little or no advantage from 'manipulating' plants while they are in the nursery.
- Composite plants (mainly cleft-grafted) are becoming more popular, a number of promising rootstocks having been identified.
- These rootstocks can impart vigour and a degree of drought tolerance to high-quality scions.
- The increase in yield comes from an increase in the number of shoots per unit area.
- There is no simple way of judging which clonal combinations will make good composite plants; field evaluation trials are still needed.
- To be successful in the nursery the cambium of the scion must be well aligned with the cambium of the rootstock and the joint must be kept moist.

Plant Density

Following a detailed review of the issues, Barua (1989) concluded: '*the spacing of tea plants has been a hotly contested subject from the middle of the nineteenth century.*' This is not a surprising conclusion, as there are many variables that impact on the answer to the question 'What is the optimum plant density at which to plant tea?' and there is no single value that can be used worldwide. The main advantages of close spacing occur during the first few years after planting, when the yield increase is proportional to the number of plants. This is when the additional revenue that is generated can help to offset the extra investment costs. The faster the crop canopy expands and covers the ground, the smaller will be the advantage of dense planting. For example, if the clone spreads naturally (and hence intercepts more light at an early stage in its development) the benefit from close spacing is likely to be less than if the natural tendency is for the branches to extend vertically (and intercept less light). Similarly, the preferred density will be influenced by the method of bringing the plants into production. 'Pegging', for example, facilitates the attainment of full ground cover sooner than frequent 'cut-across pruning'. Temperature is also important. Close spacing is more likely to be beneficial at a high-altitude cool site than under warm conditions. Drought is another interacting weather variable. The optimum density under well-watered/irrigated conditions is likely to be different from that in drought-prone areas.

In hand-harvested tea, the choice of plant spacing is limited by the need to allow pluckers to move through the field without too much hardship. Machine-harvested tea will have its own requirements, depending on the method employed.

Many experiments have been conducted over the years in attempts to determine the optimum plant density for tea plants (this assumes that there is a single value). According to Barua (1989), the first trial was initiated in India 1864, while another long-term experiment was established in Indonesia in 1927. In both examples, the plants at the closest spacing yielded most (both at $c.$ 17,000 plants ha^{-1}).

In the nineteenth century, when the tea industry in India was just beginning and planting material was scarce, square spacing at intervals of 1.5–1.8 m was common (equivalent to 4440 and 3090 plants ha^{-1}). Even in the 1930s, a density of 6940 plants ha^{-1} (1.2 m × 1.2 m) was regarded as being high, despite the experimental evidence suggesting larger populations were advantageous. Currently the 'rule of thumb' advice given to growers in most countries (China being an important exception) is to seek to establish $c.$ 10,000 plants ha^{-1} ($c.$ 1 m^2 per bush). This is despite early work in North-East India (which began in 1948/49), indicating that the optimum plant population was between 12,600 and 17,000 ha^{-1} (Barbora, 1991). In Sri Lanka, the recommended planting distance under rain-fed conditions is 1.2 m between rows and 0.6 m within a row. This equates to a plant density of 13,600 plants ha^{-1}. The wider spacing between rows allows the plucker to walk through the tea. More generally, tea in Sri Lanka is planted at densities of between 6700 and 13,600 plants ha^{-1}. Table 6.2 shows the plant density corresponding to different row spacings.

While working in Malawi, Laycock (1961) analysed the yield responses of seedling tea plants to a range of plant densities (from 2470 to 18,970 plants ha^{-1}) as recorded in three experiments conducted in contrasting locations: Malawi, Indonesia and Assam. At all three sites the yield/density relationships following planting were linear, but as the low-density plants gradually attained full crop cover, differences in yield between treatments became less. By the sixth year after planting, yields approached a plateau (or asymptotic level), at densities between 6000 and 9000 plants ha^{-1}.

By plotting the reciprocal of *yield per plant* against *density*, Kigalu (1997) confirmed that the relationships were indeed asymptotic, as the slope of each line was linear. A single annual, relative-yield response function for mature tea was also justified, since the slopes of each of the three lines based on Laycock's analysis were similar. Probably because of its relative simplicity, the results of this analysis have been used to identify an appropriate plant population in any location, For example, the *Tea Growers' Handbook*, which targets growers in Kenya, Tanzania and Uganda (TRFK, 1986),

Table 6.2 Plant population at different spacings (after Hajra, 2001)

Arrangement	Spacing (cm)	Plant density (ha^{-1})
Square	120 × 120	6,274
	135 × 135	5,315
	150 × 150	4,305
Triangular	120 × 120	7,770
	135 × 135	6,139
	145 × 150	4,353
Single hedge	90 × 60	18,518
	110 × 60	15,151
	120 × 60	13,888
	120 × 75	11,111
	120 × 90	9,259
Double hedge regular	110 × 60 × 60	19,607
	120 × 60 × 60	18,518
	120 × 75 × 75	13,675
	120 × 90 × 90	12,345
Double hedge staggered	110 × 60 × 60	20,575
	120 × 60 × 60	19,379
	120 × 75 × 75	14,414
	120 × 90 × 90	13,888

includes a tabulation of the results of such an analysis, although it was developed in Malawi.

A plant density experiment conducted at a cool, high-altitude site (2200 m) in Kenya confirmed the impact that the time taken to reach full ground cover had on the response to changes in plant density. There were only three density treatments (7690, 10,250 and 15,380 plants ha^{-1}), but three contrasting clones were compared at each density: clone TRIEA-31/8 (wide spreading frame), clone TRIEA-7/14 (narrow frame) and clone TRIEA-6/8 (intermediate), together with a heterogeneous seedling population. In all cases, the shapes of the response curves were best described by a hyperbola. Five years after planting, the relatively narrow-framed clone TRIEA-7/14 and clone TRIEA-6/8 were still yielding more at the high density than at the low density, whereas with clone TRIEA-31/8 the yield differences by that time were small, full ground cover having been reached at all three densities (Njuguna, 1977).

A follow-up experiment with a single clone (TRIEA-31/8) at the same site compared a wide range of densities (6720, 27,874 and 111,111 plants ha^{-1}) over a 19-year-period (1980 to 1998) (Bore *et al.*, 1998a). Again there were significant yield advantages from the high-density planting for the first six years, but afterwards the benefits of exceeding 28,000 plants ha^{-1} were small once full ground cover was reached in all the treatments. The response curves were again asymptotic, with a 'maximum' yield of 3900 kg ha^{-1} (Kigalu, 1997). There were also yield advantages from the higher densities in the years when the tea was pruned.

There was one severe drought year (1996/97), in which a large number of plants died. The high-density plots lost 50% of the plants, while the low-density plots lost only 20%. Examination of the roots indicated that those of high-density plants occupied a smaller volume of soil than those of plants that were more widely spaced (Bore *et al.*, 1998a).

In another plant density experiment, which was repeated at two locations in Kericho, Kenya, one at a high-altitude (1930 m) commercial estate and the other at a lower altitude (1205 m), six densities (range 5980–17,960 plants ha^{-1}) were compared. Yields were recorded for the first six years. At these two equatorial sites, the plants were brought into production by de-centring and then tipping-in at heights of 0.20, 0.30, 0.40 or 0.50 m. The graph of annual yield response to density followed a familiar pattern. Initially the responses were linear, and then they became progressively, but weakly, asymptotic in shape. A common curve represented both the high- and low-altitude sites, and the theoretical maximum yield was 4800 kg ha^{-1} (clone BBK-7) (Corley, 1997, personal communication). There was no obvious difference in yield response between the two sites despite the difference in the mean air temperature (*c.* 4 °C).

Intensification

As with many tree crops, there is interest in planting tea at high densities in order to show a return on investment as soon as possible. Several experiments have included very high plant populations in the choice of treatments for comparison. One has been referred to above (Bore *et al.*, 1998a). In North-East India, Barua (1989) compared plant densities ranging from 4444 plants ha^{-1} to 444,000 plants ha^{-1} using a systematic experimental fan design (this allows extreme treatments to be compared within a relatively small area). In practice the maximum density that could be tested, because of plant deaths at the very high densities, was 126,000 plants ha^{-1}, still at least 10 times the international practice at the time. Initially, the maximum yield obtained was from the treatment with a density of *c.* 64,000 ha^{-1}, but this value fell in the following years to 17,500 plants ha^{-1}. Here was evidence that yields could decline at high densities. In other words, the response may not always be asymptotic (Barua, 1989). This may, however, have been a statistical anomaly. In special situations, such as steep slopes, the optimum density may be higher (Figures 6.6, 6.7 and 6.8).

The one country where very high-density planting is encouraged is China. According to Yu (2012), two types of hedge planting are practised in China. One type (a) the single- or double-row system, which was introduced in the 1950s, has a spacing for the single row of either 1.5 m × 0.33 m (20,200 ha^{-1}) or 1.3 m × 0.33 m (23,300 ha^{-1}); for the double row the between-row spacing is either 0.25 or 0.50 m (range 30,000–39,000 ha^{-1}). The other type (b) multi-row was introduced in the 1970s. Within the main row space (1.5 m wide) three to six extra rows are planted to give a density of 60,000–100,000 plants ha^{-1}.

The principal advantage claimed for the double and multi-rows is the speed with which plants come into production. But if the density is too high the frames are weak with thin

Figure 6.6 When the soil is left exposed after planting, erosion can be a serious problem. Some form of protection is necessary until the crop cover reaches 60%. The closer the spacing, the sooner the soil is protected. (Tanzania, MKVC).

stems. The double-row system is recommended to give a population of between 30,000 and 39,000 plants ha^{-1}. This is substantially more than that recommended in most other countries, presumably because the plants were pure China-type with a different growth habit from Assam hybrids.

A high-density experiment implemented by the Tea Research Institute of China (31°N 120°E) consisted of four treatments, namely 60, 120, 160 and 200,000 plants ha^{-1} (cv. Longjin 43) (Guokun Yao and Tiejun Ge, 1986). Using the results of this experiment, Etherington (1990), undertook an economic analysis (a rare undertaking!) to identify the most cost-effective planting density. All the costs had been carefully recorded over a 10-year period, as well as the benefits. Etherington (1990) noted that when the experiment began (it was the time of the Cultural Revolution) there was pressure from the Chinese government to intensify agriculture. This included high-density planting and intercropping. In addition, since it was unusual to train tea bushes to obtain lateral spread, planting in hedges was a common agricultural practice (Yu, 2012). Although Etherington (1990) plotted a parabolic annual yield response curve, Kigalu (1997) showed that an asymptotic relationship was also statistically valid. After a detailed cost–benefit analysis, Etherington

Figure 6.7 Post-hole diggers being used to assist with planting tea (Tanzania, MKVC). *A black and white version of this figure will appear in some formats. For the colour version, please refer to the plate section.*

(1990) identified by interpolation a plant density of 180,000 ha^{-1} as being the most profitable.

Water Relations

In an attempt to specify the target plant population for well-watered tea, and for tea subject to drought, an experiment was initiated in southern Tanzania in 1993 (Figure 6.9). It was based on the line-source design. This allows a continuously variable quantity of water to be applied across the experiment (Chapter 15, Figures 15.10, 15.11 and 15.12) to give a range of treatments from fully irrigated to rain only (Kigalu and Nixon, 1997). Full details of the experiment and some results can be found in Box 6.3. The root systems are described in Chapter 8.

Costs and Benefits

In the first six years after planting large cumulative (extra) yield benefits were possible from high-density planting (20,000–40,000 plants ha^{-1}). For well-watered tea these were between 7500 and 9000 kg ha^{-1}, and for droughted tea between 3800 and 8100 kg ha^{-1}, depending on the clone. There were also benefits (300–500 kg ha^{-1}) in the year following the first prune.

Figure 6.8 Planting tea in Tanzania (MKVC).

Figure 6.9 This experiment was designed to compare a range of densities between 8333 and 83,333 plants ha^{-1}, at different levels of water availability (Tanzania, MKVC).

These large cumulative yield benefits, which appear to be consistent between clones and between drought treatments, demand a detailed analysis to identify the economic conditions needed to justify high-density planting (20,000–40,000 ha^{-1}), including savings in the cost of weeding (Figure 6.10). As the benefits of close spacing are so closely linked with the extra crop harvested during the first years after planting, it is surprising that so few people report these early crop yields. As with nitrogen response

Box 6.3 Plant Density × Irrigation Experiment

The responses of two contrasting clones to density and drought treatments were compared in a line-source experiment (Figure 6.9): BBK-35 is upright in nature, while AHP-S15/10 is spreading. The plant densities ranged from 8333 ha^{-1} (below the current commercial practice) to 83,333 ha^{-1} (considerably above). Drought treatments were first imposed in the third year after planting. Prior to this all the treatments were uniformly irrigated during the dry season. The young plants were brought into production by de-centring and tipping. Clone AHP-S15/10 covered the ground much more quickly than BBK-35, especially at the low densities. At the two highest densities, both clones had virtually complete ground cover within 24 months of planting.

Yields in the first cropping year increased linearly with plant density, but with the slope of the relationship three times greater for AHP-S15/10 than for BBK-35 (6.9 kg vs. 2.3 kg plant^{-1}). In the following four years, yields from well-watered treatments increased asymptotically with plant density, but with the magnitude of the responses decreasing year on year, particularly with AHP-S15/10. By year 6, there were no longer any yield benefits from increases in plant population. Annual yields peaked at about 6700 kg ha^{-1} (AHP-S15/10) and 6000 kg ha^{-1} (BBK-35).

At low densities (8300 and 12,500 ha^{-1}) clone BBK-35 withstood drought better than at a higher density (21,000 ha^{-1}). For clone AHP-S15/10, all densities were affected equally.

Inspection of the *cumulative yields* summed for the period from when regular harvesting began to the first formative prune five years later suggest that a plant density close to 40,000 ha^{-1} gives the maximum yield advantage for both clones, when fully irrigated as well as when droughted. For AHP-S15/10, the cumulative yields at this density for *well-watered crops* totalled 31,300 kg ha^{-1}. This was 7500 kg ha^{-1} (or +32%) more than from the corresponding low (current commercial) density (12,500 ha^{-1}). The benefits from high-density planting were even greater for BBK-35, the corresponding total yield being 24,700 kg ha^{-1} (+9000 kg ha^{-1}, or +57%). For *droughted crops*, the cumulative yields for AHP-S15/10 at the high density were 26,000 kg ha^{-1} (+8,100 kg ha^{-1}, or +45%) more than from the corresponding low-density treatment. The benefits of high-density planting were less for clone BBK-35 (+3800 kg ha^{-1}, or +32%).

While full crop cover was being re-established, there were further yield benefits from high-density planting in the year following the first prune. On a base yield of 1700 kg ha^{-1}, this equated to about 500 kg ha^{-1} for BBK-35 when averaged across watering treatments, and for AHP-S15/10 about 300 kg ha^{-1} (base yield only 1000 kg ha^{-1}). Clone AHP-S15/10 is well known for its slow recovery from pruning. In succeeding years, there was virtually no response to density beyond 12,500 plants ha^{-1}.

Figure 6.10 As with many tree crops, there is interest in planting tea at high densities in order to show a return on investment as soon as possible. This photograph demonstrates semi-intensive planting in Tanzania, three rows in a bed (MKVC).

curves (Chapter 12, Figure 12.2), the shape of the relationship between yield and density can be presented and interpreted in different ways.

In Kenya, Corley *et al.* (2015, personal communication) found that clones varied in their response to density. Most showed a useful response in the first five years, but not much thereafter. However, a few, including MRTM1, continued to give positive responses even 12 years after planting, and not just in pruning years. One clone gave a consistent negative response in the same trial, yielding best at 6000 plants ha^{-1}, whereas MRTM1 was best at 30,000 plants ha^{-1}.

Summary: Plant Density Experiments

- Numerous experiments have been conducted on plant density since the mid-nineteenth century.
- In the early years after planting, yield is proportional to the number of plants per unit area.
- As the low-density plants gradually attain full crop cover, so the differences in yield between low and high densities become less.
- The response curves start as linear but become asymptotic with time, with a predictable ceiling yield at the highest densities.
- Many of the experiments have failed to include a sufficiently large range of densities to allow realistic yield/plant-density curves to be plotted.

- The time taken to reach full crop cover depends partly on the method of bringing young plants into production and the natural growth habit of the clone (upright or spreading).
- The yield and financial benefits from high densities (e.g. $> 20,000$ ha^{-1}) accumulate during the early years after planting, and in the year of the prune.
- The need to intensify production will encourage tea producers (especially those using 'pluckable shoots' for mass propagation) to consider high populations.
- Very high densities are planted in China, in single, double (30,000–39,000 ha^{-1}) or multi-rows (60,000–100,000 ha^{-1}).
- These Chinese density values compare with the generally accepted plant densities outside East Asia in the range of 10,000–14,000 plants ha^{-1}.
- Clones vary in their responses to plant density.

The issues to be taken into account when deciding what plant density to select are summarised in Box 6.4, while the difficulty in undertaking a cost–benefit analysis is explained in Box 6.5.

Box 6.4 Plant Density Issues

- The annual yield response to increases in plant density is asymptotic.
- There is no clearly identified optimum plant density beyond which yields fall.
- Since the yield differences from different planting densities decline with time, the principal benefit from close spacing is the extra crop harvested in the early years after planting and in the year following a prune, before ground cover is complete.
- The time taken to achieve full ground cover is temperature-dependent (it will take longer at high-altitude sites). It also varies with the clone, whether it is naturally spreading or upright, and with the method of bringing into production (e.g. pruning, pegging or de-centring and tipping).
- The extra revenue that is generated, and associated costs and cost savings (e.g. fewer weeds) will vary depending on the answers to each of these points, but it can be substantial.
- It is not possible to specify a universal optimum plant population density. Each situation will be different, depending on the answers to the points above, and the cost and availability of plants.
- Rarely has an economic analysis been attempted, but for any new project a cost–benefit analysis for high density (usually in the range 20,000–40,000 plants ha^{-1}) should be attempted (see Box 6.5).

Box 6.5 Commentary: Close Spacing, All a Bit Fraught!

There is an advantage in terms of extra crop in the early years at close planting. It looks to me to be effectively lenticular, with no benefit at planting, more benefit as the plants grow, and eventually converging yields as the tea matures.

As for costing, this is a very sophisticated exercise! Assuming that the year of planting is year 0, extra costs arise as early as year minus 2, with nursery materials and construction, and loss of crop from mother bushes. Year minus 1 costs are obvious, as are those of year 0. Benefits start to accrue in year 1. Now subject all these extra costs to a discounted cash flow analysis, and you have to have considerable yield advantages to justify them. Ask an accountant. And don't try to offset those costs against the sale value of the extra crop – you can only include its marginal value (sales less cost of sales). All a bit fraught.

And there are other considerations. You have already touched on shortage of mother bushes as a potential constraint. If the resultant supply of plants is a constraint, does it make better sense to put your plants into a small area or a larger one? Will you get more crop from putting your 300,000 plants into 10 ha or 20 ha? If land is your constraint, go for 10, but in most cases (in Africa) it will be 20. But of course it costs more in year minus 1 to clear the 20 ha. Who ever said this was simple?

T. C. E. Congdon

Box 6.6 Commentary: Replanting

Strategies for replanting differ from those for a new estate. If you replant at, say, 2% per year, your technology will be on average 25 years old. So, in any one year, plant 100% of the best available clone, and rely on better clones coming forward over the next 50 years to give you clonal diversity.

T. C. E. Congdon

Replanting

Tea is a long-lived, perennial tree crop that can remain in production for upwards of 100 years. There comes a time, however, when a decision has to be taken about whether or not to uproot low-yielding and/or moribund tea and replant with higher-yielding and/or higher-quality plants, and at a sustainable annual rate. Uprooting tea means a loss of income while waiting for the new plants to come into full production. This cost element has to be added to the direct costs of the land clearance and rehabilitation procedures, propagating new plants and planting in the field (Box 6.6).

A core initial requirement is to try to identify the cause(s) of the low and/or declining yields, which may be completely unrelated to the age of the plants. Mwakha (1985) divided the causes of low or declining yields into two groups, pathological and non-pathological. The former includes a build-up of diseases such as *Armillaria mellea*,

> **Box 6.7** Commentary: Replanting (Low-Yielding Fields) Doesn't Work!
>
> Managers on one estate in Kenya were disappointed that yields from a replanted field averaged only 60% of yields from plantings on new areas. But they had replanted the lowest-yielding fields. These fields before replanting had an average yield 41% below the company mean. Thus, assuming that the new areas were on 'average' land, it was to be expected that the replants would yield 40% less.
>
> **R. H. V. Corley**

Phomopsis theae and *Hypoxylon serpens*, and/or pests such as nematodes (*Meloidogyne* spp.), red spider mite (*Metatetranychus bioculatus*) and tea mosquito bug (*Helopeltis* spp.). Non-pathological causes include lightning, frost, sun-scorch, drought and soil compaction. A decline in the soil organic matter (e.g. as a result of removing the prunings) may reduce soil nutrient availability.

When replanting, the aim is to achieve the shortest payback period possible. This requires good early establishment leading to sustainable high yields of quality tea achieved within a few years after replanting (Nyirenda, 1997). In southern Africa, the advice given by the Tea Research Foundation of Central Africa is that the best clones to plant (in terms of yield and quality, and drought resistance) should be chosen from within the group TRFCA-PC105 to TRFCA-PC131. Composites (using the drought-resistant rootstocks, labelled R1 to R6) should also be considered. The fields to be replanted with improved clones should be chosen on the basis that they are still producing good yields from the existing seedling tea. It is unwise to take a short-term view and to replant the low-yielding fields, since it is likely that another major (probably unknown) yield-limiting factor is present. Of course some people take the opposite view, and replant low-yielding fields first. They then complain that replanting doesn't work! See Box 6.7.

In North-East India, the recommended practice is to plant a break crop after removing the old tea plants. Guatemala grass (*Tripsacum laxum*), Napier grass (*Pennisetum purpureum*) and citronella grass (*Cymbopogon wintarianus*) are considered to be suitable species for this purpose. To provide a mulch, the grasses are cut at appropriate intervals for at least 18 months and the trimmings left *in situ*. Any soil compaction should be remedied (e.g. by using a winged tine) before the grass is planted, and nutrient deficiencies should be corrected (e.g. by applying fertiliser). In Malawi, Guatemala grass is also recommended, with a minimum of one year's rehabilitation before replanting. Experience has shown that this grass is easy to establish, grows vigorously, recovers quickly after lopping and is reasonably easy to kill at the end of the rehabilitation period (Grice, 1990a). However, a detailed analysis of the publications on this topic suggested that the effects were never large enough to compensate for the loss of more than one year's crop (Corley, 2015, personal communication) (Figures 6.11 and 6.12).

Despite these recommendations, there is little experimental evidence to support either of these proposals or on how best to evaluate the benefits and determine the payback periods necessary to cover the costs of replanting. For example, one experiment conducted over a 10-year period on Kipketer Estate in the Kericho District of Kenya

Figure 6.11 Experience has shown that Guatemala grass is easy to establish, grows vigorously, recovers quickly after lopping, and is reasonably easy to kill at the end of the rehabilitation period (MKVC).

provided no statistically valid, or even circumstantial, evidence to show that there was any benefit from rehabilitating the soil with a break crop before replanting. The cost of crop 'lost' during the period of rehabilitation had not been recovered within the (nearly) 10-year period following uprooting (Njuguna, 1993a).

In the example from Kenya, old (planted in 1926), moribund seedling tea was uprooted (with a D8 bulldozer) in 1983. In some plots, the area was replanted immediately after clearing. In other plots Guatemala grass was planted and left to rehabilitate the soil for either 12 or 24 months. A selection of clones, together with seedling tea, was then planted. The annual yields were recorded from 1984 to 1993. Responses varied between clones. Some, such as TRIEA-31/8, AHP-S15/10 and TRFCA-SFS150, performed quite well with or without rehabilitation. Others did equally badly in both situations (these included clones TRIEA-6/8, TRIEA-7/3, TRIEA-12/12 and BBK-35 together with the seedlings). In a similar experiment on the same site there were no yield advantages after rehabilitation with Guatemala grass for 12 or 24 months in the following five years (1993–1997) (Bore *et al.*, 1998b). However, there was indirect evidence of the possible benefits from replanting in neighbouring fields, where cumulative yields from newly planted clonal tea had overtaken those from moribund tea within five years from planting. It should be noted that annual yields from seedling tea are generally low at this high-altitude site close to the equator, averaging 1600–1700 kg of processed tea ha^{-1} over a pruning cycle.

In a related trial in Kenya, three methods of uprooting moribund tea plants were compared, namely with a bulldozer, using a winch, or digging up by hand. These, together with collar pruning in association with two cover crops, Guatemala grass and *Desmodium uncinatum* (a trailing perennial legume), and a control, made up the

Figure 6.12 It is important to minimise run-off in high-rainfall areas. Here we see coconut fibre (coir) being used as a mulch in a replanted field in South India (MKVC).

treatments in a complicated experiment. None of the detailed crop growth measurements, or yields of green leaf (recorded for eight years after uprooting), or the leaf nutrient contents, revealed any significant differences between the treatments (Ng'etich and Othieno, 1993).

Following 10 years of research on moribund tea, Bore (1996) reviewed the experience in Kenya. His conclusions were:

- Care must be taken when clearing moribund tea not to destroy the soil structure in the process (e.g. by compaction with heavy machinery).
- Clones should be screened for their capacity to grow well in soils where the tea has become moribund.

- Despite the lack of clear evidence, there is still a belief that nutrition (toxicity/deficiency) may be the principal cause of tea becoming moribund, either alone or in combination with other complex but unknown factors.
- Lightning, frost, or removing the prunings from the field were not the causes.
- Pests and diseases were not the cause.
- There is the possibility of a build-up of allelophatic (toxic) substances during the long period of tea monoculture (Owuor, 1996), although there is no good evidence to support this view.
- Rejuvenation (low) pruning and replacing dead bushes with clones is not an effective rehabilitation method. But dead bushes should always be replaced.
- There is no justification in Kenya for using cover crops to rehabilitate the soil before replanting.
- The need to replant increases in importance as the tea gets older and more of it becomes moribund. Research is needed to quantify the costs and benefits of replanting so that sensible decisions can be taken.

Since Bore's (1996) paper was published, Kaptich *et al.* (in preparation) have reported the results of a series of trials in which several methods of replanting old tea were compared. These trials, undertaken in Kericho, Kenya, by Unilever Tea Kenya, showed that it was unlikely that the previous tea stand had any harmful effects on the replants, and that replanting can be done either by uprooting the old tea with a tractor winch or by collar pruning, leaving the stumps and roots in place. The cumulative yields after replanting remained below those from the old stand for more than 10 years. But there was an indication that replanting with new, higher-yielding clones could be justified financially. For reasons not understood, the application of coffee pulp (1 kg per bush applied in the planting hole) gave an overall 7% yield increase. Chemical soil amendments had no effect.

In Chapter 19 (Box 19.4) a hypothetical example of an experiment designed to address some of these issues is presented. It is based on using a tractor-mounted post-hole digger to dig a 0.6–1.0 m deep hole. Experience in the Nandi Hills District of Kenya shows that this practice allows the roots to extend faster to depth and the crop canopy to cover the ground more quickly than when plants are planted in the conventional way. This suggests strongly that one important limiting factor when replanting (and indeed perhaps even when planting into virgin land) is a physical restriction to root growth associated with a dense/compact subsoil (Figure 6.7).

Summary: Replanting

There are basically two issues to address when considering the need to replant. In the case of moribund tea (and vacancies), there is little choice but to replant, providing some idea of the cause is known. The other, perhaps more difficult, decision is knowing when to uproot healthy, relatively high-yielding tea in order to replant with proven higher-yielding and higher-quality clones or composites. This is the challenge facing many long-established tea companies. Kibblewhite *et al.* (2014) have usefully reviewed this subject. The key questions are: Where, when and how much replanting is needed? How much can be afforded? – while recognising that those growers who replant now will, in 10 years' time, be those who are succeeding. By contrast, those who delayed because of the cost and loss of revenue will be struggling, in the interim, to survive. Taking important decisions like these is not helped by the failure of research to identify the causes of yield decline, and to quantify any benefits from soil rehabilitation on the following crop, despite some complex and detailed experimentation. Nevertheless, one large company in Kenya has taken the strategic decision to replant all its seedling tea by 2023. Action like this will further separate progressive, forward-thinking growers, intent on intensification, from those who take what they can from a declining resource, with minimal inputs.

Note

1. Often known by the trade name 'Speedling trays'.

7 Understanding the Growth Processes

Creating a Framework

In its natural state, the tea plant grows to become a tree of moderate size (or a shrub); under cultivation, however, it is pruned horizontally (typically at two- to five-year intervals) to form and maintain a low spreading bush. This increases the number of young tender shoots, which supply the produce, and allows them to be removed at relatively frequent intervals. The growth of a plant is therefore continually curtailed by cultural operations, which although familiar to present-day tea culture are alien to the normal growth processes of the tea plant (Carr, 1970). Unlike grass, which is also grown for its leaves, tea has its apical meristem removed at each harvest.

During the early stages of growth in the field, one of the principal objectives is to facilitate the development of a strong, resilient, multi-stemmed framework on which the crop will develop, through the processes known as 'bringing into production' (see Chapter 6). To monitor progress, we need another framework, this time a conceptual framework known as a 'framework for analysis' (Box 7.1). This is achieved by breaking down plant growth (ultimately measured as an increase in the dry weight of the plant) into three basic processes: canopy expansion, dry matter production and storage, and partitioning (Figure 7.1). This approach, based on first principles, helps us to understand how a crop grows, regardless of its location, and enables us to estimate potential and actual yields. It then becomes possible to identify possible areas for intervention (e.g. clone selection, frequency of pruning), if the differences between potential and actual yields are large.

Each of these three processes is now considered in turn.

Expansion of Leaves (and Roots)

The rate at which the leaf canopy covers the ground after planting in the field, and subsequently after pruning mature tea at regular intervals, are important determinants of yield at these times. The aim of the farmer is to facilitate full crop cover as quickly as possible in order to intercept (and utilise) as much solar radiation (sunshine) as is possible during the lifetime of the crop (Box 7.2). Of particular importance to the grower is the need to generate income and to show a return on the investment in a short time period. With young tea, canopy development is enhanced artificially by pruning and other methods of bringing tea plants into production (described below).

The other expansion process important in crop production is root development, since the size and distribution of the root system influences the volume of water that can be

---

Box 7.1 A Framework for Analysis

The yield of any crop (Y) can be considered in terms of the efficiency of conversion of solar energy (S) to the economic or useful product (young shoots in the case of tea). Thus:

$$Y = S \times f \times e \times H$$

where S is the total amount of solar energy received at the surface of the crop, f is the fraction of the energy intercepted by the leaf canopy, e is the conversion efficiency (or ratio) of solar radiation to dry matter, and H is the ratio of the dry matter present in the 'useful product' (corresponds to 'yield') to the total dry mass of the plant (normally above ground), otherwise known as the *harvest index*.

Typically, incident annual short-wave radiation totals range from c. 55 TJ ha^{-1} in the high-rainfall (cloudy) areas in Bangladesh and Assam to 63 TJ ha^{-1} in tea-growing areas of Indonesia, to 70 TJ ha^{-1} in parts of East Africa with clear skies during the long dry season (Monteith, 1972). The role of the agriculturalist is to utilise this energy as effectively as possible, as it is the primary source of wealth for a farmer.

---

extracted from the soil, and its availability. Roots expand into wet soil to extract water. Roots are less easy to describe than the crop canopy, but for comparative purposes can be expressed as root length per unit area of ground, or as root length per unit volume of soil (the most useful measurement) at different positions within the soil profile, or as rates of root extension, or simply as the maximum depth of rooting (not always easy to specify) or the maximum 'effective' depth of rooting (still needs to be defined). Roots are considered in detail in Chapter 8.

Dry Matter Production and Storage

Dry matter is produced by the process of photosynthesis, and is lost by the process of respiration, which provides the energy necessary for growth and maintenance. The conversion efficiency (e) is used to describe the proportion of solar radiation intercepted by leaves ($S \times f$), which is converted to dry matter (DM, g):

$$DM = S \times f \times e$$

Dry matter produced can be used to support current growth or, if there is a surplus, stored in the roots as starch. This reserve may then be mobilised and translocated to the shoots after pruning, extended periods of drought or low temperatures, or hail damage.

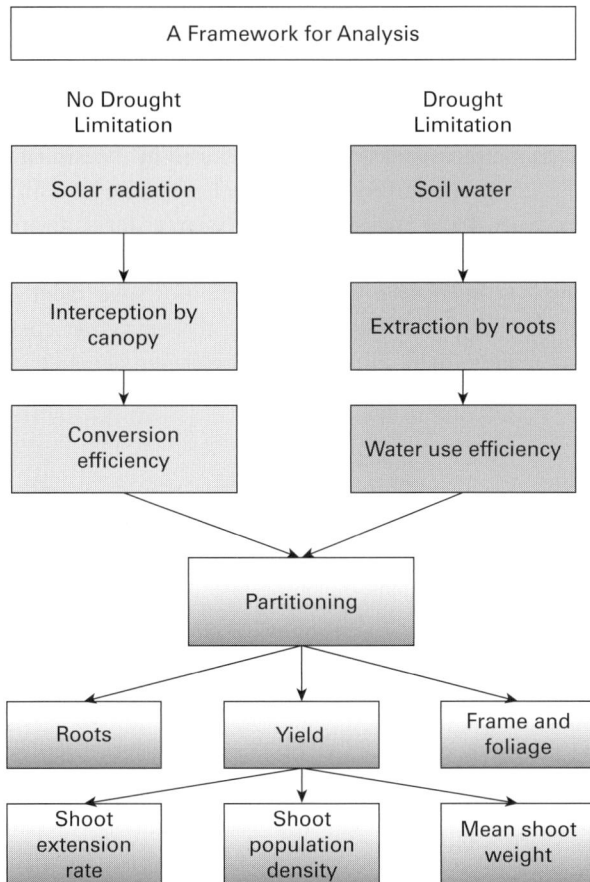

Figure 7.1 A framework for analysis.

Partitioning (Harvest Index)

One of the primary ways by which a plant breeder can improve the yield of a crop is by selecting varieties that partition a high proportion of the dry matter, or energy, to the primary product. This is how the seed yield potential of the major cereal crops has been increased. In the case of tea, it is the young shoot that has the commercial value.

Bringing Young Tea into Production

There are a number of techniques available to the tea planter that can encourage the development of a low, spreading frame and a crop canopy that covers the ground quickly. These are collectively known as methods of 'bringing into production or bearing'. They include:

Box 7.2 Estimating the Size of a Crop Canopy

This can be estimated in one of two ways: the quantity of the ground covered by foliage when viewed vertically from above (%) or, with more precision, by the leaf area index (L). This is a dimensionless quantity obtained by measuring the surface area of the leaves that cover one unit area of ground. For mature tea with full ground cover, L can vary between 4 and 10, being generally larger for cultivars with erect leaves (China-type) than for those with larger horizontal leaves (Assam-type). The leaf area required to intercept a given proportion of solar radiation depends largely on canopy geometry, which is defined by the extinction coefficient (k), a measure of how rapidly light intensity declines with depth in the canopy (Hadfield, 1974b).

$$f = 1 - \exp(k_l)$$

L is influenced by the plant density (P_d, plants m^{-2}), leaf number per plant (N_l), and the average area of each leaf, which can be estimated from two linear measurements $A \times l$ (mm^2).

$$L = P_d \times N_l \times Al$$

Rates of canopy development are largely controlled by temperature when other factors such as dry air or dry soil are not limiting. There is a base temperature (T_b) below which rates of leaf expansion are very slow, and an optimum temperature (T_o) above which rates decline before reaching zero at a maximum temperature (T_{max}). Typical values for T_b, T_o and T_{max} for tea are 12–13 °C, 25–30 °C and 35–40 °C respectively.

- Removal (or bending) of the dominant shoot (de-centring) to encourage branching
- Continuous removal of harvestable shoots above a specified height (tipping)
- Pruning of woody branches at one or more fixed heights above the ground (formative pruning)
- Spreading of lateral branches by first bending and then pegging.

Combinations include:

- De-centring, followed by pruning at a height of 350 mm above the ground
- When the plant is well developed, pruning at a height of 350–400 mm above the ground (20–26 months[1] after planting)
- De-centring (10 weeks after planting) followed by regular tipping/plucking until the first mature prune four or more years later.

In North-East India, Barua (1969), having described the available methods of bringing tea into bearing, concluded that the choice depended on the altitude, the risk of drought, the type of tea and its spacing. There is no single recommended method, since each

Figure 7.2 There are many ways of bringing young tea plants into production. The photograph shows tea plants being allowed to grow freely before they are 'tipped' at a height above ground of about 350 mm. Note the oats growing between the rows as a soil conservation measure (MKVC).

alone, or in a combination with other practices, has its proponents. The method chosen impacts on root growth, and hence on the shoot (canopy) to root ratio, as well as on the rate of development of ground cover (Figures 7.2 and 7.3).

For example, commercial growers in Sri Lanka are recommended to allow the plants to grow freely for 9–12 months in the field. The main stem is then cut at a height of 200–250 mm above the ground, leaving as many side branches as possible. A second cut is made 4–6 months later at a height of 350–400 mm across all branches. The plants are then tipped at 450–500 mm. By contrast, the practice of bending and pegging, in which lateral shoots 300–600 mm long are bent at angles of 30–45 degrees to the ground, and then held in position by wooden pegs (lateral shoots as they develop are also bent and

Figure 7.3 There are some risks associated with the practice of 'bending' and 'pegging' to bring tea into production. But, by so doing, a framework is formed on which a low-spreading bush can develop (MKVC).

pegged) is discouraged in Sri Lanka because of the risk of sun-scorch, canker (*Phomopsis theae*) and wood rot (TRISL, 2001).

Similar concerns about pegging exist in neighbouring South India, although full ground cover is achieved more quickly than with methods involving a formative prune. In South India, the main stem is cut 4–5 months after planting, which by removing apical dominance encourages branching, then tipping (the removal of young shoots) to level the crop surface (plucking table) on two occasions (two-stage tipping). This is done first at a height above the ground of 350 mm and later at 500–550 mm. This not only resulted in higher yields (+20%, when averaged over three UPASI clones) over five succeeding years compared with the conventional method of tipping once only at a height of 500–550 mm, but it also led to the formation of a (desirable) low-spreading bush frame (Sharma *et al.*, 1990).

In East Africa, three methods of bringing into production are described in the *Tea Growers' Handbook* – pruning, pegging and tipping – although it is not clear if they are all recommended. Growers are warned about the likely impact of each method on root growth, and hence on susceptibility of the immature plants to drought. The cost of pegging is greater than that of most pruning systems; it is thought that the value of the larger yields and lower weeding costs due to the more rapid attainment of full crop cover will more than compensate (TRFK, 1986), although the evidence is lacking.

Table 7.1 Choice of methods of bringing young tea plants into production after field planting

	Operations	Advantages	Disadvantages
1	De-centre to encourage lateral branching, then prune twice to create a low spreading frame, and tip	Good frame; time of pruning can match periods of dry weather	Slow to develop full ground cover: pruning will inhibit root growth
2	De-centre and then tip twice (at two heights)	Full ground cover achieved quickly	Frame less solid (but does it matter?)
3	De-centre (or nip out terminal bud) and then tip into production	Yielding soon after planting	
4	Bend lateral branches and peg to produce a frame; tip into production	Rapid ground cover, early yield	Expensive, labour intensive; risk of sun-scorch and disease; also possible temporary imbalance in shoot:root ratio
5	Any combination of the above		

In central and southern Africa, droughts occur most years and pruning of young tea is used as a way of mitigating their effect. In rain-fed tea, annual pruning is recommended for the first five years, with the successive prunes being brought forward by 10–14 days each year, beginning from the end of July. This is towards the end of the cool dry weather and prior to the hot dry season. If the tea is irrigated, the advice is to prune in alternate years for the first six years. In both rain-fed and irrigated tea, the first prune can be replaced by de-centring (Martin *et al.*, 1997; Malenga, 2001).

It is impossible to compare in meaningful ways the bringing into production methods described. Although the risks are rarely quantified, the choice will influence the susceptibility of the young plants to drought, and the time taken to come into full production (Carr, 1976). For example, annual/biennial pruning as recommended in Malawi (Martin *et al.*, 1997) will delay the attainment of full ground cover and restrict root growth during the period of defoliation. By contrast, there will be no disruption to root growth in pegged bushes. The resultant shoot:root ratios will be variable over time and difficult to describe in quantitative terms. Formal scientific evaluation is very difficult, but each system can be made to work given appropriate conditions (mainly related to rainfall) (Table 7.1). The choice is yours!

Dry Matter Production and Partitioning

Measurements made in Tanzania showed how clones differ in the amount of dry mass allocated to young shoots, as well as in their rates of canopy expansion, and hence in their capacity to intercept solar radiation (Table 7.2). Radiation-use efficiencies were similar in all four well-watered, immature clones (BBT-1, TRIEA-6/8, TRFCA-SFS150 and AHP-S15/10) with values between 0.40 and 0.66 g MJ^{-1}.

Table 7.2 The net dry matter partitioned (g plant^{-1} d^{-1}) to harvested shoots, leaves, stems, large roots and fine roots for each of four fully irrigated clones between December 1989 and November 1990, and for clone S15/10 from November 1990 to September 1992

	Component					
				Roots		
Clone	Harvested shoots	Leaves	Stems	Large	Fine	Total
1	0.32	0.67	0.99	0.86	0.12	2.96
6/8	0.35	0.49	0.82	0.65	0.09	2.40
SFS150	0.38	0.58	0.95	0.61	0.15	2.67
S15/10	0.50	0.94	1.20	0.33	0.24	3.22
SEd	0.03	0.14	0.25	0.13	0.07	0.41
n	4	3	3	3	3	3
S15/10	1.31	1.41	1.96	0.60	0.14	5.43
SE	0.23	0.25	0.44	0.10	0.02	0.71
n	4	3	3	3	3	3

SEd, standard error of the differences; SE, standard error of the mean; *n*, number of observations on which each value is based.

When drought was imposed, values fell to 0.09 g MJ^{-1} within 16 weeks (Burgess and Carr, 1996b). Clone AHP-S15/10, a high-yielding clone from Kenya, partitioned a greater proportion of dry matter to leaves and harvested shoots than the other three clones, and correspondingly less to structural roots. The harvest index was 24% (including roots) for clone AHP-S15/10. There were also differences between the seasons, with more dry matter being diverted to roots and less to shoots during the cool winter season (T_{mean} = 14–15 °C) (Figures 7.4 and 7.5). Drought had no impact on root growth, but the amount of dry matter diverted to leaves, stems and harvested shoots declined by 80–95%. In a similar comparison but with different clones, the dry weight of the foliage removed at pruning (above 0.35 m, seven and a half years after planting) was greatest for clone EPA-TN-14/3 in both well-watered (34 t ha^{-1}) and droughted (22 t ha^{-1}) treatments. The largest differential between the wet and dry plots was 13 t ha^{-1} (clone TRFCA-PC105) and the least was 6 t ha^{-1} (clone TRFCA-PC81). These differences in the dry mass of the foliage match the ranking based on the cumulative yields recorded over the first pruning cycle.

Harvest Index

Magambo and Cannell (1981) were among the first to quantify the harvest index for tea. At a high-altitude site in Kenya (2178 m), they found that over one year only 8% of the total dry matter produced (including roots) was allocated to the harvested shoots (two leaves and a bud, clone TRIEA-6/8), or 11% if roots were excluded. Harvested tea bushes produced 36% less dry matter in a year (17 t ha^{-1}) than those

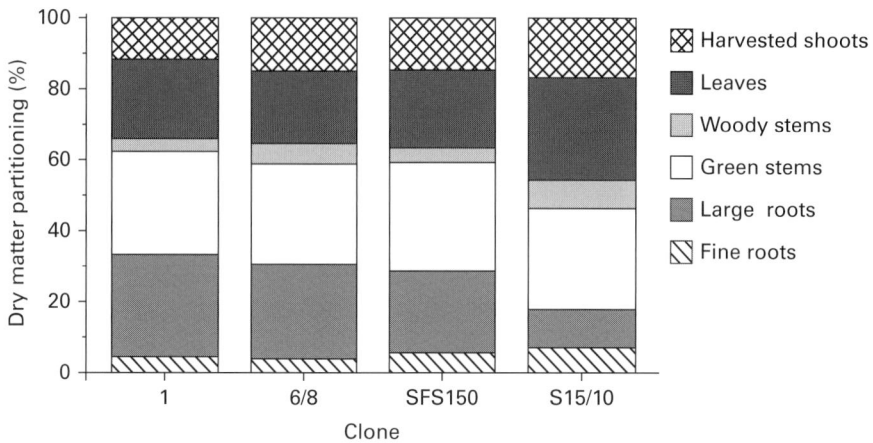

Figure 7.4 The high-yielding Kenyan clone AHP-S15/10 partitions more dry matter to leaves, and less to roots, than the other three clones shown here. An explanation, at least in part, of why this clone is high-yielding and recovers slowly and unevenly from a prune (from Burgess and Carr, 1996b).

left to grow freely (26 t ha^{-1}; plants 6–7 years old), and 64% less wood. In a follow-up study with five young clones grown in containers, Magambo and Kimani-Waithaka (1988) confirmed that the process of plucking reduced dry matter production, by up to 75% depending on the clone. Previously, Hadfield (1974b) had suggested a harvest index of about 12% or 13% for Assam-type tea in Assam, but it was not clear whether this value was with or without roots. The total annual dry-mass gain by Assam-type bushes in Assam averaged 14.7 t ha^{-1}. The total dry mass was equivalent to 49 t ha^{-1}, of which roots represented 18 t ha^{-1} (37%). With an annual yield of 2.8 t ha^{-1} this equates to a harvest index of 19%.

In Tanzania, the mean annual dry matter production from irrigated and well-fertilised mature clone TRIEA-6/8, over a five-and-a-half-year pruning cycle, was 16.9 t ha^{-1}, of which the harvested shoots comprised 3.3 t ha^{-1} (20%). This calculation was based on the realistic assumption (then) that 25% of the total dry weight was allocated to roots. The net annual increment in the total dry mass (including roots) was estimated to be 22.5 t ha^{-1}, giving a harvest index of 15% (Burgess, 1992). Over the following four years, the net annual dry matter production (including roots) was 21.5 t ha^{-1}, of which 5.1 t ha^{-1} were harvested shoots (24%). The corresponding values for young clone TRIEA-6/8 bushes (2–6 years after field planting) were 22.5 t ha^{-1}, 4.1 t ha^{-1} and 18%, respectively (Burgess and Sanga, 1994). All these harvest index values are much greater than the values (8% and 11%, with and without roots) proposed by Magambo and Cannell (1981) for the same clone in a cold location.

In a comparison of four contrasting clones at four sites in Kenya differing in altitude by 400 m (1800–2200 m), clone TN-14/3 had the lowest harvest index at 10% (assumed to include roots), and AHP-S15/10 the highest (up to 19%) (Ng'etich and Stephens, 2001). Partitioning of dry matter to coarse roots was least for clone AHP-S15/10 at all

(a)

(b)

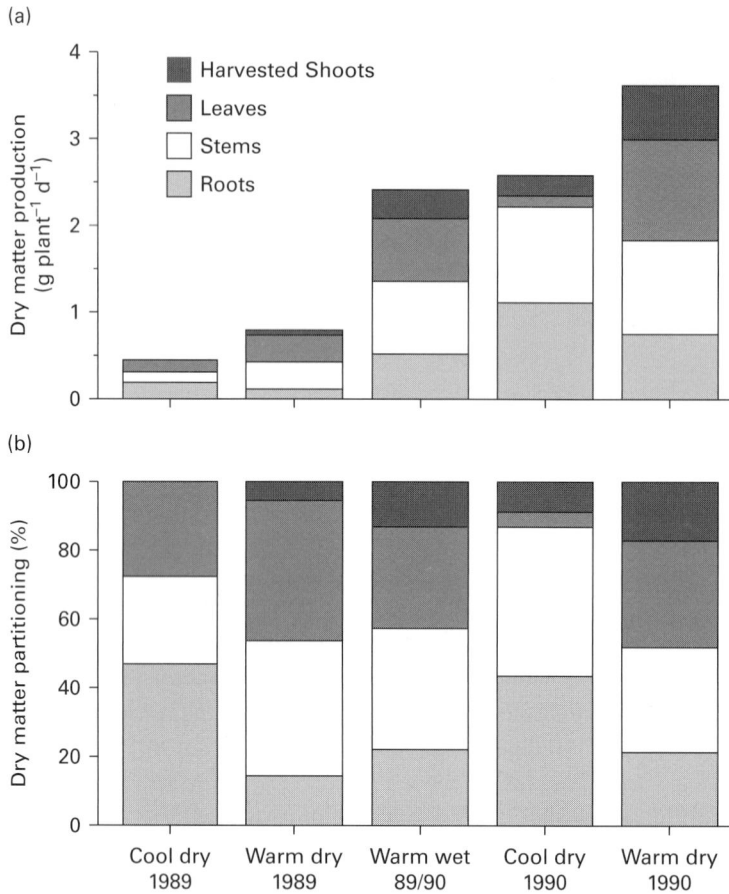

Figure 7.5 (a) Production and (b) partitioning of dry matter, averaged for four well-watered clones (BBT-1, TRIEA-6/8, TRFCA-SFS150 and AHP-S15/10) in each of five seasons, 1989–1990. More dry matter is partitioned to roots, and less to the harvested shoots, during the cool winter period than during the other seasons (from Burgess and Carr, 1996b).

four sites (*c.* 10%), and greatest for clone TRIEA-6/8 (up to 20%). The low allocation of dry matter to roots by AHP-S15/10 explains its slow and uneven recovery from a prune. Radiation-use/conversion efficiencies ranged between 0.3 and 0.45 g MJ^{-1}.

In a complementary study in Sri Lanka, partitioning of dry matter to roots increased linearly with altitude from 30 m to 1860 m. Clone TRISL-2023 was particularly sensitive to temperature (mean air temperature range from 26 °C down to 15 °C with increase in altitude) while clone TRISL-2025 was relatively stable across the four environments (Balasuriya, 2000).

To illustrate the contrasting morphology that exists within the tea population, the results of a very detailed analysis of the canopy structures, and the dry matter distribution, of two contrasting clones in South India (Assam-type and China-type) are shown in Box 7.3. The harvest indices were 7% (UPASI-9) and 16% (UPASI-3).

Box 7.3 A Comparison of Canopy Structures and Dry Matter Distribution in Two Clones in South India

Murty and Sharma (1986) compared the canopy structures of two clones, UPASI-3 (Assam-type) and UPASI-9 (China-type) in South India. The clones differed widely in their morphological appearance and in their yields (times threefold). Three bushes of each clone, chosen at random, were excavated five and a half years after planting at a spacing of 1.20 m × 0.75 m (11,111 ha^{-1}), and the foliage was separated into 250 mm strata and then divided into its components. As expected, the plant height (1.0 ± 0.05 m and 1.02 ± 0.03 m respectively) and canopy surface area were similar in both clones at the beginning of measurements, but the total leaf area was 30% greater in UPASI-3 than in UPASI-9 at all canopy depth increments except from 500 to 250 mm. The corresponding leaf area indices were 4.0 and 3.0 respectively. The leaf angle in relation to the main axis was 55 ± 5° in clone UPASI-3 compared with 70 ± 5° UPASI-9. This is contrary to what was expected: China-type bushes usually have more nearly upright leaves than Assam-type.

The total dry matter produced was also 19% greater in UPASI-3 than in UPASI-9, with the largest amount being allocated to the stem (39% by proportion), followed by the roots (28%), maintenance foliage (17%) and lastly young shoots (crop harvested over the preceding four years, 16%). By contrast, with UPASI-9 more dry matter, in total and also in proportion, was diverted to the roots (45%). This was followed in turn by the stem (32%), the foliage (17%) and then the crop (only 7%). The resultant ratio of dry matter allocated to roots compared with that stored in the stem and in the foliage was 0.9 in UPASI-9 but only 0.5 in UPASI-3. The harvest index was 16% in UPASI-3 and only 7% in UPASI-9. The cumulative yields over the four years from planting (assuming that the units are g bush^{-1} dry tea) are equivalent to about 6600 and 2400 kg ha^{-1} respectively. More detail is provided in the paper on the distribution of chlorophyll within the canopy, and on parameters such as specific leaf weight: leaves were thinner in UPASI-3 than in UPASI-9.

Pruning of Mature Tea

The duration of the immature phase in tea is open to debate, since definitions of mature and immature vary. Often it is the accountant who decides! Traditionally, immature tea was defined as plants that had been in the field for fewer than seven years. This was the time taken for widely spaced bushes to cover the ground. Now it can be as little as three to four years for a naturally spreading clone and/or clones planted at high density, or plants brought into bearing by bending and pegging. There are also other definitions of when a tea plant becomes mature, such as when its susceptibility to drought declines (linked to root:canopy ratio), or when the clone starts to produce (within nine months of planting in Kenya).

Figure 7.6 A field of pruned clonal tea, on a four-year cycle. Pruned at the end of the rains (Tanzania, MKVC).

More than 80 years ago Eden (1931) described the pruning process in mature tea as practised in Ceylon (Sri Lanka) as follows:

As a crop cultivated for tea manufacture, growth is severely controlled, the main stem being pruned back a few inches above ground level after the first few years of normal growth, with a view of inducing the plant to put out laterals and assume the bush habit. Following the primary pruning or centring, fresh growth is allowed till the bushes have filled out and obtained an overall width of 20–24 inches [500–600 mm] measured at a point 18–20 inches [450–500 mm] above the ground. They are then pruned back to a height of about 20 inches [500 mm] and taken into bearing. Hereafter at regular intervals, varying from 18 months to over three years according to altitude and locality, the bush is pruned regularly, the new cuts being made 2–3 inches [50–75 mm] above the old ones until a bush height becomes too great and necessitates down-pruning . . . Time is allowed for recovery from pruning and then the young succulent shoots are broken off or tipped so as to produce a flat or slightly convex upper surface or *plucking table*. At this stage the bush can be plucked for the manufacture of the young leaves or *flush* into normal tea.

The act of pruning tea has changed little over the years, but perhaps we are less ready to resort to the knife now than before (Figure 7.6). There are many variations on this theme, depending on the severity of the pruning act. These are reflected in the terminology used, which includes, for example, light prune, deep skiff, medium skiff, light skiff, level-off skiff, down-pruning, medium pruning and rejuvenation pruning. There are similarities with the type of haircut a man might request from a barber. Why there should be a need to have such a range of pruning practices is never clearly explained. Similarly, the evidence on which the use of any one of them is justified is not easily available.

Box 7.4 Commentary: Why Prune? A Personal View

Pruning of whatever kind generally results in reduced yield until the bush recovers. This results in reduced yield in year 1 of a pruning cycle. The MD wished to avoid this, and challenged a brain-storming group to come up with a solution.

In East Africa we are generally on a four-year cycle. Why not go to five? The mathematics are simple. If the fifth-year yield is greater than the mean of years 1–4, go to five. If lower, don't.

But the question is, why do yields decline over the cycle? My answer was to examine the three main components of yield over the cycle, and see what changes. Then at least you know what you are up against. Modern hard plucking, whether by hand or machine, results in much less table 'creep' than in the old days, so bringing the table back within reach of the pluckers is not an issue. But still the yields decline in the fifth year – unless they are on a strongly upward trend anyway, which may be the case in young tea. If you look at a running 12-month total, you may well find that the highest yield is in January to December following a June prune, and that it declines steadily thereafter.

It seems to me that we prune to get bigger shoots. Shoot size declines over the cycle until larger shoot numbers fail to compensate for smaller shoot size. This may be less important with machine harvesting, but is still a factor, as is made tea quality (lower from banjhi shoots).

After all, the objective is to maximise net revenue per hectare. This is not easy to achieve without an understanding of a whole range of related factors. The yields of various clones tend to be measured in $kg\ ha^{-1}$ green leaf, but what if moisture content differs between clones? And years in cycle? At what point does an increase in nitrogen reduce net revenue due to lower quality, higher harvesting, manufacturing and packing costs ? Have fun!

T. C. E. Congdon

Vague terms such as 'to encourage the bush to produce leaves rather than wood' or 'to ensure the development of strong new shoots with minimum die-back' (BTRI, 1986), and 'to rejuvenate the plucking table' are used to describe the process (Martin *et al.*, 1997). In scientific terms this probably describes the process by which pruning removes the apical buds from having apical dominance, thereby allowing the axillary buds below the pruning cut to develop into new shoots. The frequency of pruning is another variable over which there is little agreement. In the absence of scientific evidence to the contrary, my preferred explanation is that pruning covers over all the manager's mistakes, masking the loss in yield as a result of leaving shoots in the field that should have been in the factory (Box 7.4).

Whatever the reason, at periodic intervals it becomes necessary to reduce the height of the tea bush, as the crop canopy tends to increase in height over time. This makes it difficult for the pluckers and reduces productivity. Yields also tend to decline, but whether this is a real decline or the result of the pluckers finding it more difficult to reach the shoots is open to question. Shoots also tend to get smaller with time from the last prune, which means more hand movements to harvest 1 kg of tea. The acceptable annual rate of increase in the height of the 'table' has traditionally been specified as 200 mm (TRFK, 1986). This means that in five years the table could have risen in height by a metre! If the last prune was at 300 mm above the soil surface and 200 mm of foliage was allowed to develop above that height before the first harvest (tipping) occurred, the total above-ground height of the crop surface after five years would be 1.5 m, which makes it impossible to harvest by hand. As a result, throughout most of East Africa, and beyond, tea was often pruned on a three-year cycle, sometimes four, sometimes two. Pruning became an established convention and no one questioned the need to do it so frequently.

In central and southern Africa, comparisons have been made between two- and three-year cycles, and between three- and four-year cycles. The main advantage of a two-year cycle over a three-year one is the quicker recovery from prune, whereas the advantage of a three-year cycle is that pruning costs are less. Otherwise, according to Grice and Mkwaila (1990), there is little to choose between the three time intervals. In Malawi, pruned tea is particularly susceptible to attacks by thrips (*Scirtothrips aurantii*) during the hot dry season. The insect feeds on the new growth, particularly on those shoots exposed to sunshine. Late pruning increases the risks of attack. The mosquito bug (*Helopeltis schoutedeni*) is another pest that favours young tea and bushes recovering from pruning (Martin *et al.*, 1997).

Measurements and observations in Tanzania showed that if the tea was subjected to 'hard plucking' the rate of increase in the height of the table could be limited to 50 mm for many clones, and 100 mm for China-type clones, which are more difficult to harvest than the Assam-type. This allows the frequency of pruning to be reduced (thus reducing the loss in yield in the year of the prune when the bushes are refoliating, as well as the cost of pruning) and the interval between prunes to be extended to six or more years. A useful tool for controlling the rise in table height is a calibrated stick, which the supervisor can use to monitor the changes in table height. In the Ngwazi mechanical harvesting trials, yield was maintained for eight years with zero table rise, but larger yields and better quality were obtained if an annual rise in the table height of about 50 mm was allowed (Rutatna and Corley, in preparation).

Timing of Pruning

This is a hotly contested issue, there being no set time in the year when pruning is recommended. Factors that need to be taken into account include the time, severity and duration of the dry season, and the temperature regime (Ng'etich, 1996). For example, in eastern and southern Africa the following situations exist.

In southern Tanzania, with a unimodal rainfall distribution, and a cool dry winter followed by a warm dry spring, the best time to prune is at the end of the rains and into the dry season. When the foliage is removed, transpiration ceases and the crop then refoliates using water stored in the soil. At the same time, the prunings act as a mulch, restricting evaporation from the soil surface. Full crop cover is reached before the rains begin, followed by the peak cropping season. Irrigation can be used to bring the soil to

Figure 7.7 A 'down-prune' into thick (old) wood (Kenya, MKVC).

field capacity prior to pruning if it is necessary to delay pruning into the dry season or to extend the period available for pruning.

In one experiment in southern Tanzania, there was an 18% yield benefit as a result of pruning with the soil at field capacity compared with pruning when the soil was dry and there was already a potential soil water deficit of about 175 mm. The benefit increased to 25% when one round of irrigation was applied before the rains began. All three treatments were pruned at the same time, and the recorded yields were for the period from the start of tipping (at the beginning of the rainy season) to the end of the rains, five months later. The timing of pruning in relation to the dry season and/or an irrigation policy are important considerations in the decision-making process (Carr, 1969b).

The optimum time to prune in central/southern Africa is also at the beginning of the dry/winter season, in May, with the target to complete pruning by the middle of June. The pruning height can be increased by 20–30 mm between cycles. Down-pruning is needed when, after a number of such rises, the height after pruning is too tall to allow harvesting to continue to the end of the next pruning cycle (Figures 7.7 and 7.8). The timing of a down-prune is critical, as bush deaths observed in the hot dry season, and uneven recovery, increase exponentially for each week's delay after the end of the first week of April. This is due to attacks by thrips, since the buds in the hard wood are particularly slow to develop, especially in the winter months. The height of a down-prune is left to the manager's discretion, but the lower it occurs the smaller the yield obtained in the first year that follows, although yields are greater in subsequent years (Martin *et al.* 1997).

Figure 7.8 Prunings retained between the two centre rows (Tanzania, MKVC).

By contrast, close to the equator in Kericho, Kenya, there is one relatively short but warm dry period between December and April, variable in duration from one to three months. Potential *soil water deficits* greater than 350 mm can be expected to occur once in five years (Stephens *et al.*, 1992). Ideally the tea should be pruned at the start of the dry season but, because it is warm, this coincides with the peak crop (Othieno, 1983). Understandably, growers are reluctant to lose production. There is also a shortage of labour at this time (the workers are busy plucking tea!).

In a short-term time-of-pruning experiment in Kericho, treatments were based on plant and soil water status measurements, in order to ensure that, as far as possible, the results had wide application. The shortest time interval between pruning and the start of tipping was when the tea was pruned at the end of November, that is before the drought started (Table 7.3, treatment 1). Plants pruned in March when the bushes were under severe water stress (the shoot water potential at midday had fallen to –1.2 MPa, treatment 4) also recovered quickly, but there were fewer shoots. Tea pruned after the drought had ended (treatments 5 and 6) took up to two months longer to return to harvest than the pre-drought treatments. In this year, the maximum potential soil water deficit reached by the end of March, when the rains began, was about 300 mm (Carr and Othieno, 1972). Yields in the full calendar year following tipping were still depressed in these two treatments (Othieno, 1972).

Similar results were obtained in an adjacent experiment designed to identify the important weather variables influencing the yield of tea (Box 7.5). Reconciling these conflicting messages is not easy, but the advice given to growers in Kericho, based on this and other evidence, is to prune in the period from mid-December to mid-January,

Table 7.3 Association between date of pruning, relative to dry weather, and (a) time to the start of tipping, (b) yield in first calendar year following completion of tipping

	1	2	3	4	5	6
Treatment	Pre-drought		During drought		Post-drought	
Time taken (days)	108	116	121	107	156	185
Yield (kg ha^{-1})	2000	1967	2088	2094	1839	1806

Treatment 1: tea pruned at the end of November before the drought had started. Treatment 2: tea pruned at same time as treatment 1 but bushes left unplucked for one month before pruning to ensure that there were adequate starch reserves in the roots. Treatment 3: tea pruned when midday shoot water potential had reached –0.8 MPa (moderate stress). Treatment 4: tea pruned when midday shoot water potential had reached –1.2 MPa (severe stress). Treatment 5: tea pruned after the rains had started when the top 0.30 m of soil had been wetted to field capacity. Treatment 6: tea pruned after post-rains flush had been harvested.

Yield was measured in the first calendar year following completion of tipping, which ranged from May (treatment 1) to September (treatment 6). In the second year there were no differences in yield between treatments. Least significant difference (LSD) at $p = 0.05$ was 198 kg ha^{-1}.

Box 7.5 Time of Pruning for Mature Tea, Kericho, Kenya

One seedling bush in each 156-plant plot was pruned each month over a period of three years. All 156 plants were therefore pruned once during these three years. This process was repeated for a further three pruning cycles, making 12 years in total. The number of days taken from pruning to tipping was recorded for each bush. A cyclic pattern emerged, with the number of days varying at this high-altitude site (2180 m) from 110–120 days in the troughs, when recovery was most rapid, up to 170–180 days at the peaks, when recovery was slowest. These peaks occurred when the potential soil water deficit was in excess of 200 mm. The time taken was correlated with the peak potential soil water deficit for that year ($r = 0.615$ when $n = 12$, or $r = 0.702$ when $n = 9$, both $p < 0.05$). Annual yields, but not monthly yields, were similarly, but inversely, related to the soil water deficit ($r = 0.652$ and $r = 0.736$) (Othieno et al., 1992).

As in Malawi and Tanzania, delaying pruning into the hot dry weather exposes the branches to damage from sun-scorch (and also from hail). This can cause serious damage (covering the frame with the prunings will reduce this risk), which may in turn facilitate infestation by a stem disease, *Hypoxylon serpens*.

---

after the peak crop has been harvested and before the soil water deficit is too large (Othieno, 1983). Similar issues need to be considered in each location where tea is grown. The optimum time for pruning is nearly always a compromise.

For example, in another experiment in Kenya seedling tea was pruned at the beginning of every month (at two heights, 300 mm and 550 mm) for a year, and the cumulative yields were then recorded for 36 months from pruning. In the first year the highest yields were obtained from the plants pruned in October and November, but in subsequent years there was no consistency in response and the cumulative yields could not be correlated with any single factor (Odhiambo and Othieno, 1992). Compared with pruning at 550 mm, down-pruning to 300 mm reduced the mean annual yields obtained during the three years that followed from 2000 to 1730 kg ha^{-1}. Note the low yields obtained at this high-altitude (2180 m) site. Low temperatures have again masked the magnitude of the yield responses to pruning.

Sharma and Murty (1989) reported the results of a similar experiment conducted by UPASI in South India (alt. 1050 m). Recovery from pruning was most rapid when the tea was pruned (at 650 mm above ground level) in August and September (between the two monsoons) and from mid-March to mid-May. Rates of recovery were strongly correlated with the starch content of the roots (pencil-thin root samples were taken pre-pruning) (range 16–21%). Following pruning, starch levels declined steadily for 40 ± 10 days before starting to increase, but they had not returned to pre-pruning levels when tipping started. Bark starch levels (8–12%) did not show the same relationship with the rate of recovery as root starch.

Pruning System

Sometimes the recovery from a prune is slow and uneven across the bush, and 'partial pruning' may then be advocated and adopted. This involves leaving one or more branches with foliage on the bush so that photosynthesis can continue while the bush is in the process of refoliation. Numerous experiments have been undertaken comparing various combinations to address such questions as:

- How many of these 'lungs' should be retained?
- On which part of the bush should the foliage be left – for example, in the centre, known in Kenya as savani-lung, or on the periphery, known as rim-lung?
- For how long should the lungs remain on the bush before being removed?
- Should the lungs be removed all at once or in stages?

And so on. The number of possible treatment combinations is infinite.

It is not clear whether definitive answers have ever been obtained from experiments like these. There are so many variables to take into account, not least the practical ones of implementing a partial pruning policy (Ng'etich, 1996). For one thing, it is not easy to design an experiment, where timing is a variable, that allows like-with-like comparisons. As an example, clone AHP-S15/10 is well known in East Africa for recovering very slowly, and unevenly, from a prune. A simple short-term experiment was designed to compare a range of partial-pruning treatments in order to see whether the productivity of this clone could be improved in the year of prune (Box 7.6). The tea-making quality of leaf harvested from the lungs was of better quality than the 'tipping' leaf from the same bushes (Owuor, 1994). This difference was increased by the application of nitrogen fertiliser.

What is very clear is that, for some high-yielding clones such as AHP-S15/10, lung pruning is essential.

Box 7.6 A Partial-Pruning Experiment

An example of a partial-pruning experiment designed to identify a system of pruning that will improve the rate and uniformity of pruning clone AHPS15/10 (a clone that is slow to recover from a cut-across prune)

There were seven treatment combinations. The experiment was sited at Ngwazi Tea Research Station in southern Tanzania. The tea (spaced 1.2 m × 0.60 m) was irrigated. It was plucked at a harvest interval of two *phyllochrons*. Yields for the eight months following the initial pruning, shown in the table below, are in units of kg dry tea ha^{-1}.

Pruning treatments	Oct.	Nov.	Dec.	Jan.	Feb.	Mar.	Apr.	May	$\sum^a$
CA 100%	0	0	0	280	80	310	270	460	1400
PPB 75%	210	150	0	260	90	330	260	450	1750
PPB 50%	360	200	0	190	110	280	280	400	1820
PPB 25%	430	185	0	160	105	230	320	370	1800
PPT 75%	290	210	330	450	20	240	280	320	2140
PPT 50%	340	230	460	425	15	220	280	350	2320
PPT 25%	360	325	415	425	10	95	250	200	2080

CA = cut-across prune at a height of 0.40 m 27–30 September; tipped 15 January.
PPB = partial prune: 75%, 50% and 25% of the foliage of each bush was removed at the same time as the CA treatment; the remaining foliage was plucked as normal and pruned at *bud break* (defined as when the majority of the buds in the CA treatment had reached the 1–2 leaflet stage – 20 November).
PPT = partial prune: 75%, 50% and 25% of the foliage of each bush was removed at the same time as the CA treatment; the remaining foliage was pruned when the CA treatment was *tipped* (at 400 mm above the pruning height – 15 January). Final tipping took place on 4 April. Yield recording stopped after the end of May as yields declined with the onset of 'winter'.
[a] Least significant difference (LSD) at $p = 0.01$ is 460 kg ha^{-1}; coefficient of variation (CV) = 13.3%.

Although the differences in the cumulative yields over eight months are on the margins of statistical significance (as analysed), the consistency of response to the PPB and PPT treatments provides some confidence in the validity of the data.

In summary:

- Regardless of the degree of defoliation, partial pruning prior to bud break increased yields by 30%, compared with a single cut-across prune, and by 56% when the completion of the prune (removal of the lungs) was delayed beyond bud break to tipping.
- No die-back of shoots was observed after pruning in any of the treatments (usually a common feature with this clone).
- There are practical problems involved in implementing any of these partial-pruning treatments on a large scale.

Summary: Pruning

- The reasons for pruning need to be reviewed. Has it become a habit? Can the pruning cycle be extended if plucking control is improved?
- The timing of the prune has to be decided at a local level. It is always a compromise, depending on the sequence of seasons, the amount of crop lost, and the availability of labour.
- Other things being equal, it is best to prune when the soil is at field capacity prior to the start of the dry season.
- Most experiments in which partial removal of the canopy has been evaluated (in order to improve recovery) have failed to come up with definitive answers. For some high-yielding clones, however, lung pruning is essential.
- Rejuvenation pruning is considered elsewhere (Chapter 6).

Conclusions

A 'framework for analysis' has been presented and examples given of how this analytical approach can help us to understand and to quantify tea growth processes. The framework offers criteria for selecting outstanding clones (with a high harvest index), and can be used to guide management practices, such as how best to bring young tea into production, and when and how to prune mature tea in relation to the season. By making realistic assumptions, it is possible to predict the potential yield of a tea crop in a given location, and to compare this with actual yields (Box 7.7). If there is a large disparity between predicted and actual yields, as there usually is, the skill is then to identify the constraints to yield (from the components of the framework) and to address them systematically each in turn.

Box 7.7 Predicting Potential Annual Yields

$$\text{Yield} = S \times f \times e \times H$$

where S = annual total incoming solar radiation (TJ ha^{-1})
 f = fraction of light intercepted by crop canopy
 e = conversion efficiency (g MJ^{-1})
 H = harvest index (excluding roots)

Example 1. Kericho, Kenya (1800 m altitude)

$$\text{Yield} = 70 \times 0.85 \times 0.4 \times 0.15$$
$$= 3570 \text{ kg made tea ha}^{-1}$$

Box 7.7 Continued

Explanation: yield below potential due to:

1. incomplete interception of sunlight, and
2. low conversion efficiency (prolonged drought, December to March), despite
3. high radiation and
4. good harvest index.

Example 2. Tocklai, Assam (50 m altitude)

$$\text{Yield} = 35 \times 1.0 \times 1.2 \times 0.12$$
$$= 5040 \text{ kg ha}^{-1}$$

Explanation: yield below potential due to:

1. low light level (shade trees), but
2. full crop cover
3. good conversion efficiency
4. average harvest index.

Example 3. Compare these estimates with potential yield from existing cultivars

$$\text{Yield} = 70 \times 1.0 \times 1.3 \times 0.20$$
$$= 18,200 \text{ kg ha}^{-1}$$

Explanation:

1. high light intensity (clear skies)
2. full canopy cover (no vacancies)
3. high conversion efficiency throughout the year (irrigated, saturation deficit not limiting)
4. high harvest index (but could be increased).

Best commercial yield reported to date: 11,000 kg ha^{-1}.

There is still scope for improvement. What is your target for next year? What are your limiting factors?

Note

1. Time is not always a good measure to use. Thermal time (°C d) would be better (see Chapter 9).

8 Roots Exposed

Life Underground

Roots, being predominantly underground, are normally unseen and therefore difficult to study, and usually forgotten (Figure 8.1). They are though of critical importance to the tea crop in terms of nutrient and water uptake, the storage of starch, and as the source of various plant hormones. In this chapter, factors influencing root depth and distribution, rates of root extension with depth, shoot:root ratios and periodicity of growth are described. In particular, the effect of plant density on root distribution is considered in detail.

Root Function

Tea has two main root types:

- *Structural roots*, which are large woody structures providing support to what would be a tree if left unmanaged. These roots also store starch, which is mobilised in times of need (for example after pruning or a drought).
- *Fibrous or feeder roots* (Figures 8.2 and 8.3). These roots are thin (< 2 mm diameter), and white when young. They are important for nutrient and water uptake. Fibrous roots should be measured in units of length per unit volume (e.g. cm cm^{-3} or cm L^{-1}) of soil, but this is not easy so dry mass per unit volume of soil (e.g. g L^{-1}) is the usual descriptor. This is measured at successive depth increments down the soil profile (Box 8.1).

The greater the density of roots, the shorter the distance that liquid water has to move to a root, and the more water in the soil can be described as 'available' or indeed easily available. Similarly, the greater the root depth the larger is the total volume of water that is available to the plant, making it less susceptible to drought. The job of a tea farmer is to encourage roots to extend in depth as quickly as possible after planting by minimising soil compaction and loosening the soil where necessary (Figure 8.4), by avoiding waterlogging through effective drainage, and, where appropriate, keeping the water table at least 1 m below the soil surface.

Of particular significance in the nutrition of tea is the symbiotic association between the roots of the plant and vesicular-arbuscular mycorrhizae (VAM). These fungi, which exist partly within the roots and partly outside, play an important, although little understood, role in the nutrition of tea (Barua, 1989). By serving to increase the surface area of

Figure 8.1 Roots of tea are not easy to study (Kenya, MKVC).

absorbing tissue in contact with soil particles, and no doubt in other ways too, such as by extending into the mulch layer, VAM can improve the uptake and utilisation of nutrients, phosphate in particular. For example, in Japan, Morita and Konishi (1989) were able to confirm that VAM infection (in the presence of aluminium), even at relatively low levels, could enhance the phosphate nutrition of host plants, including tea, from soils low in nutrients.

Root Depth and Distribution

In his book on tea, Eden (1965) summarised the root system of (seedling) tea in the following way:

Tea is tap root dominant and puts out strong lateral roots. The root distribution pattern is highly variable, which is to a considerable degree a clonal characteristic. Whilst some bushes develop deep laterals, others persist in putting out almost horizontal roots, which never penetrate to a great depth.

By sampling and weighing roots at three-inch (75 mm) increments from bushes growing under varying cultural conditions, Eden (1965) found that the weight of feeding roots (thin, < 1 mm, white, unsuberised) declined, at a depth of 0.60 m, to less than 5% of the total weight. Similar evidence, showing that about two-thirds of the fibrous roots were confined to the top 300–400 mm of soil, led Harler (1966) and others to describe tea as 'a shallow rooting plant'. Data reported by Barua (1966) confirmed that in the plains of

Figure 8.2 Excavated root system of a three-year-old clone (TRIEA-6/8), roots about 1.2 m deep, Kericho, Kenya (MKVC).

Figure 8.3 Who says tea roots do not have hairs? (Tanzania, MKVC).

North-East India tea is indeed shallow-rooted, with 80% of the feeder roots (by length) from 12-year-old bushes occurring in the top 0.5 m, regardless of treatment. (This assumes that they are not part of the same data set.)

Part of the reason that tea has, historically, had this reputation of being shallow-rooted is that, during the rains, in the Brahmaputra Valley (Assam) there is a high water table.

> **Box 8.1** Exploding a Myth
>
> Before clones became generally accepted, the opinion amongst planters was that plants raised from cuttings possessed a shallower root system than those raised from seed. There is no evidence to support this view. Clones as well as seedlings can root to great depths. In a summary of trials in Assam, Barua (1989) concluded that 'the depth of root systems of clones selected with proper care will not be less than that of phenotypically similar seed progenies if the growing conditions do not differ.'

Figure 8.4 (a, b) Loosening the soil to a 1.0 m depth with a tractor-mounted post-hole digger enables the roots to go deeper and, in so-doing, allows the young plant to develop faster (Nandi Hills, Kenya, CJF).

This, together with a sticky compacted subsoil, restricts root growth to a maximum depth of about 0.7 m. This is despite field drainage (see photograph of a 14-year-old Assam-type seedling bush in Hadfield, 1974a), which is clearly not very effective.

Evidence reported since the 1960s (mainly from Africa) demonstrates convincingly that, providing there is no physical restriction to root growth, roots will extend to great depths (Figure 8.5). For example, roots of mature seedling tea in Malawi have been found at depths of at least 5.5 m, with evidence of water extraction during the dry season at that depth (Laycock and Wood, 1963). At the Tea Research Foundation of Central Africa in Malawi, the soils are deep (> 5 m), well-drained latosols, classed as clays to sandy clays with a water-holding capacity of about 130 mm m^{-1} (Willatt, 1970; Carr *et al.*, 1987). Similarly in Kenya, Kerfoot (1961, 1962) observed roots of seven-year-old

Figure 8.5 Roots go deep in Africa (Tanzania, WS).

seedling tea at depths of 3 m. He also reported that roots of tea had been exposed at depths of more than 6 m elsewhere in East and central Africa.

Later in Kericho, Kenya, Cooper (1979) traced roots of 17-year-old seedling tea to depths below 6 m. The soil at this site in Kenya is a deep, red, freely draining friable clay (humic ferrasol) with an available water-holding capacity of about 215 mm m^{-1}. In southern Tanzania, roots of eight-year-old seedling tea were found at 4.5 m (Carr, 1969a), and those of a 23-year-old clone (TRIEA-6/8) were found below 5 m (Nixon *et al.*, 2001). The crops in both of these examples were irrigated (Figure 8.5).

At this site in Tanzania the soil is classified as a xanthic ferrasol, with a brown, medium- to fine-textured topsoil over a deep, very light coloured, unmottled clay subsoil. The available soil water-holding capacity averages about 100 mm m^{-1} (Baillie and Burton, 1993; Carr, 1974). So, in Africa, where the water table is deep, tea is definitely not a shallow-rooted crop! The likely exception is in Rwanda, where tea is grown in drained swamps. Examples of the reported maximum rooting depths of tea in Africa are summarised in Table 8.1.

Table 8.1 Examples of the maximum rooting depth of tea at different locations in Africa

Site	Cultivar	Age (years)[a]	Depth (m)	Treatments	Source
Kericho, Kenya	Clone 12/12	3	0.72	Pegged; rain-fed	Carr (1976)
			0.94	Pruned; rain-fed	
	Clone 6/8		1.43	Pegged; rain-fed	
			1.17	Pruned; rain-fed	
	Seedlings	9	1.4–1.6	Rain-fed	
	Clones 6/8, 6/10, 6/11	7	3.0	Rain-fed	Carr (1977a)
	Clone 6/8	3	0.6 – > 1.2	Depending on the mulch used; rain-fed	Othieno (1977)
	Seedlings	17	> 6.0	Rain-fed	Cooper (1979)
	Clones 6/8, S15/10, TN 14/3, BBK-35	< 3	1.2–1.5	Rain-fed	Ng'etich and Stephens (2001)
Ngwazi, Tanzania	Clones BBT-1, 6/8, SFS150, S15/10	2.25	1.4–1.7	Irrigated and part irrigated	Burgess and Carr (1996b)
	Clone S15/10	4	2.8	Irrigated	
	Clone 6/8	23	> 5	Irrigated	Nixon *et al.* (2001)
		4–5	> 3		
	Seedlings	8	4.5	Part irrigated	Carr (1974)
	Clone S15/10	3	2.3	Irrigated	Kigalu (1997)
	Clone BBK-35		1.7		
	Clone S15/10	6	3.6–4.3	Plant density/water variables	Kigalu (2002)
	Clone BBK-35		2.5–3.0		
Marikatanda, Tanzania	Clone BBK-35	4	1.5	Rain-fed	Sanga and Kigalu (2005)
Mambilla, Nigeria	Clone 31/8	2	0.6–0.8	Rain-fed	Carr, personal observation, 1978
Mulanje, Malawi	Clones MT12, SFS204, SFS371	0.75	0.54–0.6	Irrigated	Willatt (1970, 1971)
			0.45	Rain-fed	

[a] Time from planting in the field.

Root Extension with Depth

For the successful establishment of young plants it is important that, in order to minimise the risk of water stress, the feeder roots extend into moist soil. This increases the volume of water available to the plant. If root growth ceases, the water has to move through the soil to a root, which can be a slow process. A limited amount of research on root extension has been reported for tea. For example, in southern Tanzania, Burgess and Carr (1996b) showed how the maximum rooting depth of four irrigated (and mulched) clones (BBT-1, TRIEA-6/8,

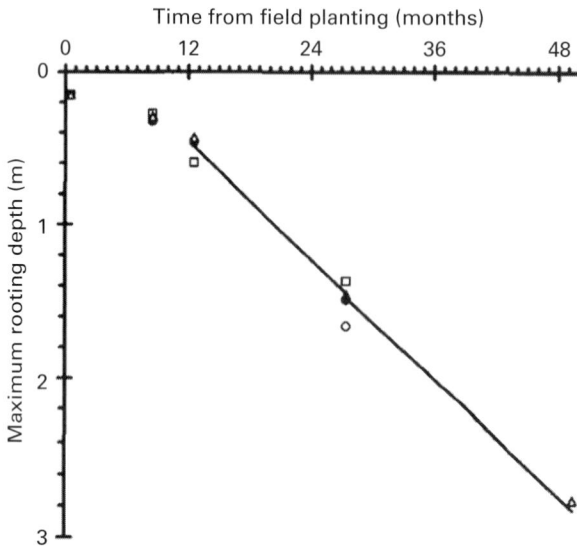

Figure 8.6 Increase in maximum rooting depth of four irrigated clones with time from planting in the field (mean rate *c.* 2.10 mm d^{-1}). This is up to twice the rate observed elsewhere in Africa (Ngwazi, Tanzania, from Burgess and Carr, 1996b).

TRFCA-SFS150 and AHP-S15/10) increased linearly with time after field planting (from 12 to 48 months) at similar rates (averaging 2.0 ± 0.11 mm d^{-1}), reaching 2.8 m within four years (clone AHP-S15/10) (Figure 8.6; Table 8.1).

By contrast, similar comparisons of four clones at four rain-fed sites in Kericho, Kenya, showed roots reaching depths of 1.0–1.5 m three years after planting (depending on the site) and averaging 1.0–1.2 mm d^{-1}, half the rate observed in Tanzania (Ng'etich and Stephens, 2001). The relationship between rooting depth and time from planting in this case was best described by an exponential equation. Clone AHP-S15/10 had the shallowest roots at each site and TN-14/3 the deepest; clones TRIEA-6/8 and BBK-35 were intermediate.

Maximum rooting depths for each clone increased linearly with the mean air temperature over the range 16–19.5 °C. Previously, Othieno and Ahn (1980) had demonstrated the positive influence of soil temperature (16–22 °C) on root mass. Low soil temperatures, as well as periodic droughts, may be the reason for the slower rates of root extension with depth observed in Kenya compared with Tanzania. A contributory factor might be that the two soil types were very different in their water-holding (pore size distribution) properties.

Root Mass

There is limited evidence of the effects of cultural practices such as irrigation, mulching and plucking, as well as the age of the bush, on root development. Such evidence as exists is summarised below.

- *Mulching* – Willatt (1970, 1971) found that grass mulches reduced the dry root mass (by 20%) and rooting depth (from 1.20 m in the control to 0.60–0.90 m, depending on the mulch used). Mulching also influenced root distribution in three-year-old plants (clone TRIEA-6/8) grown in Kenya (Othieno and Ahn, 1980).
- *Bringing into production* – From observations of the root systems of two rain-fed, three-year-old clones brought into bearing by pegging or pruning at the Tea Research Foundation of Kenya, Kericho, pruning appeared to reduce the overall size of the root system compared with pegging (Carr, 1976; Barbora *et al.*, 1982).
- *Plucking* – In Assam, Ghosh, as reported by Barua (1989), compared the depth and spread of roots of 25–30-year-old tea trees, originally grown from seed and unplucked, with similar bushes that had been plucked. On flat land with a very high water table, root depth was restricted to 0.40 m in both cases, but on small hillocks (*teela*) the trees had deeper roots (2.6 m) than the plucked bushes (1.6 m). Temporary waterlogging may have restricted root depth even on the *teela*.
- *Irrigation* – In Malawi, Willatt (1970) demonstrated the beneficial effects of irrigation on root depth and root distribution of four clones in the first year after field planting. In Tanzania, Nixon and Sanga (1995) and Burgess and Sanga (1994) studied the distribution of thick (> 1.0 mm diameter) and fine (< 1.0 mm) roots for 23-year-old clone TRIEA-6/8 plants to a depth of 3.0 m in fully irrigated and unirrigated (over the previous eight years) bushes (Figure 8.7). The mean concentration of fine roots over the whole depth in the unirrigated, rain-fed-only treatment (0.13 g L^{-1}) was only 25% of that for the fully irrigated plants (0.53 g L^{-1}), with the greatest difference being in the top 1.0 m of soil. By contrast, the concentrations of thick roots were similar in both treatments (1.67 and 1.79 g L^{-1} respectively). There is no evidence here for the often-repeated view that irrigation reduces the depth or concentration of rooting. In fact, for this example from Tanzania, the opposite is the case, unless insufficient water has been applied, as roots cannot extend into dry soil.
- *Age* – In order to help explain differences in response to irrigation by young and mature tea, a detailed comparison of root distribution of 'young' (six years after field planting) and 'mature' (23 years) clone TRIEA-6/8 plants was undertaken in Tanzania in 1994 (Nixon *et al.*, 2001). Both crops had been irrigated since planting. Sampling was restricted to the top 3.0 m, although roots of both crops went deeper than this. The total dry mass for structural roots (> 1.0 mm diameter) to 3.0 m depth was four times greater in the mature crop (5.82 kg plant^{-1}) than in the young crop (1.56 kg plant^{-1}), and for fine roots (< 1.0 mm) it was eight times greater (1.86 and 0.24 kg plant^{-1} respectively). For both crops, over 85% of the structural roots (dry weight, g L^{-1}) were found in the top 0.40 m of soil, but the absolute weights again differed by a factor of four.

Shoot:Root Ratios

In the same detailed study by Nixon *et al.* (2001), the corresponding shoot:root ratios (dry mass) were about 1:1 and 2:1 for 'old' and 'young' plants respectively. In contrast,

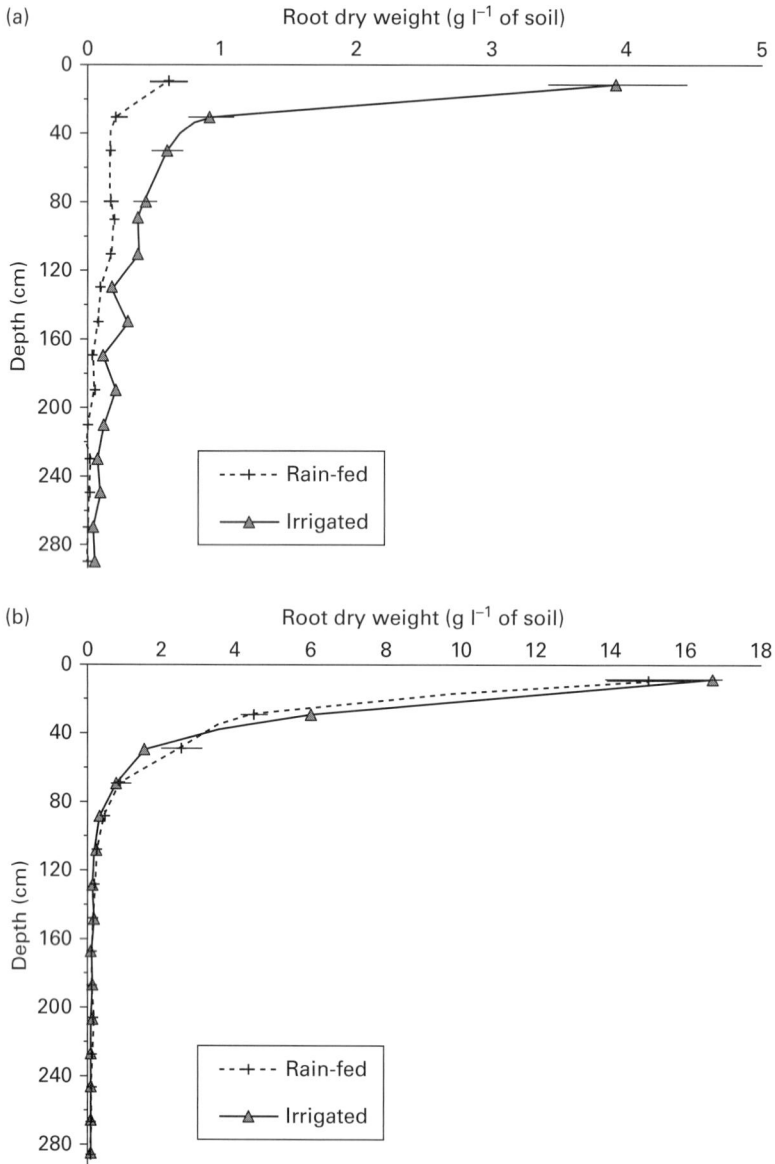

Figure 8.7 The mean root distribution of (a) fine, and (b) thick roots of 23-year-old plants of clone TRIEA-6/8 to depths of 3 m under rain-fed and fully irrigated conditions (from Burgess and Sanga, 1994; Nixon and Sanga, 1995).

Barua (1966), working in North-East India, found that the shoot:root ratio of a selection of mature clones and plants grown from seed averaged 2:1, varying between 2.78:1 (clone 19/29/13) to 1.73:1 (clones 1/7/1 and 20/23/1).

Periodicity of Root Growth

The growth of tea feeder roots (white, unsuberised) was observed and monitored against glass in several simple underground root chambers over a 21-month period between 1968 and 1970 at Ngwazi, Tanzania (Carr, 1969a, 1971a). The principal observations were:

- Roots of irrigated China-type BBT-1 grew throughout the 'winter' months (June–September) when shoot extension growth was negligible.
- Roots of previously unirrigated BBT-1 began to grow only after the first 'flush' of shoot growth was coming to an end, 6–8 weeks after water was first added to the soil.
- There was a net reduction of 27% in the length of white roots visible at the glass surface during the dry season (Carr, 1971a).
- Roots of two Assam-type clones (BBT-28 and BBT-36) grew only slowly during the 'winter' although some shoots continued to grow, albeit slowly.

In a similar study in southern Malawi, Fordham (1972) observed roots of three young clones (TRFCA-SFS204, TRFCA-MT12 and TRFCA-MFS76) and mature seedling tea under irrigated field conditions in simple root-observation trenches. Periods of maximum shoot growth were associated with minimal root growth. Pruning caused roots to stop growing for approximately three months. Herd and Squire (1976) also observed stimulation of root growth in the 'winter' months in Malawi.

In order to understand better how the products of photosynthesis are allocated between the principal plant parts, including roots, the maintenance leaves of two unpruned bushes of Assamese clones TV1 and TV25 were exposed to radioactive CO_2 at monthly intervals for a year in Assam (Barman and Saikia, 2005). A total of 31% of the dry matter produced during the year was allocated to the root system. This was divided equally between the following four categories of root: thick roots at the base of the trunk, medium-sized roots, pencil-sized roots and feeder roots. For comparison, only 11% was allocated to the young shoots (two leaves and a bud), 25% to the branches and 14% to the trunk, with 19% being retained in the maintenance foliage. The *harvest index* (that is, the proportion of the total dry mass allocated to the yield of young shoots) is considered in more detail in Chapter 7.

Plant Density

Root depth and distribution were monitored in selected treatment combinations in a plant density × drought × clone experiment at Ngwazi Tea Research Station, Tanzania. There were six spacings, giving densities ranging from 8333 (D1) to 83,333 (D6) plants ha^{-1}, of which only three are considered here (see Box 6.3 for full details of this experiment). By August 1995, 32 months after planting (and mulching), roots of well-watered clone BBK-35 had reached depths of 0.9 m (D1), and those of clone AHP-S15/

Table 8.2 Summary table showing maximum depth of rooting and mean rate of root extension with depth of two clones under well-watered and dry conditions. Averages of high and low plant densities, six years after field planting. Each value is based on data for six bushes, 12 for the mean

	Clone AHP-S15/10			Clone BBK-35		
	Well-watered	Dry	Mean	Well-watered	Dry	Mean
Depth (m)	3.6	4.3	4.0	3.0	2.5	2.8
Rate (mm d^{-1})	1.6	2.0	1.8	1.4	1.1	1.3

10 between 1.5 m (D1) and 2.1 m (D3). Roots in treatment D6 went less deep than those in D3 (Table 8.1) (Kigalu, 1997).

The largest effect of plant density on root distribution occurred in the top 0.2 m, where root density increased with plant density for both clones. Below this depth the largest root density occurred with treatment D3, followed by D6 (AHP-S15/10 only), and then D1. AHP-S15/10 had more fine roots below 0.2 m than BBK-35 at all three densities. The corresponding rates of increase in root depth equated to about 0.8–1.1 mm d^{-1} (BBK-35) and 1.6–2.2 mm d^{-1} (AHP-S15/10).

The relative differences between clones in rooting depth were not consistent between sites. Thus, roots of three-year-old BBK-35 plants were shallower than those of AHP-S15/10 in Tanzania, but deeper in Kenya.

Plant density effects on root depth and distribution were investigated again in Tanzania in early 1999, six years after field planting and four years after differential irrigation/drought treatments had first been imposed. This time the two extreme densities were compared (high, D6, and low, D1) at each of the two extreme drought treatments I0 (most droughted, a total of 40 weeks without irrigation/rain), and I6 (well irrigated since planting) for both clones. The roots of three plants from each treatment combination were excavated at 0.20 m depth increments.

The results were not easy to interpret (Table 8.2). Although, on average, roots of clone AHP-S15/10 reached depths substantially greater than those of BBK-35, the effect of water regime was not consistent between the two clones. For AHP-S15/10, dry conditions resulted in deeper rooting than in well-watered plants. For clone BBK-35 the situation was reversed, with droughted plants rooting less deeply than those that had been well watered since planting. The effect of plant density was even more complicated: for AHP-S15/10 roots of high-density plants were about 0.2–0.4 m shallower than those of plants grown at low density: by contrast, for BBK-35, although roots of droughted plants were 0.8 m shallower at low density, they were 0.2 m deeper when irrigated (Kigalu, 2002).

In terms of root distribution, there were more fine roots (< 1.0 mm diameter) at the low plant density than at the high density (expressed as g L^{-1} soil). For droughted plants this difference extended to depths of 2.0 m with both clones, but to only 0.5–1.0 m for irrigated bushes (Kigalu, 2002). In all four treatment combinations, clone BBK-35 had fewer fine roots than AHP-S15/10, low-density plants had more than those at high density and, with one exception (clone AHP-S15/10, high density), droughted plants had more than those plants that were well irrigated. The largest differences (averaged over 1.0 m depth only) in fine root densities for (a) clone BBK-35 were between low plant density, well irrigated (0.4 g L^{-1}), and low density, droughted (1.5 g L^{-1}), and for (b) clone AHP-S15/10 between high density, both droughted and well irrigated (0.9 g L^{-1}), and low density, droughted (2.0 g L^{-1}).

The shoot:root ratios (dry mass), as calculated from the slope of the linear relationship between the cumulative totals recorded in sequential whole-plant harvests over the three years following planting, were independent of plant density but were consistently lower for BBK-35 (3:1) than for AHP-S15/10 (5:1) (Kigalu, 1997). These shoot:root ratios are higher than those for the older tea mentioned above and serve to emphasise the sensitivity to drought of tea this age (under five years). There was a trend for these ratios to decline with age. It does not help crop establishment if the plastic sleeve is left on the plant when planting in the field (Figure 8.8)!

Storage of Starch

After pruning, or a severe drought that results in defoliation of the tea bush, the plants refoliate by mobilising starch stored in the roots. Before pruning, it is sometimes necessary to check that the starch reserves are adequate to ensure recovery. This is

Figure 8.8 Round and round! This bush was planted with the plastic sleeve still in place – a form of torture (MKVC).

Box 8.2 The Iodine Test

Remove a pencil-thick piece of healthy root from the tea plant and dip it into an iodine solution made up of 3 g of iodine and 6 g of potassium iodide dissolved in 1 litre of water. The development of a deep blue coloration in the cut end signifies that there are adequate starch reserves and that it may be a good time to prune. Conversely, if the blue colour is weak, it is probably best to delay pruning until the starch reserves have recovered. The test should be repeated on a minimum of three samples.

---

done using the iodine test (Box 8.2). Clones differ in their capacity to store starch, and this probably explains why some clones are slow to recover from a prune (Sharma and Murty, 1989).

Plant Hormones

The roots of tea produce cytokinins, a group of plant hormones (not specific to tea), which influence cell division elsewhere in the plant. They could possibly explain why soil temperature seems to be important in tea, as well as in other crops.

Summary

Although studying roots is not an easy task, the following broad conclusions can be reached, nearly all of which influence water availability and crop responses to drought and to irrigation:

- Vesicular-arbuscular mycorrhizae (VAM) are believed to play an important role in the nutrition of the tea plant, particularly in phosphate uptake.
- The roots of tea can extend to, and extract water from, considerable depths (5–6 m), providing that there are no physical restrictions, such as a high water table.
- Tea roots extend in depth at rates of between 1 and 2 mm d^{-1} (depending on temperature and water status), reaching 1.0 m within 18–36 months after planting in the field.
- About 85% of the structural roots (by mass) occur in the top 0.4 m of soil, regardless of the plant age.
- The density of roots continues to increase at all depths with age from planting.
- The shoot:root ratio (by dry mass) variably decreases with plant age, or stays stable at about 2:1. Clones differ in the depth and distribution of roots, but not consistently across sites.
- Methods of bringing young tea into bearing can influence the form of the resultant root system in young tea. Pruning results in the temporary cessation of root growth.

- The tea plant depends on starch stored in structural roots for recovery from pruning.
- Despite popular mythology, there is no consistent evidence that well-managed irrigation reduces the size or depth of a root system (indeed there is some evidence for the opposite response).
- Waterlogging and high water tables in the rains certainly do restrict root growth.
- Where the water table is deep, using a fence-post-hole digger to loosen the soil to a depth of 0.9–1.0 m will increase the depth and intensity of root growth after planting, and aid crop establishment.
- Grass mulches can reduce the depth and mass of root systems of young plants. It is not clear why, as there are other benefits from mulches in young tea (see Chapter 15).
- Planting density can influence root depth and distribution, but the effects are complex, and appear to vary with clone and watering regime.
- Roots can continue growing during 'winter' months when shoot extension rates are slow.
- There is evidence of periodicity of root growth, with root growth alternating with active shoot growth.

9   We Are Only Growing Leaves

Source or Sink?

To attempt to explain how the climate and its day-to-day variability (the weather) influence yield formation, and set limits on productivity, we need to understand the growth processes of a crop. This means breaking down the development of yield into its component parts, as discussed in Chapter 7, and then investigating how each in turn is influenced by external circumstances. Only in this way can one hope to identify and quantify the factors constraining yields, be they genetic, physical, chemical, biological or management-related. The alternative is a shotgun approach in the hope that something will work even if one doesn't know when or how, or whether it will be repeatable somewhere else, even if the yield component system is relatively simple, as it is with tea, or so one might think.

After all, we are only growing leaves – it's not as though it was bananas, where the useful product is a fruit and floral initiation only occurs once 30–40 leaves have been produced, and it takes up to 200 days between bunch emergence and harvest. The weather can change a lot over such extended periods of time and it becomes difficult to establish cause and effect. By comparison, tea is a relatively simple crop to understand – or is it? Read on!

In this chapter, the development of a tea shoot and its morphology are described, and the components of yield identified (Figure 9.1). The concept of thermal time is introduced and its value as a tool for predicting the duration of the shoot replacement cycle is explained. For example, it is necessary to distinguish between the base temperature for shoot extension and that for shoot development. Factors influencing crop distribution and possible yield-limiting factors, including the relative importance of the source (photosynthesis) and the sink (limited by the size of the potential crop) are examined. In the case of tea, as with grass, both the source and the sink are the same plant organs, namely the leaves. In the follow-up Chapter 10, the practical aspects of using thermal time to predict leaf appearance rates and hence to provide guidance to growers on the choice of harvest intervals, when targeting shoots of different size and stages of development, are described.

Shoot Development

In a classic study of its type Bond (1942), working in Ceylon (Sri Lanka), described in great detail the vegetative growth and anatomy of the tea plant. This description formed part of a study intended to investigate the 'phloem necrosis' disease of tea, which was of

Figure 9.1 We are only growing leaves, after all (WS).

some concern at the time. Of general interest is his description of the periodicity of growth that is a marked characteristic of the tea plant. To quote Bond (1942):

each shoot, and to a certain extent each bush, if left unplucked normally goes through a rhythmic alternation of periods of active growth, involving the successive unfolding of cataphyls and foliage leaves, and of dormancy during which growth in length is completed and a marked lignification of the tissue occurs, but in which no new leaves are unfolded. The active growth is described as the *flush* period, the succeeding interval of dormancy as the *banjhi* period.

In Ceylon, this cycle proceeds to a large extent independently of the seasonal change in climate, and about four cycles are completed each year, although there are exceptions to this rule: for example strong leader shoots can show aperiodic growth, or a terminal bud can remain banjhi for a prolonged period. A 'normal' shoot produces a fairly constant number of appendages in each flush period. On average, this consists of two bud scales, one so-called fish leaf, and four flush leaves, seven appendages in all (Figure 9.2). However, one can often find a bush with shoots at different stages of development, actively growing and

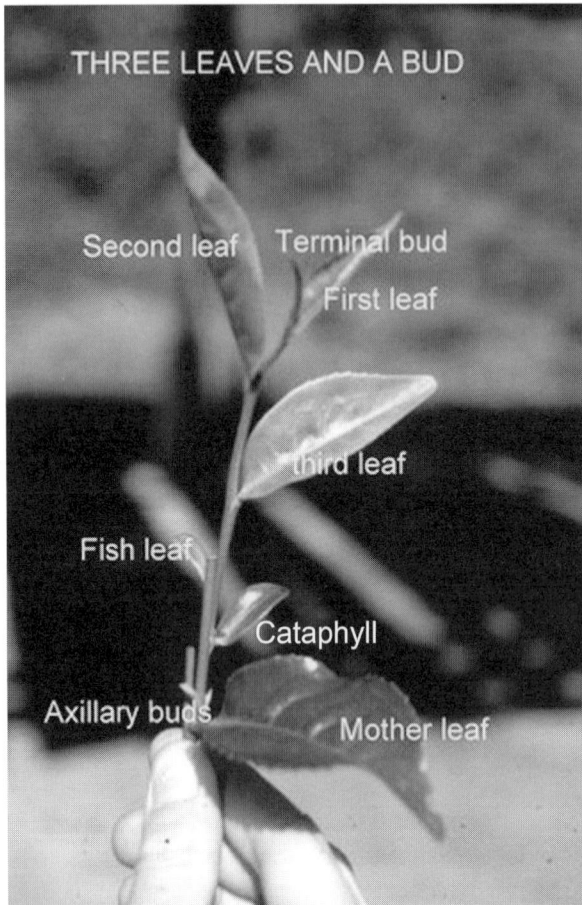

Figure 9.2 Photograph showing the sequence of leaf development on a tea shoot, including two scale leaves (or their scars), the fish leaf together with the mother leaf, flush leaves, and finally a terminal shoot in the process of going dormant (banjhi). Note that a mother leaf in parts of India is the first true leaf on a shoot, not the leaf subtending a shoot, as in this example.

dormant, indicating that shoot structure is not a constant. The degree of elongation of the internodes (the distance between two leaf nodes) is least in the vicinity of the bud scales and most in the flush-leaf region.

In a companion paper, Bond (1945) postulated that the whole cycle of events inherent in the periodic development of the flush shoot could be viewed as a self-regulating mechanism. Based on his detailed observations, which included measuring the area of vascular (and provascular) tissue in transverse section at a given distance from the growing point, Bond (1945) proposed the following four-phase mechanism by means of which the process of periodicity in tea is controlled. The onset of phase 1, active flushing, leads to a diminution in the vascular supply (of water and nutrients) to the bud and a lengthening of the plastochron (the time interval between initiation of successive leaf primordia) for the leaves in the process of formation. This in turn is followed by phase 2, the determination of the scale-leaf initials, the slower development of which, compared with the preceding flush leaves, breaks the sequence of expansion and the bud 'goes banjhi' (phase 3). While in this condition the vascular supply is again increased, the plastochron is shortened, implying that initiation of primordia continues through the banjhi period, and more young primordia receive the 'impetus' to develop into foliage leaves (phase 4) so that the cycle is renewed from the beginning.

Shoot Growth

For practical purposes, the growth of a shoot can be described as follows.

When apical dominance is removed by breaking off a shoot (e.g. by plucking), the bud in the axil of the leaf (known as the *mother leaf*[1]) nearest to the break (the *axillary bud*) begins to develop (Figure 9.3).

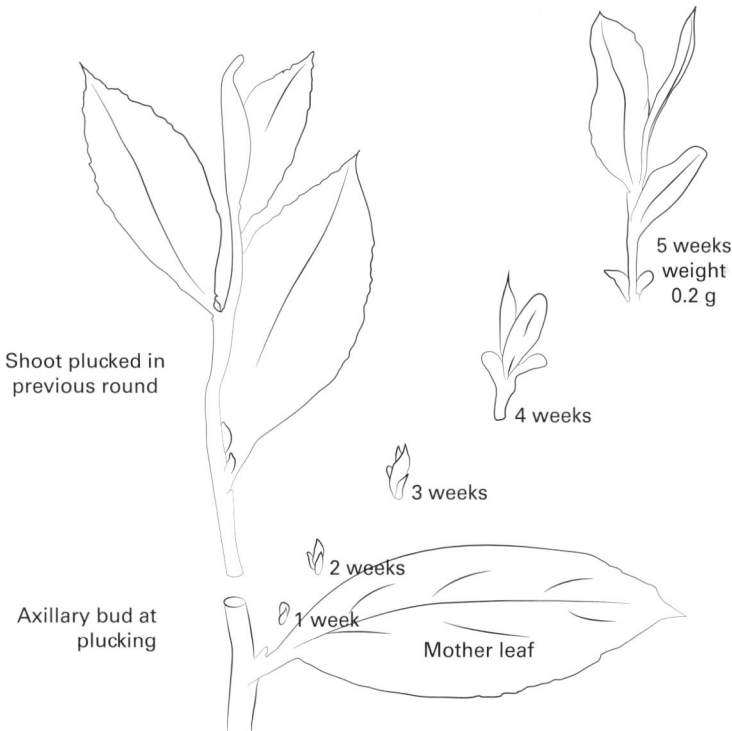

Figure 9.3 The stages of development of the axillary bud when released from apical dominance after the main shoot has been removed. The timing of emergence, the number and the mass of the developing shoots will vary depending on the situation. The numbers on this figure relate to summertime in Malawi, when the average duration of a shoot replacement cycle is 42 days (for three leaves and a bud) (Drawing by Dr Rex Ellis).

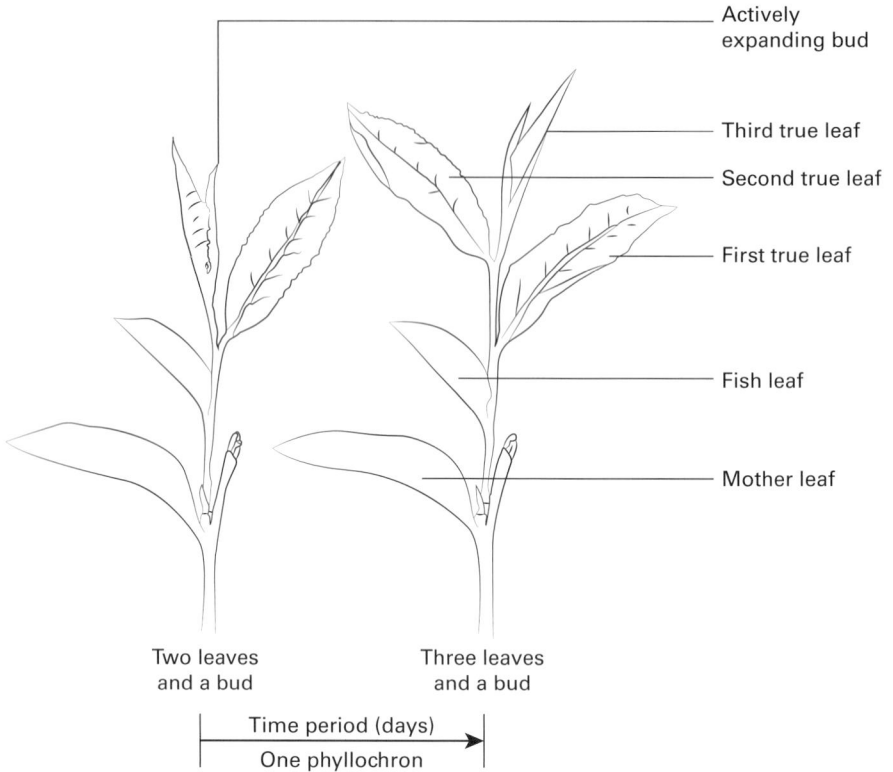

Figure 9.4 Phyllochron: a diagram depicting what is meant by one phyllochron, namely the time interval between the unfolding of one leaf and the next (between leaf two and leaf three in this example) (Diagram by Dr Paul Burgess).

This developing shoot produces, in succession, two very small *scale leaves*, which protect the bud. They usually drop off, leaving a small scar on each side of the developing shoot. There then follows a small blunt foliar structure, also a scale leaf. These scale leaves are all *cataphylls*. There is then a slightly larger unserrated blunt leaf called the *fish leaf*. Three or four (or more, up to 12) 'normal' foliage leaves then unfold. Growth of the shoot is terminated by the development of a dormant terminal bud, which is known as a *banjhi bud* (Sharma *et al.*, 1990). After a period of time (measured in weeks) the banjhi bud will begin to develop and the sequence of leaf production will repeat itself. Shoots having an actively growing terminal bud are called *flush shoots*, those with a dormant bud are *banjhi shoots*. The time interval between the opening of successive leaves on an actively growing shoot is known as a *phyllochron* (Figure 9.4) and the time for a shoot to develop, from a bud released from apical dominance to the emergence of the third true leaf, is the *shoot replacement cycle* (SRC, measured in days, *d*).

Components of Yield

There are three principal contributors to the annual yield in tea:

1. The mean *number of shoots* harvested per unit area at each harvest (N, m^{-2}).
2. The mean shoot *dry mass* at harvest (M, g).
3. The *number of harvests* (H) in a year (which depends on the duration of the shoot replacement cycles (d) but is determined by the manager).

The annual yield of tea (Y, g m^{-2}) can therefore be expressed in the following way:

$$Y = N \times M \times H$$

Much of the detailed research on factors influencing the components of yield has been done in eastern Africa. This section draws heavily on that work to provide numerical examples.

Shoot Number

This is the main determinant of yield. It varies with the clone, inputs such as nitrogen and water, temperature and stage in the pruning cycle.

The *total shoot population* is the sum of the number of *harvested shoots* and the number of *basal shoots* (small shoots remaining after harvesting that form the basis for subsequent crops) (Figure 9.5). Sometimes it is necessary to differentiate between shoots with terminal buds that are active, and those with buds that are dormant, or *banjhi* (Stephens and Carr, 1990, 1994). For example, in a fertiliser × irrigation experiment in southern Tanzania, the mean annual basal shoot population density increased over a three-year period from 310 to 560 m^{-2}, when averaged across all treatments. In well-fertilised, well-watered tea (clone TRIEA-6/8) the peak shoot density reached 850 m^{-2} over the same three-year period. In Darjeeling, Kumar *et al.* (2011) in a comparison of several China-type clones reported shoot numbers between 50 and 480 m^{-2}.

In the Tanzanian example, applying fertiliser increased both the number of shoots harvested and also the proportion that were actively growing. Irrigation had no effect on the annual mean basal shoot population density, but between seasons there were considerable differences. The main effect of water stress was to delay the peak basal shoot population density from the warm dry season to the early rains, without affecting the annual mean total as recorded in the well-irrigated plots. Over the period of the experiment, the *shoot replacement ratio* increased from 1:1.1 in the high input plots at the start to 1:1.6. In other words, by the end each harvested shoot was replaced by 1.6 new shoots (Stephens and Carr, 1994), the extra shoots having arisen from unplucked origins, or from two axillary buds replacing one plucked shoot.

Figure 9.5 The total shoot population is the sum of the number of harvested shoots and the number of basal shoots (small shoots remaining after harvesting). The peak shoot density reached 850 m^{-2} over a three-year period. (Clone TRIEA-6/8, MKVC). *A black and white version of this figure will appear in some formats. For the colour version, please refer to the plate section.*

Shoot Mass

The *fresh mass* of an individual shoot increases approximately linearly with the number of leaves on the shoot. The slope of the relationship varies with clone, season and irrigation (for example, for clone TRIEA-6/8, from 0.16 to 0.35 g for each leaf on a shoot), but fertiliser has no effect on this relationship (Stephens and Carr, 1994; Burgess *et al.*, 2006). A shoot with three leaves and a bud can weigh up to 50% more than a shoot with two leaves and a bud (clone TRIEA-6/8, Figure 9.6) or even more depending on the clone and random variation.

The dry matter content of a shoot also varies with the season and with fertiliser and irrigation (e.g. range 0.19–0.30; Burgess, 1992). The fresh mass × dry matter content gives the *dry mass*. Processed (or made) teas leaving the factory usually have water contents of 2–4%, that is, their mass is 2–4% more than the dry mass.

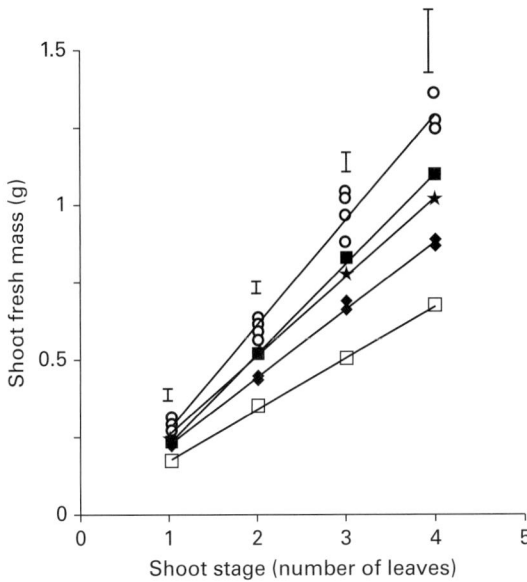

Figure 9.6 The fresh mass of an individual shoot increases linearly with the number of leaves on a shoot. The slope of the relationship varies with the clone, season and irrigation.

Shoot Replacement Cycle (SRC)

Shoot development can be considered as a three-stage process.

- *Stage 1* is a long lag phase as the axillary bud slowly expands, and as the leaf primordia within it develop.
- *Stage 2* is a period of rapid shoot extension and leaf development (the unfolding of leaves, which corresponds to phase 2 in Bond's (1945) analysis).
- *Stage 3* is when the terminal bud becomes dormant (Bond's phase 4).

The duration of the shoot replacement cycle (i.e. stage 1 and part of stage 2) is mainly a function of temperature, but it is also influenced by water stress, nutrition, and the dryness of the air, and varies among individual clones.

To aid our understanding of the processes involved, it is necessary to distinguish between shoot *extension* (increase in length) and shoot *development* (change in number of open leaves). Rates of shoot extension and development are both temperature-dependent. The base air temperature for shoot extension (T_{be}) is in the range 8–13 °C, depending on the clone. The optimum mean temperature is about 24–26 °C, with growth rates declining at temperatures above 30–35 °C. By comparison, the base temperature for shoot development (T_{bd}) is about 2–3 °C less than that for extension (Table 9.1). For this reason, internodes are shorter during the 'winter' months (or at high altitude) than in the 'summer' (or low altitudes), and tea is then more difficult to harvest by hand (Stephens and Carr, 1990, 1993; Burgess and

Table 9.1 Base temperatures for shoot development, shoot extension and internode length for each of six fully irrigated clones, with 95% confidence limits; $n = 17$. Values derived from best-fit linear relation between development rate (1/SRC, d^{-1}), relative extension rate, mean internode length (between the fish leaf and a third true leaf) and the mean daily air temperature (T_{mean}, $^{\circ}C$). Data from Burgess and Carr (1997).

Clone	Base temperature, T_b ($^{\circ}C$)		
	Development	Extension	Internode length
BBT-1	8.7 (5.1–10.6)	11.1 (9.4–12.1)	12.8 (11.5–13.6)
BBK-35	9.2 (7.4–10.5)	11.3 (10.2–12.1)	11.3 (7.4–13.1)
BBT-207	8.6 (5.7–10.4)	10.3 (9.1–11.1)	11.8 (9.6–13.1)
AHP-S15/10	7.8 (4.7–9.7)	10.1 (8.2–11.3)	10.4 (6.4–12.2)
TRIEA-6/8	6.1 (2.9–8.2)	9.5 (7.5–10.7)	10.4 (6.6–12.2)
TRFCA-SFS150	7.0 (4.4–8.7)	8.9 (6.7–10.2)	8.0 (1.4–10.6)

Carr, 1997). In Darjeeling (26°9′N 88°12′E, alt. 1240 m), Kumar *at al.* (2011) similarly found that, amongst eight clones of China origin, the base temperature for shoot extension varied between 8 and 15 °C. Rates of extension then increased linearly up to 30–35 °C. Extension stopped when the daytime maximum and minimum air temperatures fell to 18 °C and 12 °C in the autumn, and began again in the spring when they reached 20 °C and 12 °C respectively.

Thermal Time

Carr and Stephens (1992) have reviewed the concept of thermal time as an approach to understanding the effects of climate and weather on tea developmental processes. In summary, Squire (1979) demonstrated a linear relationship between shoot extension rates and mean air temperature over the range 17–25 °C. By extrapolation backwards, the base temperature for tea shoot extension was identified (about 12.5 °C for clone TRFCA-SFS204). Squire (1979) and afterwards Tanton (1982a) then applied the concept of thermal time to predict the time taken for a tea bud released from apical dominance to reach a harvestable size. Subsequently, a total of 475 day °C became the accepted value (although uncon-firmed) in Malawi, summed above a base temperature of 12.5 °C. This concept has allowed the seasonal effects of temperature on the duration of the shoot replace-ment cycle to be quantified and geographical sites to be compared (Box 9.1). A limitation to this approach occurs when temperatures exceed the optimum for growth (> 30–35 °C), and/or when the saturation deficit of the air exceeds a value of about 2.0–2.3 kPa. A field experiment in Malawi showed how intermittent misting during the day in the hot dry months from late September to early December removed the adverse effects of dry air and allowed shoots to grow at the same rate as shoots growing at the same effective mean air temperature

---

Box 9.1 Thermal Time/Day Degrees

Each growth process has a base temperature (T_b) or minimum temperature below which rates of growth or shoot extension are very slow, an optimum temperature (T_o) above which rates start to decline, and a maximum temperature (T_m) when growth ceases. For shoot extension in tea the consensus value of T_b is about 12.5 °C (although clones differ: see Chapter 7).

Thus when:

$$T > T_b < T_o$$

$$1/t = (T - T_b)/\theta$$

or

$$\theta = t(T - T_b)$$

where θ is the product of the duration in days (t) of the effective temperature ($T - T_b$). The integral of θ over time is known as *thermal time* or *day degrees*.

This concept can be used as a practical tool for predicting when to harvest tea. Thus if 475 °C day is needed for a bud to develop into a shoot with three leaves and an unopened bud, and the daily mean air temperature is 18 °C, the time taken for the shoot to develop is:

$$475 \text{ divided by } (18 - 12.5) = 475/5.5 = 86 \text{ days}$$

If the mean air temperature is 24 °C the time taken will be:

$$475/11.5 = 41 \text{ days}$$

(See exercise in Chapter 19.)

---

during the rains (Squire, 1979; Tanton, 1982b). Subsequently, Stephens and Carr (1990) showed how the apparent base temperature for shoot extension varies between clones, from 14 °C–15 °C (BBT-1) to 11.9 °C (AHP-S15/10) to 10.5 °C (clones BBT-28 and BBT-36).

Based on measurements on seven clones in Malawi, Smith *et al.* (1993a) urged caution about taking too simplistic an approach to shoot extension and over-reliance on the view that there is an inherent base temperature characteristic of each particular clone. The analysis by Stephens and Carr (1990) also showed how the results of studies like these could be biased by the shoot selection technique employed. If shoots already 30–50 mm long and actively growing are selected for measurement, there is a risk that the growth rates of only inherently fast-growing shoots are recorded. To obtain a representative sample it is essential to select axillary buds at random immediately after harvesting for subsequent measurement.

In a comparison of six clones, Burgess and Carr (1997) found that an exponential function with two constants, an initial shoot length (i) and a relative extension rate (r), provided a realistic description of the length of axillary shoots at the end of the period from the release from apical dominance to the unfurling of three true leaves. Differences between clones in the values of the base temperature for shoot extension (T_{be}) (from 8.9 °C for TRFCA-SFS150 to 11.3 °C for BBK-35) and in thermal extension rates (ρ, derived from linear relations between r and the mean air temperature) could be used to explain seasonal differences in yield in southern Tanzania. The apparent base temperatures for shoot development (T_{bd}, for the unfolding of leaves) were consistently 1.7 °C (BBT-207) to 3.4 °C (TRIEA-6/8) below those for extension, an observation supported by the effects of temperature on the length of the internodes. The value of T_{be} for young clone TRIEA-6/8 plants (9.5 °C) is close to that identified for mature plants of the same clone (10.0 °C), which is also some 2–3 °C above T_{bd} (Stephens and Carr, 1993). Note that there is considerable uncertainty in the base temperatures estimated by extrapolation (Table 9.1), and that derived mean values do vary. Allowance for diurnal changes in temperature, particularly if the night temperature falls below the base temperature, can improve the relationship.

Both sets of results from Tanzania apply to well-fertilised, well-watered plants, but nutrition and water supply influence how shoots respond to temperature. For example, shoots of well-fertilised tea were always longer (at a given stage of development) than those from unfertilised plants, while the length of shoots with three leaves and a bud ranged from 15 mm in unirrigated plots at the end of the dry season to 130 mm in high-input plots at the start of the rains. Similarly, the duration of the shoot replacement cycle varied from 65 days (warm, wet season) to 95 days (cool, dry but irrigated) for high-input plots, and from 75–180 days for unirrigated, unfertilised tea, results that all have commercial implications in terms of the choice of clones, harvesting policies and planning, and yield distribution (Stephens and Carr, 1993). The sensitivity of young tea to drought is demonstrated by the linear reduction in extension and development rates with an increase in the mean *soil water deficit* over a narrow range, 15–90 mm (Burgess and Carr, 1997).

Clones differ in many of these temperature-related attributes. For example, in Malawi, Nyirenda (2001) found that the length of the shoot replacement cycle for clone TRFCA-MFS87 to reach three leaves and a bud during the main growing season was only 37 days. This compared with the generally accepted value of 42 days deduced from shoot growth studies on seedling tea and clone TRFCA-SFS150. The average shoot length was 173 mm, and the mean shoot dry weight was 0.48 g. Similar-aged shoots of clone TRFCA-PC1 were 84 mm long and the mean shoot dry weight was 0.128 g. Measurements made at four sites in Malawi with 13 cultivars confirmed that clones can differ in their rates of shoot development. For example, the optimum shoot age for harvesting TRFCA-PC198 was 35 days, whereas for TRFCA-SFS150 it was more than 49 days (compared with 42 days cited above for the same clone) (Mphangwe and Nyirenda, 2001b).

Yield

When the simple model relating yield to shoot growth described above (see *Components of Yield*) was tested with experimental data from Tanzania, there was a very good linear relation between predicted and actual yields ($R^2 = 0.83$, $n = 9$). However, yields were consistently overestimated in the low-input plots (Stephens and Carr, 1994). It is possible that the number of shoots that reached harvestable size in these plots was overestimated. Sometimes more than one shoot can develop from a residual shoot, i.e. from secondary and tertiary leaf axils. It may also have been an anomaly associated with the reward system to pluckers operating at the time.

Source- or Sink-Limited?

As Tanton (1992) stated, crops can be divided into those which are limited by lack of photo-assimilates, i.e. they are *source-limited*, and those which are not limited by lack of sugar but instead are restricted by the size of the potential crop, i.e. they are *sink-limited*. For many years it was assumed that tea was source-limited, and research was focused on ways to increase yield by enhancing the photosynthetic process, for example the work by Barua (1970) and Hadfield (1968) in Assam on leaf size and leaf pose. But, attempts to correlate photosynthesis with yield do not work because climatic variables that increase rates of photosynthesis also increase shoot growth – the shoots increase in size faster and/or there are more of them. After reviewing the evidence, Tanton (1992) concluded that tea is clearly sink-limited. For example, tea already produces excess starch, which is stored in the stems and roots, so why should producing more lead to increased leaf yields? Of the three components of yield, only shoot number per unit area is under full genetic control; harvesting practices control shoot size, and shoot growth rates are regulated largely by the environment. Selecting clones with a large number of active shoots per unit area is likely to be the most successful way of increasing the yield potential of tea (Tanton, 1992). See also Chapter 5, Figure 5.5 in this book.

Crop Distribution

Crop yield is not uniform during the year. Large peaks in production can occur after a limiting factor, such as low temperature or drought, has ended, allowing the accumulated buds of many ages to develop together. This large peak, such as that which follows the start of the rains in southern Malawi, or for irrigated crops the rise in temperatures after the cool winter period in southern Tanzania, can cause major logistical problems for farmers and factory managers. Sometimes leaf has to be thrown away if factory capacities are exceeded at these times. Afterwards there is then a decline in production until the next generation of shoots has developed. A second, but smaller, peak in production is followed by a third and possibly a fourth until once again drought and/or low temperatures reduce rates of development and the cycle is repeated. The scale of these oscillations is proportional to the degree of synchronisation induced during the period of stress (Stephens and Carr, 1990).

When growing conditions are favourable, the terminal buds on the apical shoots are still active when they are harvested. Removal of these shoots then encourages active secondary shoots to grow from below. The terminal buds on these secondary shoots tend to become dormant when only two or three leaves have unfolded (rather than 12). This is probably because shoots such as these have developed under apical dominance (Tanton, 1992). Similarly, slow-growing shoots from unplucked origins that emerge from within the canopy also have small apical buds. When there is a large proportion of such shoots, the bush is subjectively judged to have gone *banjhi*. This condition is made worse when growing conditions are unfavourable due to drought or cold. The removal of shoots with small, dormant buds can help to reduce/dampen yield oscillations.

The initial yield peak followed by the subsequent decline is often referred to as the *Fordham effect*, after the scientist who first described and modelled the process in Malawi. It must be emphasised that the period between peaks is one of active growth and not dormancy (Fordham, 1970, 1977; Fordham and Palmer-Jones, 1977).

Summary

- Each shoot if left unplucked normally goes through a rhythmic alternation of periods of active growth (a flush) and dormancy (*banjhi*).
- A shoot normally produces a variable number of appendages (scales and leaves) during a flush period.
- By breaking off a shoot (at harvest), apical dominance is broken and the axillary bud in the axil of the leaf nearest the break begins to develop.
- The time from a bud being released from apical dominance to the emergence of the third true leaf is called the *shoot replacement cycle*.
- The time interval between the emergence of successive leaves is called a *phyllochron*.
- There are three components of yield: the number of shoots harvested per unit area, the mean shoot dry mass at harvest, and the number of harvests per year.
- Shoot number is the main determinant of yield.
- The total shoot population is the sum of the number of harvested shoots and the number of basal shoots (small shoots remaining on the bush).
- Applying fertiliser increases the number of shoots that can be harvested and the proportion that are actively growing.
- The main effect of water stress is to delay the development of the basal population of shoots from the dry season until after the start of the rains. The annual mean total shoot population is not affected by water stress.

- The fresh mass of a shoot increases linearly with the number of leaves on the shoot.
- The dry matter content of the shoot varies with the season.
- Rates of shoot extension and rates of shoot development are both temperature-dependent. The base temperature for development is 2–3 °C less than that for extension. Base temperatures vary between clones.
- The concept of thermal time, summed between the base temperature (8–15 °C) and the optimum temperature (25–30 °C) can be used to predict the duration of the shoot replacement cycle and, for example, the impact of altitude.
- The yield of tea is considered to be sink-limited.
- Large peaks of production occur after a limiting factor such as low temperature or drought has ended. This allows the accumulated buds of many ages to develop together (the so-called *Fordham effect*), causing logistical problems in the field and the factory.

Note

1. In parts of India, the term 'mother leaf' is used to describe the first true foliage leaf to unfold on the expanding shoot.

10 Plucks Shoots, and Leaves[1]

Looking After the Children

The most important job on a tea estate, or on the tea farm, is harvesting the leaf. This must be done at the right time, when the shoots are at the correct stage of development and the right size, as specified by the factory. To ensure the sustainable production of large yields of good-quality tea, the shoots (usually two or three leaves and an unopened terminal bud) must be removed with the minimum of damage to the leaves. Plucking needs to be selective, with small shoots (e.g. one leaf and an unopened bud) left on the bush (Figure 10.1). If shoots have grown too large by the time of the next harvest it may then be necessary to *break back* and discard part of these shoots, so yield is then lost. In the case of the tea crop, one harvest influences the next harvest.

There is one important difference between tea, where yield depends on the stage of shoot development at which harvesting occurs, and a fruit (or seed) crop, where the whole crop is removed when it is deemed to be mature. For example, if a large tea yield is the target, then coarse leaf is harvested. But, if the target is high quality (with a price premium), then fine leaf is harvested and a smaller yield is acceptable. This difference is often not appreciated, but one consequence is that it makes it very difficult to compare yields from one estate with another, or one treatment with another, or one clone with another, unless the same harvesting criteria are used. It is highly probable that many annual reports contain data that do not stand up to scrutiny when looked at in this way.

Harvesting tea is an activity that is repeated throughout the year at intervals that, providing there is not a drought, depend largely on the prevailing temperature. It is not a one-off operation, as it is with most other crops. There may be up to 30–40 harvests each year in equatorial areas. By contrast, at higher latitudes such as in Japan (c. 35°N), where low winter temperatures and frost limit the growing season, there may only be three or four harvests each year during the relatively short summer season. The requirements for harvesting tea are, therefore, very different depending on the location.

Growers now have a wide choice of harvesting methods available to them. These range from hand harvesting, through a number of relatively simple mechanical aids (such as blades or shears) and motorised systems (hand-held or on wheels), to large self-propelled harvesters. Hand plucking is by far the most common method of harvesting (Figure 10.2), although mechanisation is becoming more important where labour is scarce and/or expensive (this subject is covered in detail in Chapter 11). There are also opportunities to raise the productivity of pluckers through skills analysis and training (D. Evans, personal communication, 1985). In this chapter, the following topics are described in turn: the plucking process, quality assessment, scheduling harvests, and harvesting systems.

Figure 10.1 The most important job on a tea farm is harvesting the leaf. This must be done at the right time, when the shoots are at the correct stage of development and the right size (Tanzania, MKVC).

Figure 10.2 Hand harvesting, which involves 140–190 hand movements a minute, is still by far the most common method of harvesting in use today. Here the palms-up method of plucking is being used (Tanzania, MKVC). *A black and white version of this figure will appear in some formats. For the colour version, please refer to the plate section.*

The Plucking Process

Managers use many terms to describe the *plucking* process, including *selective plucking, fine/coarse plucking* and *hard plucking*. They are not all mutually exclusive. Selective plucking (cf. non-selective plucking), in which shoots are carefully chosen on the basis of their length and/or stage of development, will lead to the largest yields of high-quality leaf over the longer term. This is a skilled process, harvesting is slow and expensive, and close supervision is needed. An example of the plucking procedure as practised in Malawi is given in Box 10.1.

Fine plucking, in which only shoots with 'two leaves and a bud' together with soft banjhi shoots are harvested, should lead to a reputation for high quality, but yields can be

---

Box 10.1 Plucking Procedure

An example from Malawi based on the 'palms-up' position (from TRFCA, 1990).

1. Hold both arms in front of body with the palms of the hands facing upwards.
2. Move the hands forward and collect the shoots to be plucked individually between the first and second fingers on each hand.
3. Slide the fingers down, which pushes away the immature shoots, until the backs of the hands touch the plucking surface and the fingers meet the *janum* (fish leaf) below the second or third leaf.
4. Tighten the fingers together and, using the thumb to assist, snap off the shoots.
5. Pass the plucked shoots back into the palms of the hands with the aid of the third and fourth fingers.
6. Keeping the hands close together and near the body, move them forward to pluck more shoots as above, making sure that the palms of the hands face upwards, and continue to pass the plucked shoots back into the palms.
7. Hold the plucked shoots in the palms of the hands using the third and fourth fingers, but do not squeeze hard.
8. When the hands are comfortably full, pass the shoots back over the shoulder into the basket by moving each hand separately over the left and right shoulders.
9. Do not compress two handfuls into one.
10. Continue plucking, moving the hands forward from the body on a swath until the crossed plucking wands are reached or until the arms are fully extended. Then start from the body again, moving towards the centre of the bush and repeat until the half bush or whole bush is completed.
11. Then break back in the same area, using the thumb and forefinger with the palms facing downwards.
12. Discard the pieces of stem and leaves which are broken back.
13. Only break back hard banjhis, whole leaves and pieces of stem which are left above the plucking table after the plucking has been completed.
14. Do not grab and tear leaves.

reduced by up to 40%, compared with 'three leaves and a bud'. Labour productivity is also reduced.

Non-selective, hard plucking describes the removal of all shoots above the surface of the *maintenance foliage* (or *plucking table*), regardless of their size or stage of development. *Maintenance foliage* refers to the canopy of leaves below the level at which the tea shoots are harvested. These are the leaves that remain on the bush, intercept sunlight and, through the process of photosynthesis, manufacture carbohydrates that contribute to the growth of the tea bush. For large yields, it is essential to ensure that the leaf canopy completely covers the ground surface for as large a proportion of the *pruning cycle* as possible.

In some places, hard plucking is also referred to as *plucking black*. It has the advantage that the labour does not necessarily have to be skilled, and large yields and high labour productivity are possible. In addition, the rate of increase in height of the plucking table (referred to as *creepage* in India) will be reduced compared with less severe forms of harvesting, leading to opportunities to extend the length of the *pruning cycle*. The disadvantages include the (small) risk of reduced yields if the process is taken to the extreme, particularly on soils depleted of nutrients. In addition coarse (*mature*) and broken leaf may lead to low quality. India uses slightly different descriptors for the harvesting process.

Numerous experiments have been carried out in South India in attempts to specify how 'hard' a tea bush should be plucked, which is also known as the *intensity of harvest* (Chandra Mouli *et al.*, 2007). It describes the point at which a shoot is broken or cut. One can pluck non-selectively but lightly, by leaving a leaf on each shoot (known in India as *adding a leaf*). Equally, one can pluck highly selectively but hard: only three leaves plus bud, with no leaves remaining on the stump (Box 10.2).

For rational comparisons between treatments, it is necessary to specify how many leaves or buds are left behind after a shoot is harvested; specifying the harvest interval on its own is not sufficient. In India, hard plucking usually means breaking off the shoots

---

Box 10.2 Commentary: Selective Plucking as Practised in (South) India

In contrast to East Africa, the 'palms-down' rather than the 'palms-up' technique is used. Plucking occurs at an interval of 1.5 phyllochrons, compared with 2–2.3 P in Mufindi, southern Tanzania. By being so selective, small shoots that fail to develop into two leaves plus a bud are missed. These are left on the bush, where they harden. This represents a considerable loss of crop. The small shoots also cause the tables to rise faster than they do when 'hard' plucking is practised. This, in turn, makes plucking difficult and tiring, leading to a reduction in productivity (and hence an increase in the unit cost of the green leaf). This may then necessitate a skiff or a shortening of the pruning cycle. Much to the surprise of management, it was found in Mufindi that selective plucking, on a field scale, actually reduced crop. The plucker involuntarily leaves behind the very smallest shoots. But deliberately not harvesting the smaller shoots reduces both labour productivity and crop yield.

T. C. E. Congdon

just above the scale leaf, or close to the cutting point of the previous harvests. The advice to growers in South India is to pluck hard during the main growing season (monsoons) and to the mother leaf (Indian term, *light plucking*) during the slow-growth months of January to March and again during July and August. The intensity of harvest directly influences the rate of increase in the height of the canopy, and the depth and longevity of the maintenance foliage, and the yield (Chandra Mouli *et al.*, 2007). This integrated harvesting system maximises yield but also prevents excess maintenance foliage accumulating. Continual light plucking gives the lowest yield (Sharma *et al.*, 1990; Sharma and Satyanarayana, 1993). Previously Visser (1969) had found that hard hand plucking continuously over 18 years gave higher yields than light plucking despite a lower mass of maintenance foliage. He noted that the life of individual maintenance leaves appeared to be extended on hard-plucked bushes.

Quality Assessment

As will be discussed further in Chapter 17, the quality of processed tea is judged against many parameters. These include the appearance of the black tea, the colour of the liquor, its taste, the packing density and 'cuppage' – the number of cups of tea obtained from 1 kg of tea. Their relative importance will depend on the market, which in turn is constantly changing. No wonder it is difficult for managers to plan!

In general, quality (however it is defined) is inversely proportional to the size of the harvested shoots, but not for all clones. In order to judge the quality of the green leaf coming from the field to the factory the concept of *pieces per kilogram* is sometimes used. This is calculated from the sum of shoot number (whole shoots plus broken shoots) divided by the total fresh weight (kg) of the sample.

The usual target number is 1000–1200 pieces kg^{-1}, depending on the stage in the pruning cycle and the clone. This corresponds to about a 50:50 split between two leaves and a bud and three leaves and a bud in a sample (Box 10.3). In practice, factories tend to apply this evaluation method blindly, without taking into account differences between clones. Allowing for broken shoots is also not easy.

In South India the following composition of harvested leaf is considered to be acceptable (Sharma, 2011):

- Shoots with 'two or three open leaves and an active bud', and shoots with 'one or two open leaves and a banjhi bud' (soft banjhi): 85%
- Coarse leaf: 10%
- Immature shoots (one open leaf plus bud): 5% (implying that really small shoots should be left on the bush; in Kenya this is known as 'looking after the children').

Scheduling Harvests

Deciding when to pluck, preferably well in advance, should be linked to the actual rates of growth and development of shoots. In *programmed plucking* the intervals

---

Box 10.3 Assessing the Quality of Green Leaf Pieces per Kilogram

This index can be used as a measure both of the 'quality' of *green leaf* after plucking, and also of *plucker* productivity. In Malawi shoots taken from about five 300–500 g samples of known fresh weight, selected at the weighing point, are divided into the following categories and then counted:

- **Whole shoots:**
  1. immature (shoots with one leaf and a bud, and very small two leaves and a bud)
  2. shoots with two leaves and a bud
  3. shoots with three leaves and a bud
  4. soft banjhi
  5. coarse (shoots with more than three leaves and a bud, and hard banjhi)
- **Broken shoots:** shoots with leaves missing
- **Broken bits:** torn leaves and buds without stems.

The number of *pieces per kilogram* is then calculated as follows:

1. **Divide** the **sum** of the number of whole shoots and broken shoots by the total fresh weight of the sample, and then
2. **Multiply** by 1000. The target number of pieces kg^{-1} varies with stage in the pruning cycle, and the type of tea plant or clone. The higher the number the more small shoots in the sample.

In Kenya, the target is about 1000 pieces kg^{-1} for seedling tea in the second and third years after pruning. This is considered to represent a 50:50 split between shoots with 'two leaves and a bud' and 'three leaves and a bud'. A more detailed analysis is also possible.

From M. St. J. Clowes (1986)

---

between harvests are based on the time taken for a generation of shoots to develop to a stage suitable for plucking. In Malawi, for example, this is taken to be the equivalent of one-quarter of the duration of the shoot replacement cycle in the main growing season, that is about 42 days. Hence, in Malawi, the recommended interval between harvests is alternately 10 then 11 days. Conveniently, this takes three weeks to complete.

Other approaches to support decisions about when to harvest, from Assam, Tanzania and South India, are summarised below.

Assam

Wight (1932), working in Assam, had anticipated this concept of prediction 60 years earlier. In his report of studies of the terminal bud on a shoot he stated:

The rate at which leaves unfold determines the relation between the type of leaf plucked and the *plucking round*, and this relation is susceptible to rational analysis . . . if the value or period of leaf unfolding is known – and the great majority of the buds obey the rule prevailing at any time – then it is possible to obtain numerical values which tell us at a glance the actual leafiness of the shoot on any one day.

Referring to a publication by Cohen-Stuart published in 1919, Wight recalled the time that it takes for a leaf to develop as the *leaf-period*. He developed a table that a manager could use indicating the optimum time for harvesting. This tabular example specified that pluckers should target shoots with 'two leaves and a bud' while leaving behind on the bush shoots with 'one leaf and a bud'. In the table it was assumed that the leaf-period in Assam was four days.

Wight's ideas link closely to the more recent concepts of the *phyllochron* and *shoot replacement cycles*. The main difference is that these concepts use temperature to predict the duration of the 'leaf-period' rather than being based on an assumption about its duration depending on the season and altitude.

As mentioned in Chapter 9, the *phyllochron* is a measure of the time taken (days) for successive leaves to unfold.[2] It is influenced by temperature in particular, but is also sensitive to water stress. There are also differences between clones. The reciprocal of phyllochron is *leaf appearance rate* (leaves day^{-1}) (Burgess and Carr, 1998).

Tanzania

Burgess (1992) standardised phyllochron measurements by only selecting actively growing axillary shoots about to unfurl their second true leaf. The phyllochron was then defined as the number of days it took until it was just possible to flick out the bud by touching the tip of the third true leaf with the tip of one finger (Burgess and Myinga, 1992). Measurements like this were made on 10 shoots on each of six clones, selected on the same day each week, over a period of two years. These clones were well irrigated. Other measurements were made on part-irrigated and droughted bushes. There were differences between the clones. For example, the mean phyllochron during the main growing season was 5.8 days for clone BBT-1 but 7.9 days for clone TRIEA-6/8 and 8.3 days for BBK-35. The values for the other clones were intermediate (6.5–6.9 d). During the winter months (July and August) the phyllochron increased to 19 days for BBT-1, and 28 days for clone BBT-207, but only to 11 days for TRFCA-SFS150 (a clone well known for its capacity to continue yielding during the cold weather).

The relationship between air temperature and the phyllochron or leaf appearance rate was most simply described by a straight line, explaining up to 79% of the variability. The corresponding base temperatures for shoot development (defined by where the line crossed the *x*-axis) varied between 6 °C (for clone TRFCA-SFS150) and about 10 °C (for clones BBT-1, BBT-207 and AHP-S15/10).

Combining the data for all six clones ($n = 615$) gave the following relationship between leaf appearance rate (P, d^{-1}) and the daily mean air temperature (T, °C):

Table 10.1 Estimated effect of mean air temperature (T_{mean}, °C) on the mean phyllochron (P, d^{-1}) of six fully irrigated clones ($P = 0.0162 (T_{mean} - 9)$)

T_{mean}	P	T_{mean}	P	T_{mean}	P	T_{mean}	P
10.0	0.017	10.25	0.021	10.5	0.025	10.75	0.029
11.0	0.033	11.25	0.037	11.5	0.041	11.75	0.045
12.0	0.049	12.25	0.053	12.5	0.057	12.75	0.061
13.0	0.065	13.25	0.069	13.5	0.074	13.75	0.078
14.0	0.082	14.25	0.086	14.5	0.090	14.75	0.094
15.0	0.098	15.25	0.102	15.5	0.106	15.75	0.110
16.0	0.114	16.25	0.118	16.5	0.122	16.75	0.126
17.0	0.130	17.25	0.134	17.5	0.138	17.75	0.142
18.0	0.146	18.25	0.150	18.5	0.155	18.75	0.159
19.0	0.163	19.25	0.167	19.5	0.171	19.75	0.175
20.0	0.179	20.25	0.183	20.5	0.187	20.75	0.191
21.0	0.195	21.25	0.199	21.5	0.203	21.75	0.207

Note: A well-sited maximum and minimum thermometer in a standard Stevenson screen is essential for this technique to work.

$$P = 0.0162 \ (T_{mean} - 9.0)$$

It is assumed that, because of the diversity between the six clones, this equation represents a population of heterogeneous seedling tea.

When the potential *soil water deficit* was less than 150 mm the relationship was the same as that for well-watered tea, but when it reached 250 mm the leaf appearance rate declined to a value less than half that predicted from temperature data alone.

This detailed study has been put to good use. On the assumption that after a harvest the shoots remaining on the bush have, on average, one unfolded leaf and that the target shoot at the next harvest is 'three leaves and a bud', the interval to the next harvest is two phyllochrons. Using the mean of the daily maximum and minimum temperatures, it is possible, by referring to tables, to determine the time to the next harvest and, by so doing, to plan ahead (Burgess, 1992). Tables 10.1, 10.2 and 10.3 show how this is done. The commentary in Box 10.4 shows how this concept has been adapted for use on a commercial estate.

South India

Following the work in Malawi reported by Grice and Clowes (TRFCA, 1990), in which they used the concept of day degrees or thermal time to predict how long it would take an axillary bud released from apical dominance to develop into a shoot suitable for harvesting, Murty and Sharma (1989) evaluated the same methodology at Anamala in South India (alt. 1050 m), where it is known as the *leaf expansion concept*. On average, over a year, about 625 °C d (above a base temperature of 12.5 °C) were needed for the axillary bud to grow into a harvestable shoot (three leaves and a bud). This value for thermal time was obtained by tagging six axillary buds after removing the shoot (at the fish leaf) each week over 31 weeks, and then monitoring the axillary buds' development (clone UPASI-

Table 10.2 Phyllochron calculator

ESTATE:
MONTH:

Day	Max. temp. (°C)	Min. temp. (°C)	Mean temp. (°C)	Phyllochron	Cumulative phyllochron			
---	---	---	---	---	Field 1	Field 2	Field 3	Field 4
1								
2								
3								
4								
5								
6								
7								
8								
9								
10								
11								
12								
13								
14								
15								
16								
17								
18								
19								
20								
21								
22								
23								
24								
25								
26								
27								
28								
29								
30								
31								

Table 10.3 An example of how the leaf appearance rate (LAR) or phyllochron can be used to predict the date of harvest as derived from measurements of the mean air temperature (taken inside a Stevenson screen) (from Burgess and Carr, 1998)

Day	Air temperature (°C)			LAR (leaf d^{-1})	Cumulative total (phyllochrons)	
	Maximum	Minimum	Mean			
					0.00	
1	19.0	9.5	14.25	0.085	0.085	
2	19.5	10.0	14.75	0.098	0.183	
3	18.0	10.0	14.00	0.078	0.261	
4	18.0	10.0	14.00	0.078	0.339	
5	17.5	9.0	13.25	0.054	0.393	
	———				1.737	
	———					
19	21.5	10.0	15.75	0.118	1.855	
20	21.5	8.5	15.00	0.103	1.958	
21	21.0	9.5	15.25	0.109	**2.067**	**Harvest**
22	22.5	8.5	15.50	0.114	0.114	
23	24.5	11.0	17.75	0.146	0.260	

Box 10.4 Commentary: Rule of Thumb

Go into a field that you assess as just right for plucking. Back-calculate the number of phyllochrons to the previous harvest date. The total will be the phyllochron interval that is right for your style of plucking. Now keep a rolling total, and monitor changes as they happen, and then speed up or slow down as temperatures change, instead of being caught out yet again.

T. C. E. Congdon

---

10). It compares with 475 °C d recorded in Malawi. The time durations corresponding to thermal time were from 52 to 68 days. In a parallel study, the number of degree-days for the period of unfolding and expansion of the third leaf were: (1) similar for each clone tested, (2) independent of the year in the pruning cycle, and (3) independent of the intensity of harvest. This surprising set of results is probably an anomaly, since at that time there was no accurate way of measuring a phyllochron.

The difference in the value for the thermal time for a shoot to develop between Malawi and South India could have been due to a number of factors, one of which is the assumption about the base temperature, which, as research elsewhere has shown, can differ quite considerably (by up to 5 °C) between clones. (To reduce the thermal time from 625 to 475 °C d the base temperature needs to be increased, over say 60 days, from 12.5 to 15 °C, which is a high value.) Murty and Sharma (1989) proposed the use of thermal time to plan a commercial harvest schedule, rather than reliance on fixed time intervals. An analysis based on data recorded in the field suggested that there would be a yield increase of *c.*

200 kg ha^{-1} if this approach was followed, the composition of the harvested shoots would not change, and there would be an increase of 1.5 kg green leaf in the daily plucking average. The value of the benefits, although they appear to be small, more than compensated for two additional rounds of plucking and the deployment of an extra 32 pluckers per hectare during the course of the trial.

Harvesting Systems

There are several ways in which plucking is organised and scheduled. The traditional way is by so-called *gang plucking*, in which a group of people, the number depending on labour availability and the size of the crop, moves across the field together (Figure 10.3). *Scheme plucking* by contrast involves individuals or families taking responsibility for harvesting small areas of tea throughout the season (Figure 10.4). The criteria to take into account when assessing how best to manage the harvesting process include labour productivity, skill and reliability, the prescribed quality of the green leaf, the importance of a sense of ownership, and ease of day-to-day management. Some system of *programmed plucking* is essential if managers are to take the best decisions based on realistic predictions of the future crop. Responding only to events is no longer enough.

Programmed Plucking

The aim of programmed plucking is to reduce the number of shoot generations on a bush. The fewer the plucking rounds the more pronounced is the difference between generations (the ones to be plucked now and those that will be left on the bush) thereby

Figure 10.3 *Gang plucking*: a group of pluckers move across a field together (MKVC). *A black and white version of this figure will appear in some formats. For the colour version, please refer to the plate section.*

Figure 10.4 (a, b) Palms up, or down? *Scheme plucking* is a family business. Growers now have a wide range of harvesting methods available to them (Tanzania, MKVC).

minimising the degree of selectivity needed when plucking. Thermal time is used to specify the harvest intervals in advance. In cool areas when rates of shoot development are slow it is not so easy to separate the generations. Non-selective plucking is essential if the generations are to be separated. In Kenya it has not been found possible to persuade the pluckers to remove immature, one-leaf-plus-bud, shoots because it would 'compromise their future incomes'.

Scheme Plucking

This management system was developed in Malawi (Box 10.5). Each plucker (or a family) is given an area of tea for which he or she is responsible throughout the season, the area of which represents the number of bushes that can be harvested comfortably in one day. Typically, this is about 0.05–0.1 ha depending on the yield and the harvesting interval. Sometimes two (half-size) plots are allocated each day so that it is easier to accommodate a five-and-a-half-day working week, and to introduce flexibility into the system. Plots are demarcated by stakes at each corner or by allowing a shoot of tea to grow to a length of about 0.5 m in place of a stake. The responsibility for fertiliser application, weeding and pruning of each plot can also be delegated to individual pluckers. Systems that suit local practices need to be developed to suit individual estates. *Programmed scheme plucking* is the term used to describe the combination of *programmed plucking* with *scheme plucking*. The procedure is well described by Grice and Clowes (TRFCA, 1990). Absenteeism can make this a difficult practice to operate commercially.

Plucking Intensity

Ergonomics need to be considered when deciding how high to allow the leaf canopy to rise before pruning. For example, in India, an average worker is around 1.5 m tall, and their elbows are about 0.9 m above the ground. The maximum bush height at which s/he could work efficiently is about 1.0 m. The height of a bush is governed by the height at which

Box 10.5 Scheme Plucking: a Case Study

Scheme plucking was pioneered in Kenya and Malawi. It was intended to give pluckers a sense of ownership by allocating permanent plots for which they were responsible and which they returned to at each plucking round. They would not be tempted to remove small shoots, as they were aware that a small shoot would be larger and heavier at the next round. It was intended that plucking of each plot would be completed in one day. Family and other help could be used when needed, but one person had the responsibility for the six plots (assuming a six-day week).

Field number	Area (ha)	Plant spacing (m)	Plant density (ha^{-1})	Bushes per plot	Bushes per field	Plots per field
1	6.0	1.2 × 0.6	13900	425	83400	196
2	3.5	1.2 × 0.75	11100	425	38850	91
3	7.1	1.2 × 1.2	6950	425	49345	116
4	8.9	1.2 × 0.9	9300	425	82770	194
5	4.6	0.9 × 0.9	12300	425	56580	133
6	5.0	1.2 × 1.2	6950	425	34750	81
Total	35.1			2550		811

Number of plots to be included = 811
Harvest interval = say 12 days
Number of working days out of 12 = 10
Number of plots to be plucked per day = 81
Plot size is designed to be plucked in one day
THEREFORE 81 PLUCKERS ARE NEEDED

a bush is *tipped in* after pruning. For bushes pruned at or below 0.65 m, the upper limit for tipping is 0.75 m. *Tipping* is the name given to the first and second harvests after pruning when a new *leaf canopy* is being established and a level *plucking surface*, running parallel to the ground and following the natural slope of the land, is being formed at a specified height. This means that there is 0.25 m between the tipping height and the maximum height at which a plucker can work efficiently. Once the height has been established, a plucking wand 2–3 m long helps the plucker to ensure that the plucking table level is maintained. The only shoots to be plucked are those of the right size, preferably 'three leaves and a bud' and all soft banjhi shoots, which appear above the plucking table (Sharma, 2011).

If *creepage* occurs at a rate of 0.10 m a year, then there are only two and a half years before it is necessary to prune again, which means a three-year pruning cycle. If, however, creepage can be kept at 0.05 m a year (or taller pluckers are employed!) the pruning cycle can be extended to five years, but only if the yield in year 5 is likely to be at least as great as the mean yield obtained in the first four years.

In some countries, a process known as *skiffing* is used to control the height of the canopy. This is a very light prune intended to level the plucking surface if it has become uneven (known in India as a *level-off skiff*), or sometimes to extend the length of the

(a)

(b)

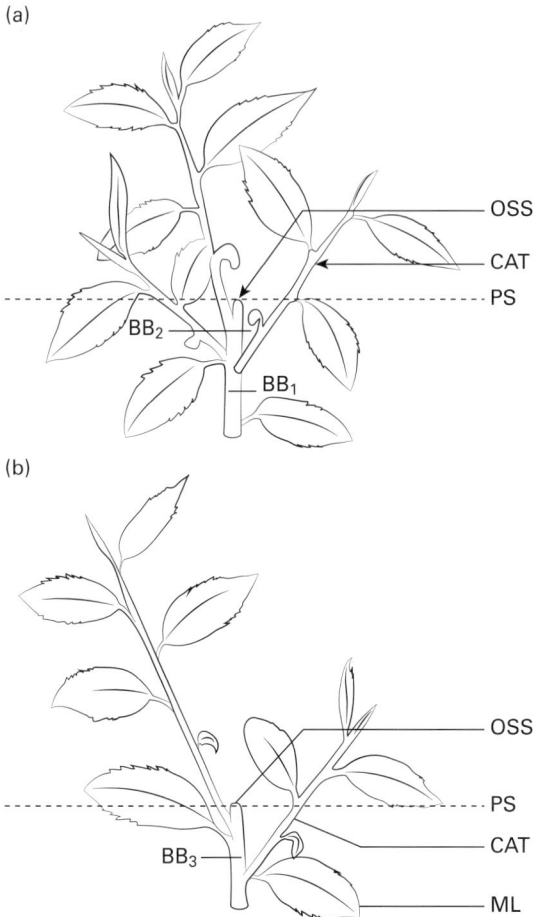

Figure 10.5 Breaking back a shoot because harvest was delayed is not to be encouraged – it should be avoided at all costs (Dr Rex Ellis). BB, break-back; CAT, cataphyll; ML, mother leaf; OSS, old shoot stalk; PS, plucking surface.

pruning cycle in a mature crop. Skiffing is usually an indicator of management's failure to control the harvesting process.

In North-East India, skiffing has replaced the previous practice of an annual prune. Whether or not it is needed or beneficial is not clear, but it is claimed to result in an early, high-quality first flush. There are different levels of skiffing: a *deep skiff* (a cut is given midway between the *pruning* and *tipping* levels); a *medium skiff* (the cut is made 0.15 m above the pruning level or 0.05 m below the previous year's tipping level); and a *light skiff*, which is made 0.20 m above the pruning level (Barua, 1989).

There is also a debate about the need for the practice known as *breaking back*. Sharma and Murty (1989) argue that it should be used as an essential measure 'to maintain the vigour of the bush at the plucking table level'. *Breaking back* in their view involves breaking off the old shoot stalk at the same time as plucking the new shoot from amongst a cluster of shoots (Figure 10.5). At the same time, any clusters left behind from previous harvests (commonly known as 'crows' feet') can be broken back into green wood below. Other benefits, such as the control of bush height, are merely incidental and not the primary reason for breaking back. Templer (1978) presents a contrary view, arguing that it is unnecessary if the 'correct' plucking interval is maintained, and that it also reduces the productivity of a plucker by about 10% over a day, with no proven benefit other than a cosmetic one due to the 'compulsive habit of growing tea like a lawn'! Breaking back is only necessary if one is late getting into a field to harvest, which indicates poor management – and, yes, it costs!

In Japan and elsewhere in eastern Asia, green tea is grown as a hedgerow with a convex plucking surface. In most other areas, where black tea predominates, the plucking surface is indeed like a flat lawn. It is not clear why there are these two distinct methods, or what the advantages of one over the other might be in each context. The hedgerow with clear pathways between the rows is certainly easier to harvest compared with the continuous ground cover, which is awkward to walk through, especially with a full basket of leaf. This must have an effect on labour productivity. The level surface will intercept all the incoming solar radiation compared with probably about 80% interception by the curved surface and the wide inter-row spacing (where weeds will grow). At high latitudes there may be advantages in terms of light interception in having the hedgerows running north/south rather than east/west.

Templer (1978) summarised experience in East Africa as follows: curved surfaces as used in Japan, in comparison with flat surfaces, were at times: more laborious to form; lower-yielding; more laborious to pluck; but easier to walk through. Similar comments applied to sloped plucking surfaces.

Leaf Handling

A further important, but sometimes overlooked, aspect of management is *leaf handling*. This term covers the movement of green leaf within the field, weighing and storage, and transport from the field to the factory. The positioning of weighing points and the ease of access from the field affects the productivity of the pluckers. Similarly, bad handling of the green leaf in the plucking basket and poor storage conditions after weighing and during transport can have adverse effects on the condition and tea-making quality of the leaf entering the factory. Since leaf handling extends from the time that an individual shoot is plucked in the field to the beginning of the formal withering process in the factory, it is often a neglected part of the total tea production process, being the full responsibility of neither the field nor the factory manager. Why has no one taken this 'missing link' in the tea processing chain seriously? Perhaps they have . . .

Conclusion

As Ellis and Grice (1976) once stated, '*the greatest crime as far as yield is concerned is to pluck shoots that are too small.*' The next biggest crime is unnecessary *breaking back*,

---

> **Box 10.6** Actions to Take to Improve Harvesting Practices
>
> What actions can be taken as a manager to improve profitability through better plucking and leaf handling?
>
> - Evaluate all your harvesting procedures systematically.
> - Look to improve the timeliness of harvests, using a knowledge of shoot development rates (phyllochrons) in order to be able to plan ahead.
> - Monitor regularly the rate of increase in the height of the table (creepage); aim to keep it below 100 mm a year, preferably closer to 50 mm; set this as a target for the field supervisor – but remember, breaking back is a sin!
> - Monitor the quality of the harvested leaf at the leaf shed (pieces kg^{-1}).
> - Review the handling and storage of the leaf between the bush (yes, the bush) and the factory.
> - Undertake a skills analysis of the best pluckers, transfer these skills to those who are less able, and by so doing raise the general level of ability.
> - Train, involve and reward labour. Explain to them, for example, the concept of the shoot replacement cycle and the importance of not removing immature leaf. It is in their personal interests as well as those of the owner.
> - Determine the potential yields that can be achieved in the locality and set these as your target. Remember, not so long ago, 1000 kg ha^{-1} was considered to be a large yield. Now it is perhaps 4000–5000 kg ha^{-1}, depending on location. What will it be in 20 years' time?
> - Identify those factors that are preventing the potential yields from being achieved.

which again represents a loss of yield (Figure 10.5). A plucker needs to exercise a high degree of skill to meet these exacting standards, although once the generations have been established selectivity is no longer an issue. Many pluckers are inexperienced when they begin. To improve productivity skills, training is essential (Box 10.6).

Summary

- The following topics are described in turn in this chapter: harvesting criteria, quality assessment, scheduling harvests, and harvesting systems.
- Managers use many terms to describe the plucking process, including *selective plucking, fine/coarse plucking* and *hard plucking*. They are not all mutually exclusive.
- The advice to growers in South India is to pluck hard during the main growing season (monsoons) and to pluck to the mother leaf (Indian term, *light plucking*) during the slow-growing months.
- The intensity of harvest directly influences the rate of increase in height of the canopy, and the depth and longevity of the maintenance foliage, and the yield.

- In general, quality (however it is defined) is inversely proportional to the size of the harvested shoots.
- In order to judge the quality of the green leaf coming from the field to the factory the concept of *pieces per kilogram* is sometimes used.
- In *programmed plucking* the intervals between harvests are based on the time taken for a generation of shoots to develop to a stage suitable for plucking.
- The *phyllochron* is a measure of the time taken (days) for successive leaves to unfold. It is influenced by temperature in particular.
- The relationship between air temperature and the phyllochron or leaf appearance rate is most simply described by a straight line.
- The corresponding base temperature for shoot development varies between 6 °C and about 10 °C.
- The traditional way to organise harvesting is by *gang plucking*, in which a group of people moves across the field together. *Scheme plucking* by contrast involves individuals or families taking responsibility for harvesting small areas of tea.
- In North-East India, *skiffing* has replaced the previous practice of an annual prune. Whether or not it is needed or beneficial is not clear.
- There are also strong views about the practice known as *breaking back*, which has no proven benefit other than a cosmetic one due to the 'compulsive habit of growing tea like a lawn'.
- There are no clear benefits from curved crop surfaces as used in Japan and elsewhere.

Notes

1. With apologies to Lynne Truss, author of *Eats, Shoots & Leaves: the Zero Tolerance Approach to Punctuation.*
2. In India this is known as the *leaf expansion time* or *concept* (Dharmaraj, 2012), which is not strictly a correct term.

11 Machine-Assisted Harvesting

The Need of the Hour

M. K. V. Carr and C. J. Flowers

Almost regardless of the industry represented, economic pressures are forcing organisations to increase the productivity of their staff in order to reduce the unit costs of production. Only those businesses that are competitive in the world are likely to survive unless they attract government subsidies. These pressures have always been there. But, as the world shrinks in commercial terms, they are perhaps greater now than they were. The tea industry is not exempt from these pressures. We are all having to run faster to stay in the same place.

As productivity increases, the real price of tea continues to decline and labour costs continue to rise. Notwithstanding, tea production remains a labour-intensive industry. A thousand-hectare estate for example may employ from 1000–1500 people, and perhaps even more when harvesting is at its peak. A comparable-sized wheat farm in Australia would be run by one person. The difference between a modern industrial company and a plantation is also highlighted by the fact that, within one well-known multinational company, the plantation sector represents only 2–3% of the total turnover yet that sector employs more people than the rest of the company combined. The difference is 'mechanisation'.

Harvesting tea by hand has evolved with the tea industy. It is a highly skilful process involving between 140 and 190 individual hand actions per minute (D. Evans, personal communication, 1993). It is also labour-intensive. The skills are, however, transferable through systematic training. In this way, the weaker pluckers can become as good as the average, and the average as good as the best (but who has ever seen pluckers trained in a formal way?). In Tanzania, pluckers typically harvest between 4 and 9 kg of fresh shoots per hour. For a six-and-a-half-hour day this equates to about 30–60 kg d^{-1} depending on the season, and represents the normal range of competence (Squirrell, 1995). The best pluckers can harvest in excess of 100 kg green leaf each day (it used to be 20 kg).

Improving plucking speed is one way of increasing productivity. Other opportunities include: the sensible siting of weighing points, scheme plucking, programmed plucking, health and welfare, and appropriate incentives (see Chapter 10).

Although labour is cheap in absolute terms, in many so-called developing countries, for example in Africa and the Indian subcontinent, when the provision of housing, medical and other social benefits is taken into account it becomes expensive. On commercial tea estates in these countries labour costs can represent from 40% to

70% of the total field costs. Although from the outside the job of a plucker looks picturesque, in reality it is repetitive and boring (see Figure 10.1). To attract people in sufficient numbers on a regular and reliable basis, at a cost the industry can afford, is likely to become increasingly difficult. In the same way, for a smallholder, managing as little as 0.3 ha of tea is a full-time job. On larger farms, there can often be a shortage of family and hired labour, especially at critical times. Shortages occur, for example, after the start of the rains when tea production peaks at the same time as other labour-intensive farming activities, such as weeding the maize and other food crops (Burgess *et al.*, 2006). In some areas, smallholders compete with neighbouring estates for labour, while conversely smallholders who employ labour (and it is a misconception that they do not) are often unable to compete with the large companies.

In addition, there has been a strategic need in some places to plan for a reduction in the workforce because of the incidence of HIV/AIDS, although this is reported to be declining. As a result of all these factors, it is prudent to seek to identify appropriate methods of machine-assisted harvesting. It is not wise to wait. Since the 1980s, there has been a great deal of commercial interest in methods for mechanising tea harvesting. This search has not always been sustained for long periods but rather it has been intermittent, often opportunistic and short-term, depending on the labour situation prevailing at a particular time. Many of the developments have been initiated in workshops on the estates. Such initiatives were sometimes not supported by trade unions or by government for fear of increasing unemployment.

Mechanical harvesting of tea, however, is not new. The use of shears is referred to in Assam and Japan in the late nineteenth century, while tea was being harvested by self-propelled machines in the former Soviet Union (now Georgia) during the 1950s. Japan has been the leader in developing machines for harvesting (green) tea for many years, although at a very different scale from elsewhere in the world. Machine harvesting was also pioneered in Australia (it was introduced in Queensland in 1968), Papua New Guinea and Argentina, but it has been slow to gain acceptance in some of the other principal tea-growing areas of the world where one might expect a lead to be taken, such as North-East India.[1] Conversion from hand plucking to machine is currently happening in Taiwan.

During the 1980s shortages of labour (and its cost in South Africa) for different reasons, and increased yields, forced commercial companies in Africa and South India to begin to consider machine-aided tea harvesting. Their first attempt was to purchase off-the-shelf shears and other harvesting machines from Japan. Although conceptually valuable, many of these tools failed to withstand the intensive use and the very different physical conditions common in Africa. It therefore became necessary to develop harvesting aids/machines appropriate to where they were to be used. The focus for much of this chapter is therefore on experiences in South India and Africa. This is where much of the empirical research on and evaluation of machine-assisted harvesting has been undertaken and reported in the international literature in recent years, on machines designed and developed elsewhere (Japan, Argentina, Australia, UK).

Options

The industry is fortunate in that many options exist to challenge the supremacy of hand harvesting. These range in cost and complexity from the simple (the 'blade' – a blunt knife), through modified garden shears, various hand-held 'hedge-cutter' type devices, lightweight wheeled, hand-pushed or self-propelled harvesters, large self-propelled tea combines capable of harvesting over 100 ha a year, to fully automated machines riding on rails. The choice is large, and each has its place depending on the economic circumstance, the cost and availability of labour, the supporting infrastructure and the value of the final product (Carr, 1996).

Chris Flowers, who with others is responsible for much of the development work in East Africa, has recently reviewed the evolution of mechanical harvesting in eastern and southern Africa. His report (Flowers, 2013), together with the papers by John Kilgour and others, form the basis of this chapter. In addition, there are examples/case studies of the evaluation of mechanical harvesting from two different regions, South India and Africa. Bore and Ng'etich (2000) reviewed the role of mechanical harvesting in the context of Kenya, stressing the need for an integrated, systems approach.

It is difficult to extrapolate from one region to another, since conditions vary so much. For example in Japan, where many of the harvesting machines under test in Africa have been developed, the growing season is short and there are only two or three harvests a year. The farm size is very small (0.5 ha) and there is virtually no labour available. It is an industrial society and the processed tea sells at a high price in a protected market. In Australia large self-propelled machines are used. The tea farms are relatively large (100 ha) and labour is scarce and expensive but skilled. Tea has to compete as a high-cost beverage on a local market. In this case, for tea to be marketed as Australian it must contain at least 10% of tea grown and processed in Australia, with the balance made up of imports.

Compare all these situations with those in most tea-producing countries in eastern, central and southern Africa (and elsewhere). Here tea is produced either on large estates (100–600 ha), or on smallholdings (0.5–3.0 ha). The topography can be severe, with steep slopes and undulating land. The rows are short and may follow the contour. The tea is harvested for up to 12 months a year, with perhaps 15–25 individual harvests. There is a local resource of skills and technical services, and this is still developing. Here tea is a low-value commodity, which has to compete on the world market.

In India, Saikia and Sarma (2011) reported the results of what appears to be one of the first experiments in Assam on the use of shears for harvesting tea. Based on the results of two years of experimental treatments (clones TV1 and TV23 under light *Albizia* shade) they, like UPASI, recommended using shears intermittently, namely during the peak cropping period (July to September), alternating with hand plucking early and late in the season. The largest yield (in the second year) was obtained using shears with a 10 mm step (the space between the blade and the leaf tray) intermittently at 11-day intervals (3080 kg ha^{-1}). Flat shears, also used intermittently at the same 11-day interval, came next (2760 kg ha^{-1}). Continuous hand plucking at seven- or nine-day intervals yielded 2550 kg ha^{-1}. A flat shear used continuously at nine-day intervals, or hand plucking at the

Table 11.1 Decision matrix for the choice of machine tea harvester (from Flowers, 2013)

Harvesting system	Management	Maintenance	Machine	Most appropriate field layout	Cost structure
Complete mechanical solution	Highly technical, focused on planning and performance monitoring	Complex hydraulics and highly skilled artisans	Tracked, stable, large units	Flat, long, straight runs	Expensive capital, needs high productivity
Harvest only	Technical ability required, poor productivity can be overcome by spare machines	Simple, but owing to light construction not durable	Single-person, hand-held	Steep hills, short runs, difficult headlands	Low capital, only requires low productivity

same interval, produced the lowest yields (1860 kg ha^{-1}). Although the yield differences are meant to be statistically significant, and the coefficient of variation is very small at 2.8%, it is not easy to reconcile all the data presented, including those on green leaf and made tea quality. Surprisingly, there was little difference in productivity of the harvesters in terms of green leaf harvested per day ($c.$ 35 kg d^{-1}) regardless of treatment.

There are therefore a number of options available to the tea grower, but great care must be taken when considering how best to proceed with the introduction of mechanical harvesting. Direct technology transfer, from one location to another, however tempting, is rarely successful with any agricultural mechanisation system unless due notice is taken of differences in cultural, economic and technical conditions in each location.

The designer needs to optimise the design and operation of the cutting mechanism in order to minimise the removal of small tea shoots and buds (and to avoid cutting maintenance leaves). This means precise control of the height of the cutter bar as it moves across the field and, for motorised machines, matching the cutter bar speed with the forward speed. There are two principal cutting mechanisms: the reciprocating A-blade and the revolving drum. Both are entirely non-selective.

Leaf handling after harvest should always be considered as a component of the total harvesting system. The type of harvesting unit selected depends on which components of the harvesting process are to be mechanised. For example, will the leaf be transported by machine to the headland or will it be carried? Similarly, how will it be loaded into the transporter for transfer to the tea factory? There is a continuum of choices, each with a cost as indicated in Table 11.1. The advantages and disadvantages of the different methods are summarised in Table 11.2.

Each of the plucking aids will now be considered in turn, beginning with the simplest. Box 11.1 shows the topics covered.

Table 11.2 Advantages and disadvantages of different machines

Machine	Advantages	Disadvantages
Hand-held single- or two-man	No field preparation required, highly adaptable, relatively cheap to run	Leaf quality highly dependent on operator desire, needs strong operators. Productivity limited due to nature of physical work. No forward speed control
Modified hand-held with skids	No field preparation required, highly adaptable, relatively cheap to run	Leaf quality less dependent on operator desire, needs strong operators. Productivity limited due to nature of physical work. No forward speed control
Rickshaw – manually pulled	Removes a lot of operator influence on leaf quality, depending on design. Highly manoeuvrable, requiring limited turning circles. Low capital costs. Running costs the same as hand-held	Requires a good degree of field preparation, although not too much in terms of field access if a six-foot (1.8 m) wide row is sacrificed for headland turning. Very tiring in undulating conditions, limits productivity. No forward speed control
Rickshaw – mechanical drive system	Removes all of operator influence on leaf quality. Very good manoeuvrability, requiring limited turning circles. Modest capital costs. Running costs slightly higher than hand-held	Requires a good degree of field preparation, although not too much in terms of field access if a six-foot (1.8 m) wide row is sacrificed for headland turning. Not as manoeuvrable in and out of field as the manually pulled unit due to increased machine weight. Must also be more productive to cover additional capital costs
Four wheel, self-propelled units	Removes all of operator influence on leaf quality. Can have good manoeuvrability if using skid steering. Can carry a reasonable quantity of green leaf to the field headland, good for long runs	Without four-wheel drive and some form of floating beam axle, not good on undulating conditions as loses traction. A considerable level of field and headland preparation required. Unit costs can be high if time in work is not maximised, difficult to do unless fields are ideally planted for mechanisation
Large-capacity units	A complete solution to harvest, forward speed control and loading of transporter	Can only be used on flat land, generally specifically planted for mechanised operations due to high capital costs and low capacity to cope with tight headlands or steep banks. Upwards of 15,000 kg of green leaf per day would be required to justify such a unit. Because of high capital costs, often not affordable to have any redundancy in the system, so maintenance is a considerable challenge

---

> **Box 11.1** Systems Covered in This Chapter
>
> - Plucking aids
>   - Blades
>   - Shears
> - Comparisons with hand plucking
>   - South India
>   - East Africa
>   - Central and southern Africa
> - Portable machines
>   - Single-person
>   - Two or more people
> - Wheeled lightweight machines
>   - Southern and eastern Africa
> - Large self-propelled machines
>   - Kenya
>   - Papua New Guinea
>   - Uganda
> - Recent developments

Plucking Aids

Blades

Brooke Bond Kenya Ltd (now Unilever Kenya Tea Ltd) developed the blade harvesting technique in Kenya during the 1980s. The blade consists of a blunt knife made from galvanised steel, positioned 15 mm above a horizontal aluminium platform or base plate. The blade is held in one hand so that the platform rests on the bush surface. The blade is then moved towards the operator, who uses the other hand to guide shoots into the path of the knife, where they are broken (Figure 11.1).

Shears

Harvesting tea with modified garden shears, which are usually used for hedge trimming, has been practised in South India for many years (Sharma and Satyanarayana, 1993), while renewed interest by researchers in this technique has been shown in East Africa (Burgess et al., 2006), and central and southern Africa (Wilkie, 1995; Nyasulu, 2001) following the lead taken by the tea industry. Figure 11.2 shows an example of the shears being used in Tanzania by Burgess et al. (2006):

- A tray is attached to the lower blade for collecting the harvested shoots.
- The upper blade supports a plate that pushes the cut shoots into the tray.
- The base of the tray rests on top of the crop canopy.

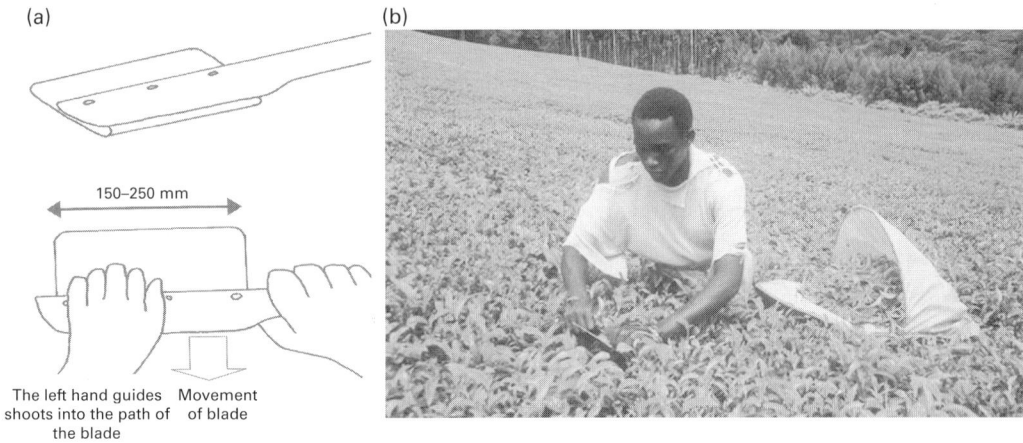

(a)

150–250 mm

The left hand guides   Movement
shoots into the path of   of blade
the blade

(b)

Figure 11.1 (a, b) The industry is fortunate in that many options exist to challenge the supremacy of hand harvesting. These range in cost and complexity from the simple (the 'blade' – a blunt knife, as shown here, Kericho, Kenya. MKVC) through modified garden shears (Figure 11.2), various hand-held 'hedge-cutter' type devices, lightweight wheeled, hand-pushed or self-propelled harvesters (Figure 11.3), large self-propelled tea combines (Figure 11.4), to fully automated machines riding on rails (Figure 11.5).

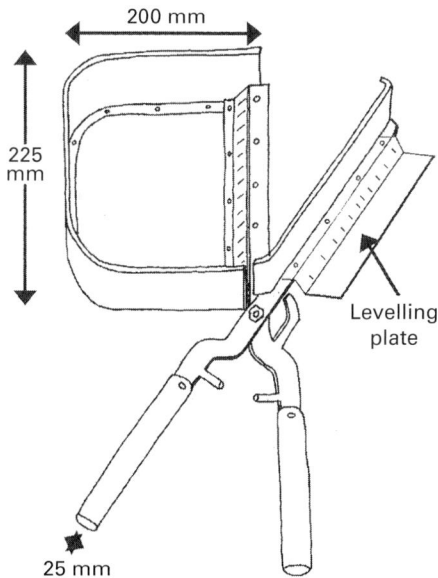

200 mm

225
mm

Levelling
plate

25 mm

Figure 11.2 Harvesting with modified garden shears (diagram courtesy of Paul Burgess).

- The height at which the shoots are cut, relative to the effective surface of the canopy, is a measure of the *intensity of harvesting*. With shears, this can be changed by varying the space (or step) between the base of the tray and the top of the lower blade (Suryanarayanan and Hegde, 1993).
- In turn, the step controls the height of the crop canopy above the ground.

Comparisons with Hand Plucking

South India

In an experiment in South India lasting five years, Sharma and Satyanarayana (1993) compared shear harvesting (without a step) at different times during the cropping season with hand plucking. They concluded that shear harvesting when judiciously applied in combination with selective hand plucking did not affect either the yield (which averaged about 4500 kg made tea ha^{-1}) or the bush 'vigour'. Shears were particularly helpful during the yield peaks when labour was at a premium. Continuous hard plucking reduced yields by about 9%. Yields from this treatment continued to be depressed after the cessation of treatments and the resumption of hand harvesting.

Similar experiments were carried out at several sites across South India (at Anamala, central Travancore, High Range and Nilgiris). These lasted for one or two complete pruning cycles. Satyanarayana and Sharma (1994) found that continual shear harvesting (again with no step) depressed yields in the short term and, in the longer term, by cutting into the maintenance layer. They too recommended an integrated approach to harvesting, with shears being used during the peak cropping periods from April to June and again from mid-September to mid-December. During the intervening low-yielding periods, hand plucking was advised. In the case of China-type plants they said this schedule should begin 15 months after pruning, and with Assam-type bushes 18 months after pruning.

In the mid-1990s, 60% of the estates in South India that responded to a questionnaire ($n = 59$ estates) were practising shear harvesting, with half of these using integrated methods. The majority were using shears only in the third and fourth years of the pruning cycle. In some cases the use of shears allowed the pruning cycle to be extended from four to five years. No adverse effects on the quality of the processed tea were reported. There was an increase in labour productivity at all yield levels (from 2000 to 3500 kg ha^{-1}) with integrated harvesting as well as with continuous shearing.

Again in South India, research at UPASI found that the chemical quality parameters and sensory evaluation of black teas were all influenced by the method of harvesting. Hand-plucked teas were very rich in their green-leaf biochemical precursors (e.g. catechins) and had higher contents of made-tea quality constituents (e.g. theaflavins) than tea harvested with shears. This loss of quality from shear harvesting was thought to be due to mechanical injury to the shoots with the shears and also to the lack of shoot selection. Where tea had been mechanically harvested for a number of years (five in this case), the quality of the tea was improved compared with tea that was being harvested with shears for the first time after many years of hand harvesting. In South India, the one change noticed during a recent visit was the click-click-click sound made by the numerous shears now operated by lady pluckers. This was particularly noticeable in the early morning when all else was quiet.

East Africa

In a series of experiments over an eight-year period, simple mechanical aids (blades and shears) were evaluated against hand harvesting on mature morphologically contrasting tea clones in southern Tanzania (Burgess *et al.*, 2006). The details of the experiments are shown in Box 11.2, while the principal conclusions that emerged from this very detailed study were:

- The use of blades, with an appropriate plate and good supervision, can result in higher yields than hand harvesting, particularly for clones with a large number of small shoots that are difficult to harvest by hand (yield +13%).
- Using shears, at the same harvest interval as hand harvesting (two phyllochrons, see Chapter 10), results in lower yields and a larger proportion of broken shoots. However, because the mean shoot size is smaller than that harvested

---

Box 11.2 Comparing Blade with Shear and Hand Plucking

In a series of experiments over an eight-year period, simple mechanical aids (blade and shear) were evaluated against hand harvesting on mature morphologically contrasting tea clones in southern Tanzania (Burgess *et al.*, 2006). The effects of shear step height (5–32 mm) and the harvest interval (1.8–4.2 phyllochrons) were also examined. Except in the year following pruning, large annual yields (5700–7900 kg dry tea ha^{-1}) were obtained by hand harvesting at intervals of two phyllochrons.

For clones K35 (large shoots) and T207 (small shoots), the mean harvested shoot weights were equivalent to three unfurled leaves and a terminal bud. The proportions of broken shoots (40–48%) and coarse material (4–6%) were both relatively high. Using a blade resulted in similar yields to hand harvesting from clone K35, but larger yields from T207 (+13%).

The yield increase from clone T207 was associated with the harvest of more shoots and heavier shoots, smaller increases in canopy height, and a higher proportion (7–9%) of coarse material compared to hand harvesting. On bushes which had been harvested by hand for two years following pruning, using flat shears (no step) supported by the tea canopy resulted, over a three-year period, in yields 8–14% lower than those obtained by hand harvesting and, for clone K35, a reduction in the leaf area index to below 5.

The development of a larger leaf area index is possible by adding a step to the shear. However, since annual yields were reduced by 40–50 kg ha^{-1} per mm increase in step height, the step should be the minimum necessary to maintain long-term bush productivity. As mean shoot weights following shear harvesting were about 13% below those obtained by hand harvesting, there is scope, when using shears, to extend the harvest interval from 2 to 2.5 phyllochrons.

by hand, the harvest interval with shears could be extended slightly, up to 2.5 phyllochrons.

- When shears are used so that the base of the tray rests on the crop canopy, the step height should be the smallest possible to sustain long-term productivity. Where the leaf area index (see Chapter 7) was maintained above about 4, each millimetre increase in step height (over the range 5–32 mm) reduced mean annual yields of dry tea by 40–50 kg ha^{-1}.

- In the 10 months following pruning the plots were all plucked by hand. There was little influence of the previous harvesting method on yield or on shoot quality. Shoot numbers per kg of fresh leaf declined from 900 kg^{-1} in the year before pruning to 610 kg^{-1} in the 10 months following for clone BBK-35, and from 1600 to 1170 kg^{-1} for clone BBT-207. The pruning weights from bushes previously harvested by a blade or by shears without a step were about 25% less than those from bushes harvested with shears with steps.

Central and Southern Africa

The use of mechanical methods of harvesting tea has been investigated in Malawi since the 1970s, and more recently in Zimbabwe and South Africa, where machine plucking was introduced in 1980. Detailed evaluation of shears commenced in 1988 and continued until 2000. This was primarily in response to shortages of labour in Zimbabwe, which was made worse in the early 1990s when peace was agreed in Mozambique and refugees who had been employed on tea estates in Zimbabwe returned home. In Malawi, there were shortages of labour in the north of the country but not in the south, where the bulk of the tea is grown. In South Africa labour had become expensive following revaluation of the currency (Martin, 1999). As a result, it became so serious that several tea estates became unprofitable and were abandoned.

A meeting was held in Zimbabwe in 1995 to review experiences of using shears in these three southern African countries. In the summary of the meeting the points in favour of shears were listed (Nixon, 1995):

- Good shear harvesting is as good as hand plucking.
- Major increases in productivity are achievable.
- Periods of peak production could be accommodated, and plucking rounds maintained without additional staff.

Subsequently a further series of experiments with shears was conducted across the region (Nyasulu, 2001). Yields and quality (based on evaluation) were similar from the shear-harvested bushes and from those plucked by hand. There was no difference in yields between flat and 10 mm stepped shears. The annual rise in the height of the plucking surface was kept below 50 mm with shears, and 100 mm with hands. These results suggested that it might be possible with shears to extend the length of the pruning cycle to more than three years. Based on a 42-day shoot replacement cycle, the optimum harvest intervals were either 10/11 (four generations of shoots on a bush) or 14/14 days (three generations). Both gave similar results. Nyasulu (2001) recommended that harvesting with shears should begin soon after the start of the main growing season.

Portable Machines

The Japanese are very keen to export their tea-harvesting technology. For example, well-known motorised reciprocating-blade harvesters include the Ochiai and Kawasaki brands. However, in Japan there are typically only up to four harvesting rounds in a season/year, compared with up to 30 in equatorial Africa. To cope with rough-terrain conditions, tea-harvesting machines imported into Africa from Japan have to be modified and strengthened in order to improve their reliability. Kawasaki developed a unit (H140D) in Papua New Guinea in the 1990s and then sold it in Africa.

Since the 1970s (Dale, 1974), the search for a tea harvesting machine suitable for African conditions has continued. It is still continuing, as the twin issues of labour shortages and labour costs intensify. This search has become like that for the holy grail, with enthusiastic individuals (amateur engineers) seeking to build the perfect machine.

Single-Person Machines

These consist of a 600 mm wide reciprocating A-blade cutter with air jets to blow the cut shoots into a small container. They can be powered by a two-stroke petrol engine carried on the operator's back (or by battery). The accuracy of the cut is entirely dependent on the skill of the operator.

Two or More People

This unit is carried by two people walking either side of a double row. The bushes either side of the path are side-trimmed to facilitate access. This machine is particularly suited to hilly terrain. The width of cut is between 600 and 1600 mm and the unit is powered by a 49 cc two-stroke engine. The cutting and transporting mechanisms are similar to those on a one-person machine. An alternative cutting mechanism is a horizontal cylinder with spiral blades cutting against a ledger plate (like a cylinder-type lawn mower). This provides the air movement to push the leaves into the collecting chamber. A skid that rides on the surface of the bush can be added to improve cutting height control. The New Century Corporation H140 and H140D (stronger frame) cutting head and a similar unit made by Ochai in Japan have now become the industry standard and are used on the wheeled machines described below. Operators tend to carry the portable machines at a convenient height (to them) rather than at the required table height. When they are carried too low, the maintenance leaf is cut, thus reducing quality.

Figure 11.3 (a, b, c) A selection of mechanical harvesters, at various stages of development (Tanzania and Kenya, MKVC). *A black and white version of this figure will appear in some formats. For the colour version, please refer to the plate section.*

During the mid-1990s similar machines were in use in Uganda in tea that was being rehabilitated after being abandoned during a period of civil strife. The same machines were also evaluated in Kenya, where they were modified in several ways:

- Height control (skids or 'gliders' were added to allow the machine to slide over the top of two adjacent tea bushes).
- Ease of movement through the tea, by extending the span of the handles.
- Adjustable height control on the handles, to allow compensation for the weight of engine.
- Bag-straps removed and large zip fasteners incorporated at rear, to improve leaf distribution in collecting bag to prevent it dropping down between rows.

Wheeled Lightweight Machines

Southern and Eastern Africa

The first small lightweight wheeled machines were introduced in Argentina during the 1990s. Known as the Jachacha machine (Brazilian type), the rickshaw was further developed and tested in Zimbabwe before being widely used in Uganda, where labour was scarce and the Japanese two-person machines were not appropriate at the time. The basic wheeled machine is pushed/pulled through the tea by the two/three/four operators (Figure 11.3), one of whom carries the cut leaf to the headland.

While the rickshaw units are relatively manoeuvrable in the field and on narrow headlands, they do not eliminate all the quality problems associated with poor table-height control, mainly due to the machines' tendency to tilt depending upon the operators' movements.

The need to cut pathways or wheelings through the tea also limits the use of such machines to tea which is planted in parallel rows and where these terminate at suitable turning areas.

Although the machines are simple and lightweight, the resulting tea quality can be very poor. The smaller two-person machines have to do as much work in Africa in one year as the same machine would do in its entire working life in Japan. The physical and mechanical properties of these machines when used in South Africa and Zimbabwe, including operating costs, have been described. For example, mechanical harvesting with a two-wheeled machine on a 10/11-day round in Chipinge, Zimbabwe, over a three-year period reduced yields of seedling and clonal (TRFCA-SFS150) tea by 10–20% (on a base yield of about 4000 kg ha^{-1}) compared with hand harvesting at the same interval (Mukumbarezah, 2001). Yield reductions of 23% and 10% for seedling and clonal tea respectively occurred when the harvest interval was increased to 14 days. These harvest intervals were based on an assumed shoot replacement cycle of 42 days at this site in Zimbabwe. The yield from the clonal tea when harvested on 21-day rounds was slightly higher when mechanically harvested than when it was hand plucked, but less than 60% of the green leaf was considered to be of an acceptable standard. On this basis, a 14-day plucking round during the main growing season was recommended for machine-harvested tea. This confirms other research reported here. The effect of the cutting height (or intensity of harvest) on tea yield was very variable. With seedling tea, yields only declined in the third year of treatments; the higher the cutting height, the greater the yield loss (Mukumbarezah, 2001).

A three-wheeled machine was modified in 2004 and then tested extensively on undulating land in western Uganda, at Ankole Tea Estate. The machine has three small (300 mm diameter only) wheels, the front two of which are hydraulically driven, as are the air blower and drum blade. The operator stands at the rear of the machine, steering by means of a tiller and adusting the forward speed. The unit is powered by an air-cooled 10 hp diesel engine, and spans one row of tea only. Its main disadvantages are that it is impossible to steer in reverse, it bounces in the field because of the small wheels, it cannot climb banks, and the hydraulics are often ripped by tea bushes, leaving the unit stranded in the middle of the field!

Design improvements to the two-man hand-held harvester were made in Malawi, with further improvements in Tanzania. These included:

- The H140D cutting head was suspended from a purpose-built frame running on motorcycle wheels.
- A large aluminium leaf carrier situated at the rear helped to set the cutting height so that the cutting head could pivot.
- The unit spanned two rows but the cutting head only cut one: this prevented the operator from cutting the tea in the wrong direction and also reduced the number of pathways or wheeling required.
- Two operators pulled from the front. One pushed from the back.
- Simple frame construction: this unit could cover 1 ha in a day with three operators.

A meeting to review the results of three commercially managed trials at three sites in southern Tanzania was held in 1996. These trials were designed to assess the performance of several mechanical-harvesting techniques, including blade and shear, one- and two-person portable machines, and a wheeled power-driven machine. At that time, it was agreed that more work was needed to determine long-term effects on yield and bush health and on tea quality of clones with contrasting morphologies. Wheeled and two-person portable machines offered the possiblity of increasing labour productivity by between four- and sevenfold (Nixon, 1996b).

During the 1990s a green field tea estate was planted in Njombe by the then Commonwealth Development Corporation. Kibena Tea was laid out specifically for mechanical tea harvesting, with each field being planted to a set pattern of 500 × 180 m and with a 5 m road around each field's perimeter.

This layout allowed further development of small, lightweight machines. Initally the Assam 02 and Assam 02A were trialled in the mid-1990s with some success, but leaf quality and machine reliability remained a challenge. Both of these units employed a drum blade which, if not correctly synchronised to the ground speed, resulted in a single leaf being sliced into small sections.

With the objective of consistent production of good-quality tea, a number of design considerations had to be taken into account during the selection of the most appropriate unit. These were condensed into three key attributes that the machine and its operators had to fulfil. These were to enable the user to:

1. Accurately set and repeat a cutting height. Repeatability was considered a key attribute, as without it skiffing or excessive table creep would occur. Plucking rounds and consequently shoot length and yield would therefore be impossible to optimise.
2. Control the ground speed so as not to exceed the capacity of the blades to cut and discharge leaf from the cutting face prior to another shoot being presented. It was noted that many quality issues associated with mechanical tea harvesting can be

Figure 1.1 A tea estate in Kenya, west of the Great Rift Valley. The best estates in Kenya still lead the way in terms of productivity (MKVC).

Figure 1.3 Where did all the shade trees go? During the 1970s it was recognised that in the cool high-altitude areas of Africa shade was no longer necessary (MKVC).

Figure 1.4 Shortages of labour in some places have meant that mechanical harvesting is essential. Many systems are now available and being tested (Uganda, MKVC).

Figure 2.5 Green tea is very popular in Japan, Taiwan and China (MKVC).

Figure 2.13 Tea plants growing into multi-stemmed trees in western Uganda, about 15 years after Idi Amin took over power and tea estates were abandoned. Tea is very forgiving and soon comes back into production when normal management practices are resumed (MKVC).

Figure 3.1 An example of a poorly sited weather station. For example, the Stevenson screen (contains thermometers) and the evaporation pan are both far too close to the tree. Ideally, the measurements should be made over a short, well-watered, grass surface. We depend upon good weather data to judge the suitability of a site for tea (MKVC).

Figure 3.3 In South India, tea is grown on steep slopes in the Nilgiri range of mountains at altitudes from 300 m to 2500 m (MKVC).

Figure 3.5 Numerous small earth dams have been created in Mufindi, southern Tanzania, to store water in order to irrigate the tea during the dry season. Note also the *Hakea salicifolia* hedges planted as shelter belts across the direction of the prevailing wind. The benefits of shelter are questionable (MKVC).

Figure 3.7 Tea is grown in the Chipinge and Honde Valley districts in the Eastern Highlands of Zimbabwe. Drought is a recurring challenge, as shown here (MKVC).

Figure 4.1 Tea shoots (*Camellia sinensis* var. *assamica*). Paintings by Mrs W. Wight, 1950.

Figure 5.1 A tea flower (RHVC).

Figure 5.4 (a) Close-up of tagged tea fruit, (b) with cotton bag in place and (c) after removal of cotton bag (Kenya, RHVC).

Figure 6.3 (a) No one appears to have attempted to quantify the degree of shade needed at different times and to specify this in terms of the light intensity. (b) It is important to move the plants at regular intervals to stop them getting too big and to limit the degree that the roots escape from the polythene sleeve. (c, d) Plants have been graded in size prior to the larger ones going to the field (East Africa, MKVC).

Figure 6.7 Post-hole diggers being used to assist with planting tea (Tanzania, MKVC).

Figure 9.5 The total shoot population is the sum of the number of harvested shoots and the number of basal shoots (small shoots remaining after harvesting). The peak shoot density reached 850 m^{-2} over a three-year period. (Clone TRIEA-6/8, MKVC).

Figure 10.2 Hand harvesting, which involves 140–190 hand movements a minute, is still by far the most common method of harvesting in use today. Here the palms-up method of plucking is being used (Tanzania, MKVC).

Figure 10.3 *Gang plucking*: a group of pluckers move across a field together (MKVC).

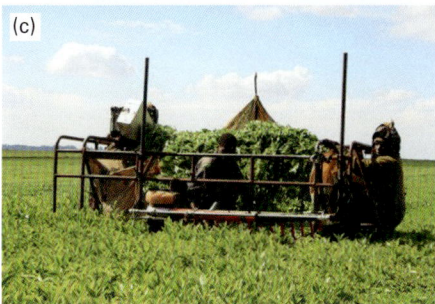

Figure 11.3 (a, b, c) A selection of mechanical harvesters, at various stages of development (Tanzania and Kenya, MKVC).

Figure 14.2 On some large estates fertiliser is applied from the air. This photograph shows the aerial application of zinc in Malawi – aeroplane vintage unknown (WS).

Figure 15.4 (a, b) Rehabilitation of eroded tea land in Sri Lanka: construction of soil conservation terraces (MKVC).

Figure 15.5 Soil and water conservation has become an important topic in Sri Lanka because of the degradation of soil over many years (Save Our Soil project). The sign lists the cover crops being tested (MKVC).

Figure 15.6 Disposing of excess water in areas where annual rainfall can exceed 4000 mm is an important aspect of estate management. Note the down-drains (South India, MKVC).

Figure 15.8 Line-source experiment at Ngwazi Tea Research Station. The crop closest to the line-source is fully irrigated, but the amount of water applied declines with distance in both directions so that at the extremities the crop receives only rainfall. To be effective, a line-source irrigation system should only be used when there is no wind. This usually occurs very early in the morning or late in the evening (Tanzania, MKVC).

(a)

(b)

Figure 17.7 (a, b) Oxidation is allowed to continue for 1–3 hours, depending on conditions. During this time, the aroma and flavour develop and the colour of the macerated leaf becomes darker. The leaf is then dried. The quantity and proportions of theaflavins and thearubigins determine the colour and taste of the tea liquor (MKVC).

Figure 18.1 A successful business needs to attract and retain good staff. This means providing services that would be expected living in an urban area. Managing a tea estate is like running a small town, with all of the responsibility that goes with it (Sri Lanka, MKVC).

Figure 18.2 The provision of housing on tea estates is improving, in some cases (Tanzania, MKVC).

Figure 18.4 Sustainability is everywhere, but how sustainable is sustainability? (Tanzania, MKVC).

Figure 19.1 (a, b, c, d) A prerequisite for establishing a successful tea estate includes an assessment of the indigenous vegetation, a soil survey, assessment of the water resources, a topographic survey, and reliable weather data (MKVC).

Figure 20.1 A potential new site for mechanised irrigated clonal tea (MKVC).

Figure 20.2 Research stations have an obligation to communicate the results of their research in the most effective way to all stakeholders. The line-source irrigation system provided an excellent visual picture of the effect of drought and irrigation on the yield of different clones. (Ngwazi Tea Research Station, Tanzania, MKVC).

Figure 20.3 Some tea research institutes have their priorities right, with their own cricket ground (Sri Lanka, MKVC).

Figure 20.5 Tea revives – it is the sovereign drink of pleasure and health. It is time for tea.

attributed to leaf which is left uncut during the operation becoming excessively long by the next round.

3. Control the direction of travel so as not to harvest the same line in opposite directions on different occasions. Cutting in alternate directions was found to increase the percentage of maintenance foliage that is harvested.

In later years the estate was completely converted to machine tea harvesting, using a hybrid of the Japanese reciprocating-blade cutting head mounted on a purpose-built self-propelled tool carrier that to a large extent met the criteria above. However, further work is still required. The unit is capable of harvesting up to 40 ha within a plucking round, operated by a team of five people. On a 12-day round this is equivalent to 1.2 ha d^{-1}.

Problems still remain, although these are currently being addressed in Kenya (Box 11.3). In particular, the ability to control the cutting height accurately still needs to be achieved.

Box 11.3 Some Factors That Influence the Quality of Mechanically Harvested Leaf

-   **Type of tea**. The physical attributes of the tea shoot vary with the stage of growth (e.g. within the pruning cycle), and the clone (e.g. clone TRFCA-PC81, from Malawi, has a vertical leaf pose; this results in more half-cut leaves than occurs with clones TRFCA-PC168, TRFCA-PC165 and TRFCA-SFS150 with a more horizontal angle of orientation). Leaf pose is infinitely variable within a population of seedling tea. The degree of coarseness within the stem of the tea shoot also varies (e.g. clone TN-14/3 has relatively hard stalks, so that yields can be 30% higher when this clone is machine harvested than when it is harvested by hand).
-   **Setting the cutting height**. It is important that the machine blade is set at the correct height. Hand-held units operate on the basis that side skids provide some degree of support on the bush, but 'digging-in' to the maintenace foliage can occur if the operator is not vigilant. Wheeled and tracked vehicles use the ground as the reference point, while one innovative machine (the 'Magic Carpet') floats on the surface of the bush.
-   **Control of forward speed**. This is influenced not so much by the ground conditions, but by the rate at which the blade can cut and discharge shoots from the cutting surface. If this rate is too slow, shoots are not cut and remain on the bush; by the time of the subsequent round they will be overgrown. High forward speeds also lead to 'bouncing' in the field, and this impacts on the control of the cutting height, means larger and heavier machines, and increases fuel consumption.
-   **Control of direction of travel**. It is now accepted that the only way to prevent a large amount of maintenance leaf being cut is always to harvest a row of tea in the same direction.

Large Self-Propelled Machines

Kenya

In trials of a large self-propelled machine in equatorial Kenya (alt. 2140 m), Mwakha (1986) compared the yield responses of clonal tea (AHP-S15/10) to three round lengths (21, 42 and 63 days) and two cutting heights (10 and 20 mm above the last cut). At the start of the experiment (1985), which lasted 12 months, the tea was eight years old. The harvester was a French-made Bobard Jejune (type Bob M55), which had been modified in Kericho (Kenya) for tea harvesting (it's not clear how). One driver operated the machine with the help of two assistants. The cutting mechanisms were reciprocating A-blades. The plots were long (493 m) and thin (2.44 m wide). The control treatment was hand plucked at intervals determined by the estate. This treatment out-yielded all the others by up to three times (surprisingly?). The only treatment that produced a comparable yield was the one harvested at 63-day intervals at a height 10 mm above the previous cutting height. Cutting of immature shoots was the explanation proposed for the low yields from the mechanically harvested bushes.

A similar experiment, but this time with seedling tea conducted over a 27-month period (beginning in 1987), gave rather different results. This time hand plucking resulted in the lowest yields (about half those from the machine-harvested plots: no explanation was offered for the difference from the first experiment), but the orga-noleptic scores were slightly higher, and the proportions of various tea shoot grades (using the method described by Templer, 1978) were reported. As would have been expected, the proportions of larger shoots (four or more leaves and a bud) and broken pieces were much greater in the machine-harvested tea (longer harvest intervals than in plots that had been hand plucked (at irregular intervals) (Mwakha, 1990).

It is impossible to draw meaningful conclusions from either of these two experiments, based on this evidence. No attempt appears to have been made to understand or to explain how the cutting mechanism and/or forward speed of the harvester or the intensity of harvest (was it the same for all treatments?) were controlled and how they interacted with the seedling tea or clone. This is fundamental if our understanding of the processes involved in mechanical harvesting of tea is to be advanced and machine design improved. Very different results might have been obtained if either of these variables had been changed.

Based on the evidence from these two experiments, no clear advice on mechanical harvesting for the tea industry in Kenya was forthcoming, except to state that it is much quicker than hand plucking (up to 50 times). In fact, the principal conclusion from a paper by Owuor *et al.* (1991) describing essentially the same experiment as Mwakha (1990), but with the emphasis on the quality of the processed (black) tea was that '*mechanical plucking will lead to a reduction in quality.*' But is this always true, and by how much? It can only be claimed if we are comparing like with like. Does it matter? It will depend on the market at the time when the tea is sold. There are examples when machine-assisted harvesting resulted in the same or occasionally better quality teas (see below). This is an example of how misinformation is spread, and of how all mechanically harvested teas can get a bad name. If the rounds had been extended in the hand-plucked tea, as happens when there is a shortage of labour, the quality of this tea would have been reduced (and/or yield would have been lost because of the need to 'break back' the long shoots).

Papua New Guinea

Tea was harvested mechanically with self-propelled machines in the Soviet Union (Georgia) in the 1950s. In 1975, a tracked machine was developed for use in Papua New Guinea by W. R. Carpenter & Company Estates Ltd (a tea and coffee company). Based in part on this design, a prototype Valiant harvester was built in Japan in 1981, and later several machines were supplied (see New Century Corporation website: www.newcentury-japan.com). Maintenance costs were found to be high, and there were weaknesses in the frame. In 1985 the first Williames T5000 harvester, which had been designed in Australia, was operating in Papua New Guinea. This number was increased to seven over the next three years. These Australian machines were found to be much more fuel-efficient than the Valiant (158 kg, compared with 85 kg green leaf per litre of fuel) and six times more productive (G. A. Williames, personal communication, 2013). The land is almost flat in Papua New Guinea where the tea is grown, and there is a tendency for the soil to become waterlogged and drainage is needed. For these reasons the emphasis has been on track machines. In contrast, Australia has concentrated on wheeled machines, as the soils there are better drained (Clowes, 1991).

Uganda

At the same time as these initiatives were being taken in Papua New Guinea, civil strife had followed the overthrow of Idi Amin in 1979 in Uganda. This forced companies in western Uganda to look at machine harvesting as an option, as there was a severe labour shortage. Tea that had been abandoned 10 years earlier had grown into 7 m tall (polled) trees and was in the process of being rehabilitated. In 1981, Mitchell Cotts (a tea company) commissioned Silsoe College, Cranfield University, in the UK, to design and build a tea harvesting machine to meet the company's own specifications (Figure 11.4a). These included the need to have adjustable widths to suit the existing plant spacings (although tea planted in Uganda during the 1950s and 1960s was at a spacing that facilitated the use of machines), the capacity to harvest 110 ha in a year (equivalent to the output of 150 manual workers), and the ability to work safely on a 10-degree side slope.

The half-track, straddle-type prototype machine was shipped to Uganda in 1983, stopping en route in the Nandi Hills in Kenya for initial field trials. Before the first year of operation had been completed, four new, modified machines (Mark 1), designed to be more reliable than the prototype and with improved performance, were built. The first two were operational in 1983 and the two others in 1984. To facilitate movement of harvesters through the tea it is essential that tracks are

Figure 11.4 The continuing evolution of self-propelled mechanical tea harvesters, manufactured and tested in countries around the world, including Australia, Japan, Kenya, Tanzania and Uganda: (a) a Silsoe machine; (b) a Williames.

opened (side pruning of bushes adjacent to the wheel track). To minimise excess damage to the cut leaf and to the maintenace foliage, it was found necessary to harvest bushes in one direction only. There were again problems with reliability and the availability of spare parts, and this resulted in a Mark 2 version being produced. It became operational in 1987. The oldest machine by then had operated for more than 4000 hours. Training was provided for the mechanics, drivers and support staff. Full details of the designs were provided by Kilgour (1990). One major problem was the size of the harvester. It weighed 10.5 t, and sometimes got stuck in the mud. Another constraint was the transport system (tractors and trailers) taking the leaf from the field to the factory. The only other comparable machines at that time were those of Russian origin (then operating in Georgia, USSR), the Valiant and the Williames T5000 (Figure 11.4b).

Further development of the Mark 2 was curtailed by a company takeover. Instead the decision was taken to adopt the Williames T5000 machine from Australia (Figure 11.4b). By then, this model had been operating successfully for several years in Papua New Guinea. Five machines were purchased for use in western Uganda, which were immediately upgraded, and two more for estates in the east. There were many similarities in the design and operation of the two machines (see below), but the Williames was faster, lighter, cheaper to run, more fuel-efficient and more manoeuvrable than the Mitchell Cotts machine (Mark 2). Full details of the design and relative performance of the two machines have been described by Kilgour and Burley (1991). A video of the most recent Williames tea harvesters (namely the T150 Ground Supported model) can be seen on the Williames-Tea website (www .williamestea.com).

Recent Developments

On the same website, the Selective Tea Harvester can be seen operating. This machine is based on an ingenious concept first tested with the so-called 'Magic Carpet'. The Selective Tea Harvester literally moves across the surface of the bush,

Figure 11.5 (a) A harvesting machine designed specifically for Japanese conditions. (b) To achieve optimum height control some machines are designed to operate on rails.

with no ground contact. It is best suited to level ground (as are most self-propelled harvesters). It is apparently capable of differentiating between shoots that are ready to be harvested and those that are too small. It operates at 4 km h^{-1} on flat land in India. A prototype (T1000) was tested by the Tea Research Institute of Tanzania in 2000, and at Mufindi Tea Company on a commercial basis, but a number of inherent mechanical problems were encountered. This machine cannot be used on young tea.

Another recent innovation is based on harvesting mechanisms used by the salad crop industry. A company by the name of Jilanjo, in association with UK-based Nicholson Engineering, has used a bandsaw blade for cutting the tea shoots. When tested in Tanzania on flat land, the cut was apparently 'perfect'. Version 3 (V3) used four skid steer wheels on each side, powered by hydraulic hubs, with electronic controls. The skid steering made the machine highly manoeuvrable despite its large bulk. It is powered by an 80 hp engine and weighs nearly 5 t.

Bulky self-propelled machines like the ones described above are only suitable for large areas of flat land. Ideally future plantings of tea should be done with mechanisation in mind, with long rows on flat land as in Tanzania, where three estates have recently been developed on flat land in non-traditional tea areas (Figure 11.6). They are all irrigated (or will be) and are harvested mechanically.

Existing tea usually requires something different, and here the machine has to be designed and built for the tea (as opposed to the tea being planted to suit the machine). In other words, harvesters must be capable of coping with varying topography, row widths and lengths, turning areas, access to the field, and types of tea (Box 11.4).

Conclusions

When evaluating the effectiveness of different mechanical harvesting systems, several parameters must be assessed, including the composition of the harvested shoots, and the yield and value of the processed tea. These can all vary with the frequency and intensity of harvesting and, in turn, with the stages of bush and shoot development. Long-term effects of the machine on bush productivity are also important to monitor. Similarly, the ease of use and manoeuvrability of the harvester, together

Figure 11.6 (a, b, c) Views of the tracks used by large motorised harvesting machines. Note the long, straight rows required for their efficient use (MKVC).

Box 11.4 Mechanical Harvesting: Issues to Consider

1. The tea industry (in a given location) is faced with possible shortages of labour and/or relatively high labour costs. To remain competitive in the world market it needs to give serious consideration to the development and use of appropriate mechanical harvesting systems within the foreseeable future.

2. For technology transfer from one country to another to be successful, due consideration must be given to local physical, biological and socioeconomic conditions. Techniques developed elsewhere for mechanical harvesting of tea will, however, form the basis for initial trials and associated development work.

3. Tea is unusual amongst cultivated crops in that the way that it is harvested impacts on the number, size and type of shoots that remain on the bush. This

Box 11.4 Continued

    has a direct effect on the yield and quality of the shoots available at the next and subsequent harvests.

4.   Mechanical harvesting systems for tea must not be designed in isolation from the crop. They need to be designed, developed and operated in ways that match the developmental biology of the tea bush, at different stages in the production or pruning cycle, and not in mechanical isolation of the plant.

5.   Any mechanical harvesting system must take into account the skills and support services available for operation and maintenance. It is sensible to consider leaf handling and transport as integral components of such a system.

6.   It is prudent for the tea industry to be prepared for the time when mechanical harvesting becomes a cost-effective option. This includes ensuring that any new planting is done with this in mind.

Box 11.5 Mechanical Harvesting: Design Principles

These are some of the lessons learned during 20 years of experience of designing, building and operating tea harvesting machines for use in a wide range of conditions (largely through trial and error), in four East African countries (Kenya, Malawi, Tanzania and Uganda) together with South Africa (from Flowers, 2013).

    Harvesters must be:

- Stable and able to repeat the same cutting profile at each harvesting event
- Highly manoeuvrable when entering or exiting fields, as well as turning in limited space on the headland
- Simple to construct and maintain
- Reliable and thus well engineered
- Able to produce quality leaf
- Able to function without large tracks being cut in the tea
- Able to cope with hills, undulating field conditions and steep slopes while remaining stable
- Commercially viable
- Limited to a maximum forward speed of 3 km h^{-1} in order to maintain a perfect cut when using double reciprocating blades
- Able to function on wet soils without causing soil compaction.

---

with the costs of machine maintenance, must all be considered (Box 11.5). The aim is to improve productivity and welfare.

    To be effective, mechanical harvesting systems should be robust, with precise height control of the cutter bar (the speed of which should match the forward speed

Table 11.3 A case study of a machinery selection procedure in East Africa (after Flowers, 2013)

Item	Challenge	Solution
1	Short runs, tea not planted with mechanisation in mind	Productivity will be reduced; costs must account for this
2	Access in and out of fields involves negotiating banks and drains	May be necessary to uproot some tea to allow easy access
3	High cost of labour	Daily wage rate divided by 33 kg gives the piece rate (shillings/kg) for hand plucking; with mechanisation, increase daily task up to 500+ kg without compromising quality; calculate new piece rate
4	Harvest-only concept	Terrain constraints, short runs and difficult access in and out of fields reduces the time in work considerably
5	Quality of green leaf harvesting to match or exceed hand plucking	Slow forward speed and excellent height control is needed to achieve this
6	Maintenance capacity	If staff are well trained and provided with right tools, this should be OK
7	Managerial capacity	Again depends on training; keep machine as simple as possible
8	Large area under tea but fragmented	Better to have a large number of small units rather than a few large ones
9	Undulating terrain	Uphill work is tiring; some kind of self-propelled unit is required

of the machine) and some shoot selectivity, and the operators need to be well trained. Mechanisation of harvesting offers some advantages, including increases in labour productivity and associated cost savings. Disadvantages can include limited selectivity of shoots, poor reliability and extended periods of downtime during repair if maintenance is not robust, and the poor, perhaps sometimes unwarranted, reputation of mechanically harvested tea in the market place. A case study of the ongoing selection procedure adopted by commercial growers in eastern Africa is shown in Table 11.3.

Too often mechanical harvesting of tea is seen as a panacea, an easy way of solving a range of problems associated with shortages of skilled staff. Machines are sometimes introduced in an opportunistic way (e.g. because someone has donated a tea harvester) or because there is a short-term problem with labour availability. Unless great care is taken to ensure that the machines are used effectively, neither of these approaches to innovation can be justified, *except* in emergencies. Sometimes, in trials, the intensity of harvest has differed between treatments, resulting in misleading results, because like-for-like comparisons were not being made. These then enter into the folklore of tea production. Mistakes can also be very expensive over the longer term as a result of lost crop and/or poor-quality, low-value teas. Reputations too can be quickly lost. Forward planning is essential, for example, to ensure that new areas of tea are planted in rows that facilitate

---

Box 11.6 Commentary: Machine Harvesting

It strikes me as strange that TRFCA are selecting clones for machine harvestability. There is considerable political pressure against mechanical harvesting in Malawi, for reasons you can well imagine. In any case, I'm not sure how much difference it makes. Once tea becomes conditioned to machine harvesting there is little maintenance leaf in the bag. And the amount of premature fermentation caused by cut leaf is tiny compared with the crushing caused by hand plucking. One nearby company has been machine harvesting for a while. I had a chat with the CEO of a neighbouring company, and suggested they should be moving in the same direction. He was concerned that it would result in loss of quality. But better quality was precisely why his neighbour was machine harvesting – much less crushed leaf! And much better control over the harvest interval (although that could result in sharper peaks in yield). And of course huge savings in wages, wage-related benefits, housing maintenance, medical expenses – you name it – all of which vastly outweighed the cost of the machinery.

In Tanzania, I worked on the principle that while productivity in the tea industry was increasing faster than in the economy as a whole, the real cost of plucking was declining. Once productivity in the tea industry plateaued, and that of the economy as a whole took off, the real cost of plucking increased, and mechanical harvesting became a necessity.

T. C. E. Congdon

---

the efficient use of machines in the future, that 'machine-friendly' clones are planted, and that operators are suitably trained and rewarded.

The search for the perfect machine continues. But perhaps we cannot improve, or even match, what well-trained human beings can achieve when plucking tea with their hands. Engineers on the other hand are pre-programmed to believe that every problem has an engineering solution. In short, the success of a machine-harvester intervention largely depends on the level of effort that is applied. It is a fallacy to believe that machine harvest operations are easier to manage than hand plucking. It is also a fallacy that machines can be successfully deployed in a previously hand-plucked field without any field preparation. Unfortunately all too often once a machine is purchased the need for regular maintenance is overlooked, again contributing to the belief that machines produce worse tea than a hand plucker (Box 11.6).

Once the fields have been prepared for mechanised tea harvesting, the three 'Ms' apply:

* **Machine** – Choose the correct unit for the objective and terrain.
* **Management** – Machines require perhaps more management than hand plucking.
* **Maintenance** – Without sensible maintenance and replacement policies, machines will be unreliable, plucking rounds will be compromised, productivity will be reduced, costs will increase and quality will deteriorate.

In addition, there is prejudice to overcome, as witnessed in this statement:

We recently swapped an invoice of machine harvested tea with an invoice of hand plucked, and guess what . . . The machine harvest invoice got the same price as all the other hand-plucked invoices, while the hand-plucked tea was discounted to machine prices.

So much for the discerning tea buyer/taster!

Note

1. The results of research with shears in Assam have recently been reported by Saikia and Sarma (2011).

12 Hidden Hunger and Intelligent Guesswork

We Can Only Build on What Has Gone Before

In this chapter, the soils in which tea is grown in the principal tea-producing regions of the world are briefly described as a prelude to discussing the major nutrient requirements (nitrogen, potassium and phosphorus) of tea, based on the results of research in many countries. The problems associated with specifying how much nitrogen to apply (or, increasingly, how little) are considered in detail, in particular the difficulties associated with scaling-up the results obtained from field experiments to field, estate, regional and national levels. The roles of sulphur, calcium and magnesium in the productivity of the tea crop, together with those of the two most important micronutrients, zinc and copper, are also considered, albeit briefly. Finally, to show how this information is interpreted in terms of advice to tea growers, fertiliser application rates as recommended in a number of countries are compared. Issues relating to the sustainability of soil fertility are considered in Chapter 14.

Soils

Tea is grown on a wide range of soil types that have developed from diverse parent rock materials and under different climatic conditions. The properties of soils formed, for example, in tropical climates, or at low altitude, will differ from those formed in the subtropics, or at high altitudes. The common factor is that all these soils will have been formed, or they exist, under high-rainfall conditions, which leads to intense leaching of the products of weathering. This has resulted in acid soils, a prerequisite for successful tea production. The capacity of soils to support (sustainable) tea production is variable. For example, in many eastern Africa countries, tea soils derived from volcanic ash tend to be deeper and better drained than the soils found in Assam in North-East India, the bulk of which are alluvial in origin. The alluvial soils are of fairly recent geological origin, having been washed down from the Himalayas and other mountain ranges. They are found on both banks of the Brahmaputra River and on the banks of its tributaries. The land here is flat or gently sloping and supports 200,000 ha of tea. In the Cachar District of Assam, tea is also planted in alluvial soils, but these are derived from sandstones which were deposited in the Barak Valley. In neighbouring Bangladesh, in the important tea-growing region of Sylhet, tea is cultivated on small hillocks of sandstone, known as *teelahs*. Similar soils are found next door in Cachar. Tea is also planted in drained peats, known as *bheels*.

The *bheels* are very rich in organic matter and are similar to the peaty soils found in Rwanda (East Africa). In Bangladesh, China, Sri Lanka and Taiwan, and in some areas of Malawi, tea is grown mostly on sedimentary soils derived from gneiss or granite rocks (Barua, 1989). The tea soils in South India (latosols) are also derived from gneissic rocks, and contain a lot of mica as well as sesquioxides. The predominant clay mineral is kaolinite. Leaching of potash and nitrogen, in particular by excess rain, together with erosion of the topsoil, means that the fertility of these soils is poor (Ranganathan, 1976).

In the Kericho District of Kenya (where the Tea Research Foundation of Kenya is sited at 0°22'S 35°21'E, alt. 2180 m) the soils are very deep, red, freely draining humic ferrasols.[1] These also form the dominant soils in southern Tanzania on the middle and lower slopes of the hills, with humic acrisols on the ridge tops and the upper slopes. In other areas of Kenya, and parts of Tanzania and Uganda, most of the soils where tea is grown are classified as nitisols, soils that are derived from recent volcanic deposits.

The consensus view is that the optimum pH for tea is between 5.0 and 5.6 (in water). Values at or above 6.5 or at or below 4.0 are unsuitable. In Africa, areas occur within fields where the pH is high and it is difficult to establish tea plants. Commonly known as 'hut-sites', these areas were previously homesteads or cattle kraals. Remedial action includes mixing sulphur in the soil from the planting hole at rates dependent on the pH, or applying aluminium sulphate (Othieno, 1992).

Indigenous vegetation is used to indicate the suitability of areas for tea, the most reliable species being aluminium accumulators. *Albizia* species are particularly good indicators. Tea plants can root to considerable depths (> 5 m, see below). Deep (at least 2 m), well-drained soils with a high water-holding capacity are to be preferred, although tea grows well in locations with a water table maintained through drainage at or below a depth of one metre, as in Assam and Rwanda (Othieno, 1992). For more information on the characteristics of tea soils, please see the publications by Barua (1989) and Othieno (1992), both of whom refer to the pioneering work of Mann (1935).

Nutrition

As tea is grown on a diversity of soils in tropical, subtropical and temperate areas, there is no easy way of specifying the fertiliser requirements except through field experiments, and these do not always give clear, unambiguous answers.

One of the first statistically designed fertiliser field experiments undertaken with tea was reported from Sri Lanka by Eden (1944). Yield responses to N at three levels (initially 0, 22 and 45 kg N ha^{-1}, later 45, 67 and 90 kg N ha^{-1}, combined with different levels of phosphate and potash), were recorded over four pruning cycles (a total of 12 years). In this pioneering work, the data were statistically analysed and presented graphically. The results showed that the yield response increased with increasing nitrogen up to 90 kg N ha^{-1}, with the efficiency being low in the first year following a prune, but increasing from one pruning cycle to the next, reaching a peak of 6.4 kg made tea for each kg of N applied. The levels of N tested are low compared with today's recommendations, but so were the yields, at only 500–800 kg made tea ha^{-1}. As is commonly found in fertiliser tea experiments, the shapes of the response curves varied from one pruning cycle to the next. A response to potash (45 kg K_2O ha^{-1}) emerged only in the last year of the fourth cycle (year 12), and that may have been a statistical freak. There was a small response to phosphate at all three levels tested (0, 22 and 45 kg P_2O_5 ha^{-1}) in the second and subsequent cycles.

This example published in 1944 has been presented in order to highlight:

- The work of our predecessors, and to ensure that it is not forgotten.
- The contribution that science and statistics have made to tropical agricultural research in the last 100 years.
- The low yields obtained, which at that time were considered to be acceptable. (Are we guilty of being equally complacent?)
- The need to identify and overcome limiting factors in a systematic way.
- Our limited ability to think beyond current conventional wisdom (e.g. the low rates of N applied in this experiment).
- The need to include at least one experimental treatment that is seen as extreme, in order, for example, to establish the potential yield. A control treatment with zero inputs sets the lower yield limit so that the benefits of added inputs can be calculated.

At about the same time (the 1930s) similar work was under way in Assam. Cooper (1946) reported the results of a series of trials comparing different forms of N, both inorganic (a new concept) and organic (traditional). The results of these early organic experiments are considered in Chapter 14. They probably still have relevance today, given the renewed debates surrounding healthy and sustainable food production systems.

Nitrogen

The important role that nitrogen plays in the productivity of tea is summarised in Box 12.1, but judging how much nitrogen to apply to a crop is one of the most difficult decisions to take, especially as it is a very expensive decision if you get it wrong. This dilemma is not helped by the different responses obtained in experiments from one year to the next or from one field to another, which are partly due to different weather conditions, partly the result of soil variability, and partly (mainly) unknown.

Response Curves to Nitrogen: a Warning!

Growers want a clear unambiguous answer, but this is unfortunately not possible. Scientists do not always make it any easier. Indeed they often confound the problem by allowing the computer to provide the answer, and failing to engage their own brains (Box 12.2). The difficulty arises in determining the *optimum* level of N for yield and/or for profit (as opposed to the level of N giving the *maximum* yield), for green leaf and/or for made tea, and this depends on the shape of the response curve (Boyd and Needham, 1976). With a limited number of nitrogen treatments in an experiment (usually from four

---

Box 12.1 Nitrogen and Tea

Nitrogen (N) is a key constituent of chlorophyll, proteins, nucleic acid and proto-plasm. It therefore plays an important role in the physiology of the tea plant. Nitrogen deficiency retards the overall growth of the plant, with a general yellowing of the leaves, formation of smaller leaves, and shorter internodes. In India, the 'replacement theory' operates. Nutrients, especially nitrogen, are added to replace those lost through the removal of the harvested crop, with some allowance for fixation, leaching and gaseous losses. This nutrient management regime may clearly limit any yield increases, as the amount of N applied is limited to the current yield level. To allow for yield increases, additional N can be applied in incremental quantities until the law of diminishing returns means that it is no longer profitable to apply additional N. In this so-called 'expansion ratio' approach, the amount of N to apply is based on the organic matter status of the soil and the expected/target yield. The altitude influences the contribution of N from the organic matter, together with rainfall. The greater the altitude, the cooler it becomes and the slower the activity of the soil microorganisms. The amount of N released as a result of decomposition of the organic matter then declines.

Box 12.2 Commentary: More or Less?

As tea growers, we do not make the tea produce more crop. What we do is identify constraints to yield, and by relieving them we enable the tea more nearly to achieve its full crop potential.

One factor constraining yield is lack of nutrients. Once this basic concept is understood, the rest follows naturally.

Take two estates:

- Estate 1 was carved out of a forest reserve. The tea was planted in deep forest soil, rich in organic matter.
- Estate 2 was mostly planted on old cultivated land, which had previously been under maize, potatoes and latterly pyrethrum. The soils were eroded, and depleted of organic matter.

Naturally, the base (unfertilised) yield of Estate 1 was higher than that of Estate 2, because the nutrient deficit on Estate 2 was greater than that on Estate 1.

But perversely, under a fertiliser policy based on historic or anticipated crop, it was Estate 1 that received the most nitrogen.

When I took over the running of the estates in 1973, I would have liked to reverse this policy, and give Estate 2 a higher rate than Estate 1, but this was too much for The Powers That Be to accept, so I had to settle for a blanket rate across all estates.

The effect on the low-yielding estates was immediate and large, to the extent that in my first year in office we achieved a 30% crop increase – helped, I will admit, by good weather, but in spite of factory constraints.

Box 12.2 Continued

I do hope this serves to explain why a policy based on yield should not be followed. But I am reminded of the words of Leo Tolstoy:

I know that most men, including those at ease with problems of the greatest complexity, can seldom accept even the simplest and most obvious truth if it be such as would oblige them to admit the falsity of conclusions which they have delighted in explaining to colleagues, which they have proudly taught to others, and which they have woven, thread by thread, into the fabric of their lives.

I rest my case.

T. C. E. Congdon

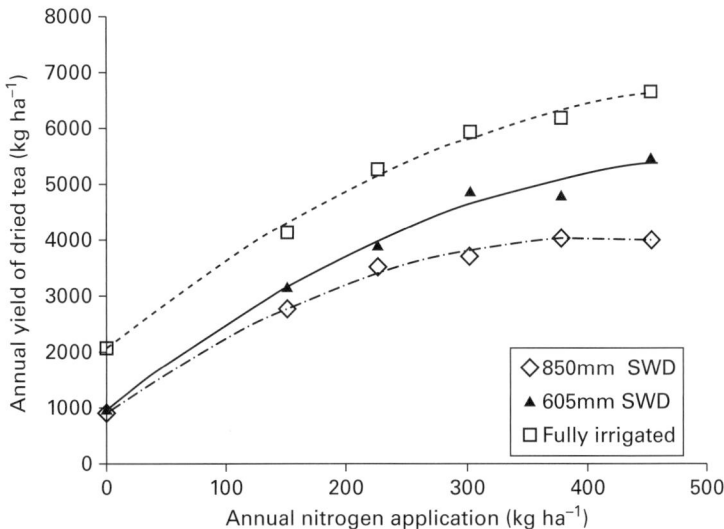

Figure 12.1 An example of the yield (dried tea) response to nitrogen at three levels of water availability: fully irrigated, partially irrigated, rainfall only (clone TRIEA-6/8, Ngwazi Tea Research Station, Tanzania, 1993/94). SWD = potential soil water deficit at the end of the dry season (from Burgess, 1996).

to six, which must include a zero N, and an excessively high treatment) it is statistically possible to create a number of different curves that fit the data points equally well.

Consider first Figure 12.1. This shows the response to nitrogen at three levels of water availability: fully irrigated, partially irrigated, and rain only. The three curves all have a similar shape (asymptotic) and judging 'by eye' the optimum quantity of nitrogen to apply (depending on what criteria you use) is probably about 400 kg N ha^{-1} for the well-watered treatment compared with about 200 kg N ha^{-1} for the rainfall only crop.

But now consider Figure 12.2: again it shows yield responses to nitrogen, this time for two years, with five different curves fitted to the data. All the relationships shown

(a)

(b)

(c)

(d)

(e)

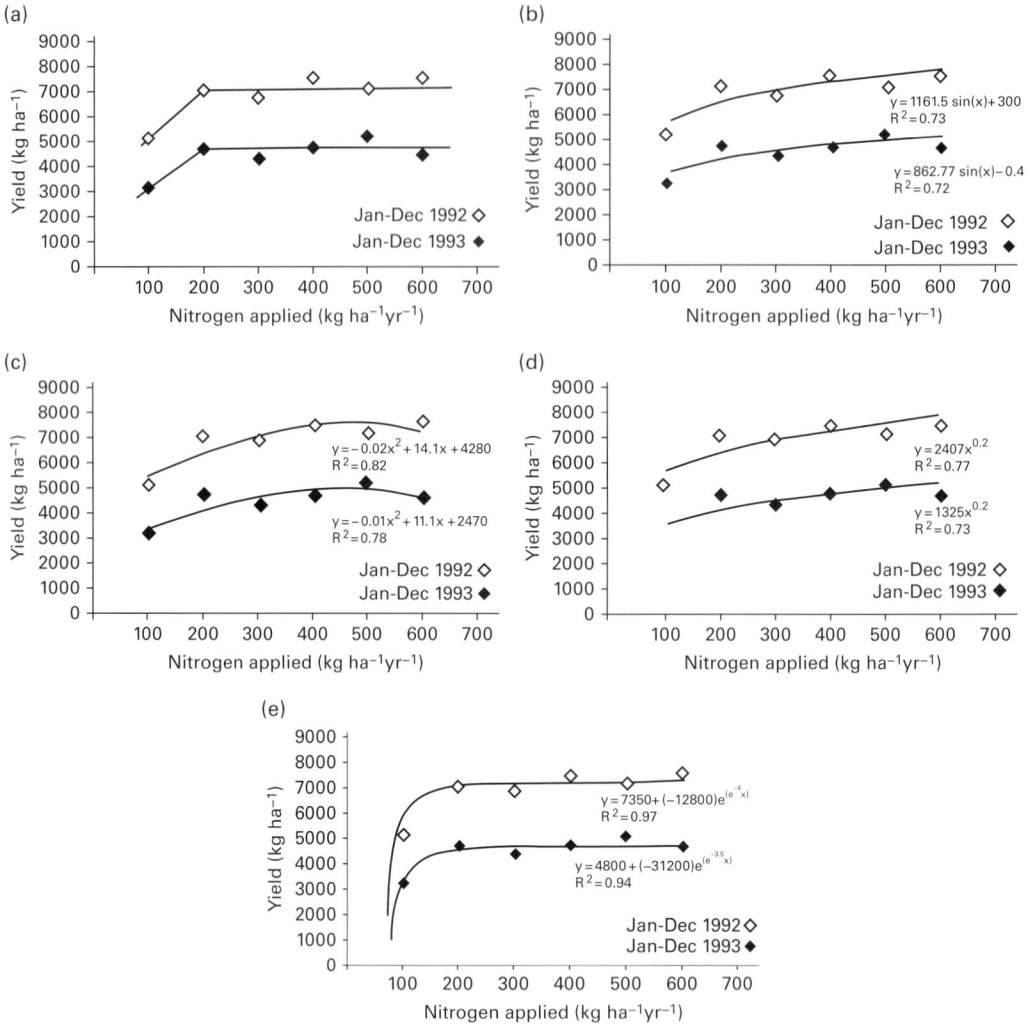

Figure 12.2 Examples of five ways of showing the relationships between yield (dried tea, clone AHP-S15/10) and annual nitrogen fertiliser application rates for two periods of time (January to December 1992, and January to December 1993), Kericho, Kenya (alt. 1860 m): (a) two-stage linear; (b) asymptotic; (c) quadratic; (d) logarithmic; (e) hyperbolic. Data (seven-day harvesting intervals) from Owuor *et al.* (1997a).

are statistically significant, but each would suggest a different optimum level of N. For example:

1. Linear segments (Figure 12.2a): typically, yield at first increases proportionally (or nearly so) to the amount of nitrogen applied, until a turning point is reached where the response ceases or slowly decreases; this horizontal line is the second segment. There may be a third segment if there is clear evidence that yields decline in a linear fashion once a certain level of N has been exceeded. Recommendation: 200 kg N ha^{-1}.

2. A law of 'diminishing returns' function, whereby each additional increment of nitrogen produces a progressively smaller increase in crop yield (and profit) (Figure 12.2b). Recommendation for highest yield is about 500–600 kg N ha^{-1}.
3. A quadratic function (Figure 12.2c). This curve implies that yields decline once a threshold level of N has been exceeded. Recommendation: 400–500 kg N ha^{-1}.

The advantage of the linear segment model (Figure 12.2a) to the farmer and adviser is its simplicity. The optimal dressing remains at or near the turning point so long as the value of the crop exceeds the cost of applying the fertiliser. In addition, the results can be expressed meaningfully in terms of the rate of increase in crop yield, i.e. how many kilograms of extra crop result from each additional kilogram of N?

The data to be wary of are graphs with no actual data points, when only computer-generated points, derived from the function used to test the data, are shown. Such graphs give a very misleading impression of the level of precision in the advice being given!

An NPKS fertiliser experiment in Kenya provides an example of the problem. It was situated in a field of the high-yielding clone AHP-S15/10 planted in 1970 on Kaproret Estate in Kericho at an altitude of 1860 m (Owuor et al., 1997b). The soil was classified as a well-drained humic nitisol and had the properties of a clay loam near the surface but of (kaolinite) clay to great depths below. Six rates of NPKS 25:5:5:5, namely 100, 200, 300, 400, 500 and 600 kg N ha^{-1} per annum, were compared. The plots were split to allow three harvesting intervals to be compared, namely 7, 14 and 21 days. The target shoot to harvest was 'two leaves and a bud' (fine plucking). It was claimed that the response curves were quadratic ($p < 0.05$), with the annual N input corresponding to the maximum yield (recorded over 20 months only) cited as being 466, 487 and 491 kg N ha^{-1} for each of the three harvest intervals respectively. Even if a quadratic function was the best way of representing the data, the precision of these estimates of the N level giving maximum yields is completely unjustified and misleading. A better value would be a collective c. 480 kg N ha^{-1}, or somewhere between 450 and 500 kg N ha^{-1}, as it is possible that the differences, although statistically significant, were not real. However, it is likely that functions other than a quadratic curve would have fitted the data just as well. There was no significant yield response to N inputs above 200 kg ha^{-1}. This was also the most profitable N level under the production costs and revenue conditions in 1993.

Subsequently, Owuor et al. (2008) reported the results of the same experiment, this time over a period of 18 years. The differences in the response curves from year to year are clear to see. Quadratic curves were developed which indicated maximum yields at 400 kg N ha^{-1}. Wisely, the authors looked at the data again and came to the conclusion that, despite the quadratic curve fitting the data at a statistically significant probability level of $p < 0.05$,[2] there was in fact no yield benefit at application rates above 300 kg N ha^{-1}. They then adjusted the figure down again by stating (without formal justification) that the most economical application rate was 250 kg N ha^{-1}. This is 50 kg more than they recommended in 1993. Looked at over one four-year pruning cycle (1986–1989), the yield response to 1 kg N declined from 20 kg made tea ha^{-1} at 0–100 kg N ha^{-1}, through 14 kg ha^{-1} at 100–250 kg N ha^{-1}, to 11 kg ha^{-1} at 250–300 kg N ha^{-1}, 7 kg ha^{-1} at

300–450 kg N ha^{-1}, and 5 kg ha^{-1} at 450–600 kg N ha^{-1}. The average annual yields of made tea over this four-year period in the 'best' treatments were in excess of 6000 kg ha^{-1}, and in the best years (e.g. 1987) yields from this outstanding clone (AHP-S15/10) exceeded 9000 kg ha^{-1} in the high N plots. The nitrogen was applied both as a compound NPK 20:10:10 fertiliser and as NPKS 25:5:5:5 as a single annual application.[3] There were no differences in yield response between the two formulations. When the fertiliser applications were split, there were no additional yield benefits. Later in the paper it is stated that during six years of one of the experiments (comparing split applications) there were no significant yield benefits at N rates above 200 kg ha^{-1}. So no clear answers, and for managers difficult decisions indeed!

These results apply to a very high-yielding clone, with yields reaching 8000–9000 kg ha^{-1} in peak years. This corresponds to an annual N loss of 320–360 kg N ha^{-1}, which for several of the low-input treatments is greater than the N input from fertiliser. This is especially the case for the zero-input control, which consistently yielded about 3000 kg ha^{-1} over the 18 years without any fertiliser. This is substantially more than the yields obtained from seedling tea, and more than the average yields obtained by smallholders. In other words, the N reserves in the soil were being mined. It is worth remembering that additional P and K were being applied; this was really a series of NPK/ NPKS experiments, not N alone.

In both these examples the benefits to the farmer or fertiliser distributor were indirect. For example, NPKS 25:5:5:5 is easier to handle than NPK 20:10:10 (needs fewer bags per hectare), minimises storage, and improves cash flow. The uniformity of application is likely to be better if the dose is in two applications rather than one.

The problem of specifying a level of nitrogen fertiliser for general application across or within a country has been highlighted by a detailed study undertaken by Owuor et al. (2010a) in Kenya. The fertiliser trial described above (Owuor et al., 2008) was replicated in five major tea-growing regions of the country,[4] with the same clone in each case (BBK-35). NPKS fertiliser was applied annually at rates of 0, 75, 150, 225 and 300 kg N ha^{-1} for 10 years from 1998 to 2007 and yields of green leaf recorded every seven days. In the last year, 1 kg of green leaf from each plot was processed by a miniature CTC method (TeaCraft), and unsorted black teas were subjected to plain tea quality evaluation, both chemical and sensory.

The yield response to fertiliser varied from site to site, and from year to year. The incremental response to additional N also varied between sites. For example, from 0 to 50 kg N ha^{-1} the annual increase in yield of made tea ranged between 2.0 kg tea kg^{-1} N at one site and 14.2 kg tea kg^{-1} N at another. Similarly, from 225 to 250 kg N ha^{-1} the corresponding yield responses were close to zero at the first site and as high as 5.1 kg tea kg^{-1} N at the other. There was no obvious explanation such as altitude (range 1800–2620 m) for these differences in yields and yield responses (Figure 12.3).

These figures tell only half the story. There is a difference between maximum yield, maximum return and maximum profit. Other variables that influence the decision on how much nitrogen to apply include the cost of fertiliser, including its application, and the value of the tea. Remember that any change in the value of the processed tea (up or down) applies to the total crop and not just to the yield increase. These costs and returns

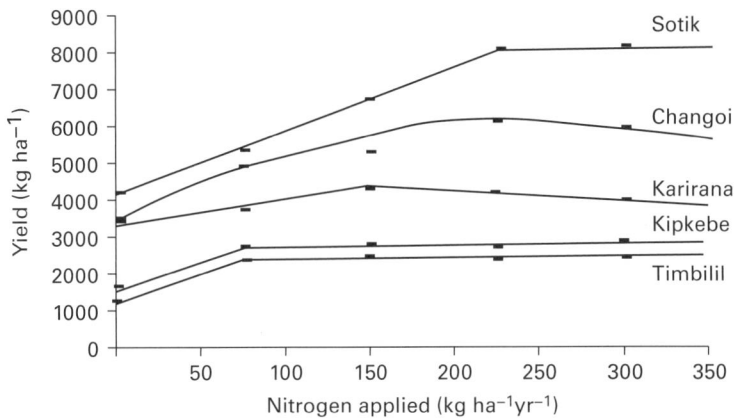

Figure 12.3 Contrasting yield (dried tea) response to nitrogen at five sites in Kenya: clone BBK-35, year 2004. Data from Owuor *et al.* (2010a).

are not constant, but changing day by day, whereas the decision on how much nitrogen to apply has to be made up to 18 months in advance of its application. So not only is the advice usually based on the results of one experiment, but the cost of fertiliser, and the future price of tea (green leaf and/or made tea) have to be predicted. No wonder it is not possible to provide site-specific advice that will maximise profits. The chances of the advice to a grower in one specific site being 'correct' are low, and if they happen to be correct one year (and we never will know the answer to the question) there is no guarantee that the same advice will hold true for the next year. A blanket recommendation for a country, a region or even neighbouring estates is likely to be the wrong recommendation for most people, in most years, at most places. Why then do we persist in this delusion of precision? Answer: the paymasters expect it, often not realising there is no simple single answer.

And indeed there isn't! Since this chapter was first drafted, Msomba *et al.* (2014) have reported the results of a multi-location nitrogen experiment repeated at eight sites in eastern Africa, namely Kenya (three sites), Rwanda (two sites) and Tanzania (three sites). There were five rates of N compared (0, 75, 150, 225 and 300 kg N ha^{-1}) all applied as NPKS 25:5:5:5. The tea was all clone TRIEA-6/8, the most popular clone in the region. Yields differed considerably between sites, averaged over the two years and the five N treatments, from 1900 kg made tea ha^{-1} in Mulindi (Rwanda) to 6200 kg ha^{-1} in Sotik (Kenya). This was despite every attempt being made to standardise tea management practices, including the stage in the pruning cycle and the harvest intervals. The target shoot to harvest was two leaves and a bud (fine plucking) at all the sites. Three different plucking rounds were compared (7, 14 and 21 days) within the same experiment. Yield responses to N varied between sites. The conclusion was that the 'most suitable application rate was between 150 and 225 kg N ha^{-1}', but that 'blanket recommendations for the region were not appropriate' and that 'location-specific advice was needed'.

Returning to the previous experiment (Owuor *et al.*, 2010a), all the plain tea quality parameters (thearubigins, theaflavins, total colour and brightness, as well as the sensory

Table 12.1 Classification of soil organic matter status, as used in Uttarakhand, northern India

	Organic matter content (%)		
Classification	Low altitude (500–1500 m)	Medium altitude (1500–2000 m)	High altitude (2000–2500 m)
Low	< 2.6	< 5.2	< 7.8
Medium	2.6–7.8	5.2–10.4	7.8–13.0
High	> 7.8	> 10.4	> 13.0

Table 12.2 Table for estimating annual application of nitrogen depending on target yield and the organic matter status of the soil – as recommended for use in the northern Indian state of Uttarakhand. Note: the response to N is assumed to be linear across the whole range of N application levels.

	Soil organic matter status		
	Low	Medium	High
Yield of made tea (kg ha^{-1})	Nitrogen application (kg ha^{-1})		
Minimum	160	120	100
2000	240	200	180
2500	265	225	200
3000	290	250	220
3500	315	275	240
4000	340	300	260
4500	365	325	280
5000	390	350	300
5500	415	375	320
6000	440	400	340
6500	465	425	360
7000	490	450	380

evaluations by two professional tasters) declined with increases in the amount of nitrogen applied (the relationship was linear in the case of theaflavins) and, with the exception of thearubigins, they also varied with the geographical location of the experiment.

The N response of a crop is influenced not only by the amount of applied N but also by the organic matter content of the soil. This is taken into account in the hilly state of Uttarakhand in northern India (28–31°N). The soil organic matter status is simply estimated as low, medium or high using the data shown in Table 12.1. This estimate is then carried forward into a table that lists the rates of nitrogen to apply depending on the expected yield (Table 12.2). Note that this assumes a linear response to N across the whole range of N applications, from 100 up to 490 kg ha^{-1}, depending on the organic status of the soil. These recommendations, summarised by Verma and Palani (1997), are based on the results of research undertaken at the UPASI Tea Research Foundation in South India.

Kamau *et al.* (2008) investigated the effects of the age of the tea plant on its response to nitrogen in Kericho, Kenya. They imposed five N rates (from 0 up to 400 kg N ha^{-1}) on tea fields 14, 29, 43 and 76 years after planting. The two youngest fields were clones and the other two were derived from seedlings (age and genotype were therefore confounded). Plant densities varied. The experiments ran for three years from November 2002, although yields were recorded for two years only. In a good growing year the clones out-yielded the seedling tea, but not in a poor year. The yields from the clones increased with the level of nitrogen applied, whereas the older seedling tea failed to respond to N. Within the clonal tea, made tea yield and N uptake (based on an analysis of the whole plant) were closely correlated. The effect of genotype on productivity was greater than the effect of age. Age on its own did not affect the yielding ability within the same genotype. The 'apparent recovery' of N was greater in the clonal tea than in the seedling plants. Nitrogen management strategies should therefore be based on the yield potential of the tea bushes in the target environment as defined by the genotype and the age of the plantation.

As a follow-up to this study Kamau (2011) developed a simple decision-support model for managing ageing plantations of tea. The model, known by its acronym, MAP-Tea, calculates the net economic returns over time (up to 100 years) for (a) low-yielding seedling tea and (b) high-yielding clonal tea. Four phases of development were identified with the associated costs of inputs and tea prices. In the simulations it was assumed that seedling genotypes remain productive for 35 years, after which the degradation phase begins.

The cost of nitrogen continually increases, while the price paid for processed black tea is always fluctuating, as is the premium paid for quality. Farmers and managers have to monitor and to justify their expenditure on fertiliser, particularly nitrogen. When the financial situation is really uncertain, it is sometimes necessary to take extreme action, namely not to apply any fertiliser at all for one or more years. The question that then needs to be answered is how much yield will be lost, in year 1, year 2 etc., or are there sufficient reserves in the soil to maintain production? Or, indeed, is it false economy to seek to save money in this way? The answer will be different depending on whether you are only selling green leaf to a factory or whether you are a processor selling made tea, as well as a grower.

Few experiments to answer this question appear to have been reported, although some large tea companies have been forced to take action similar to that described above, while smallholders with limited cash are unable to purchase fertiliser every year. As Gokhale (1955) stated, 'if the rate of manuring is changed in and from a given year then the manner in which the yield changes in the years immediately following is of great practical interest and deserves investigation.' Gokhale (1960) subsequently developed a generic relationship, based on data from field experiments conducted in North-East India, to estimate the probable change in yield with time when the level of manuring (organic or inorganic) is either increased or decreased from a given year. It would be of interest to know if the same curve-fitting approach applied to the much higher yields obtained now. Verma and Pund (2014) also referred to an old experiment (1940–1959) in which NPK fertiliser application ceased after 14 years. It took only three years for yields to decline back to the threshold level. But again, the yield levels were very low (300–700 kg ha^{-1}), and one questions whether this conclusion would be valid today.

Summary: Responses to Nitrogen

If you are looking for an easy answer to the question of how much nitrogen to apply, you will not find one here. When listening to an adviser:

- Beware computer-generated graphs that do not show genuine data points, as they can be very misleading.
- Yield/nitrogen response curves can be based on different assumptions: on the law of diminishing returns (asymptotic); on yields declining once a threshold has been exceeded (quadratic); or on linear segments. Check to see whether another type of response curve would fit the data equally well. Ask awkward questions.
- The shape and position of the response curves vary from site to site and from year to year, making it difficult to plan ahead.
- The 'optimum' amount of nitrogen to apply will vary depending on the costs of application and the value of the tea (green leaf or made tea).
- Tea quality declines with the amount of nitrogen applied. It also varies with the geographical location.

Potassium: Hidden Hunger

Planters' saying:

If you look down through the foliage of a bush and can see your boots then you have a potash problem!

Plants take up potassium mainly from the soil solution. To a much lesser extent potassium uptake can occur by direct contact exchange between root hairs and mycorrhiza with soil particles, but tea roots do not have root hairs (or do they? see Figure 8.3). Although potassium is the second most important element in the nutrition of the tea crop, surprisingly little is known about tea yield responses to potassium and potassium fertiliser requirements. This is despite numerous field experiments in many diverse locations, and attempts to develop diagnostic techniques based on soil and foliar analysis (Ranganathan and Natesan, 1985).

In South India, the important role of potassium in tea culture has been recognised since the 1940s. In this region's kaolinite clay soils, there is a limited number of fixation sites for potassium ions (i.e. the soils have a low cation exchange capacity). As a result, in the high-rainfall areas of South India, potassium fertiliser is leached down through the root zone, especially in acid soils. Verma and Pund (2014) have reviewed in great detail 70 years of research on potassium in tea in South India.

Potassium is very mobile in the plant and circulates freely, particularly towards younger, meristematic tissues. When potassium is deficient, it moves from older leaves

---

Box 12.3 Potassium Deficiency Symptoms in Mature Tea

- New leaf production ceases at the perimeter of the bush; normal growth continues in the centre of the bush, making it columnar in appearance.
- Leaves turn darker green.
- Pronounced necrosis is seen on the margins of mature leaves, which continues until the leaf dies and falls off, leaving a crown of young leaves at the top of the stem.
- New leaves become progressively smaller.
- Freshly fallen leaves accumulate under the bush.
- Development of a 'thin and twiggy frame'.

---

to younger tissues, which is why, in part, attempts to base fertiliser recommendations on leaf analysis have not been successful.

Until the 1960s, the tea crop's requirement for potassium was largely met from soil reserves. These were adequate to sustain the low yields at the time. Visual symptoms of potassium deficiency only appear in mature tea when the requirement is large. The crop passes through a stage known as 'hidden hunger' (Verma and Pund, 2014) (Box 12.3). Similarly, foliar analysis is only useful as a diagnostic tool when remedial applications of potassium are needed, not to help guide routine applications (see Chapter 14). As a result of this uncertainty, annual applications of potassium are generally based on the quantities needed to replace that lost by removal of the harvested crop, together with some estimate of the potassium status of the soil. Special consideration is given to applications in the year of pruning, when it is important to ensure that there is sufficient potassium to allow rapid and even recovery of the foliage. Tea responds quickly, when there is a need, to applications of potassium fertiliser (Willson, 1969).

As yields have increased, so has the amount of potassium removed with the crop, and the capacity of the soil to mobilise and supply enough potassium to meet the needs of the crop has become a yield-limiting factor. In round figures, the young shoots contain about 2.0% potassium (dry mass). For a tea crop yielding 5000 kg ha^{-1} dry tea, this represents a loss (in other words, export to the consuming country) of 100 kg K from each hectare of tea.

In a long-term nutrition experiment in Mufindi, Tanzania, a soil analysis seven years after treatments (rates of NPK fertiliser × irrigation) had first been imposed indicated a potassium content in the 0–150 mm soil layer of 0.53 (for the low level of NPK) to 0.57 mEq 100 g^{-1} (for the high level of NPK; Burgess and Sanga, 1993). Both these values are well above the critical level signifying a deficiency of 0.4 mEq 100g^{-1} quoted by Landon (1984). Leaf analysis (based on the 'mother' leaf) produced values ranging from 1.75% (low NPK) to 2.19% (high NPK; Burgess, 1992). Both values exceed the critical level of 1.5% as specified in Kenya by Owuor and Wanyoko (1983). The potassium content of the leaf decreased (significantly) with increased nitrogen application. Similar results in Kenya have been associated with competition between ammonium and potassium ions for cation exchange sites in the soil. There is a sensitive balance between nitrogen and available potassium in the soil, especially in soils with a low cation exchange capacity, such as those at the Ngwazi Tea Research Station in Tanzania.

For example, in Kenya, increasing the annual nitrogen application rates (as urea) from 0 to 200 kg N ha^{-1} reduced the potassium nutrient concentration in mature leaves from 2.6% to 2.1%. Similarly, increasing the potassium level (as potassium chloride, KCl, also known as muriate of potash) from zero to 80 kg K$_2$O ha^{-1} reduced the nitrogen concentration in maintenance leaves from 2.95 to 2.54% (Sitienei et al., 2013). Nitrogen ions are highly mobile, and N tends to be concentrated in the younger leaves.

Summary: Responses to Potash

Although potassium is the second most important element in the nutrition of the tea crop, surprisingly little is known about tea yield responses to potassium and the fertiliser requirements.

Phosphorus

Although phosphorus is an essential element in growth processes, even less is known about the yield response of tea to phosphate-containing fertilisers than is known for potassium. The amount of phosphorus removed in the harvested crop is relatively small. A dry tea shoot contains 0.20–0.25% phosphorus, so that for a crop yielding 5000 kg ha^{-1} of made tea between 10 and 12.5 kg of phosphorus are removed in the dry leaves in a year. This comes from soil reserves, replaced by applications of compound fertilisers such as NPKS 25:5:5:5. This application of P is a safety-first approach, with little experimental evidence to justify this as a management practice (although consistent responses to P have been obtained in Zimbabwe; Grice, 1990c). On the other hand, there is much evidence to support the widespread practice of mixing phosphate (60 g single superphosphate; 30 g triple superphosphate) with the soil from the planting hole to aid crop establishment (Grice, 1990b).

The availability of phosphorus in the soil is very much influenced by the pH of the soil. Acid soils tend to fix phosphorus, making it unavailable to the tea plant, except in the surface mulch zone where the pH is marginally higher than in the mineral soil. Leaving the soil undisturbed allows the roots to proliferate in the soil–mulch interface and extract phosphorus. This explanation is still a working hypothesis yet to be completely proven by experiment (Willson et al., 1975).

Symptoms of P deficiency have been widely seen in Indonesia, but only when yields exceed 2000 kg ha^{-1}. Although andosols (soils derived from volcanic ash) have been shown to have the capacity to fix phosphorus, to a greater extent than latosols or podzols, experiments designed to test the effects of rate and frequency of P application on all three of these soils have failed to show any responses in terms of yield. In andosols the ionic form of phosphorus readily reacts with oxides and hydroxides of aluminium (abundant in acid soils) to form insoluble compounds (Othieno, 1992). In latosols and podzols iron is the principal metallic element responsible for the fixation of P.

Phosphorus deficiency symptoms show as dull bronzing of recently mature leaves and stems becoming thinner, with internodes reducing in length. Wanyoko *et al.* (1996) suggested that the baseline soil value for a deficiency of available phosphorus to become apparent is 14 mg P kg^{-1} dry soil, while Owuor and Wanyoko (1983) cite 0.17 mg P kg^{-1} as being the critical (mature) leaf concentration associated with a deficiency. There is still a great deal more that we need to know about phosphorus nutrition of tea grown on acid soils. In the meantime, or until phosphate becomes too scarce (and expensive), expect the application of P in the form of NPKS or NPK compound fertilisers to continue on a 'just in case' basis (Cordell *et al.*, 2009). For a crop yielding 4000–5000 kg ha^{-1} made tea, an annual application of 20–25 kg P ha^{-1} will replace that being removed in the crop.

Summary: Responses to Phosphate

Until phosphate becomes too scarce (and expensive), the application of P in the form of NPKS or NPK compound fertilisers is likely to continue to be on a 'just in case' basis.

Sulphur

Sulphur deficiency symptoms (commonly known as tea yellows) were first seen, and the cause identified, in Malawi in the 1930s. The symptoms develop in the following sequence (Storey and Leach, 1933):

- Yellow mottling of the leaf, with veins remaining green in younger leaves (net-veining).
- Leaves turning yellow, becoming brittle and reduced in size.
- Scorching on leaf tips and margins.
- Leaves shedding, leaving small leaves at tip of shoot.
- Shortening of the internodes; axillary buds growing into small shoots with yellow leaves.
- Bush becoming totally defoliated, with progressive die-back of shoots leading to the death of the bush.

Reports of sulphur deficiency are still relatively few, but have come from the following countries: Malawi, Tanzania, Kenya and Uganda in Africa, and India, Bangladesh and Indonesia in Asia. The symptoms have largely been offset in the past by the sulphur contained in sulphate of ammonia (primarily a source of nitrogen), and single super-phosphate (a source of phosphorus) fertilisers.

In contrast, where sulphate of ammonia has been replaced by urea as a source of nitrogen, some tea soils in North-East India have since become depleted of sulphur. The current recommendation (Anon., 2012) is to apply 20 kg S ha^{-1} when the soil analysis indicates that the available S content is below the critical limit (40 ppm). Ghosh *et al.* (2006), in an analysis of the results of tests on more than 5000 soil samples taken in the tea-growing areas of Dooars (West Bengal), found that the S contents of more than half the soils were less than 40 ppm. There was a positive correlation between organic carbon status and available sulphur, but (surprisingly) a negative correlation with pH. See Rao and Sharma (2001) for a detailed review of this topic.

Calcium

In terms of the amount removed from the soil, calcium is the third most important element after nitrogen and potassium. On average the harvestable crop contains between 0.5% and 1.0% calcium, which for a crop yielding 5000 kg ha^{-1} means an annual loss of 5–10 kg Ca ha^{-1}. Even in acid soils, deficiency symptoms are rare.

Magnesium

Magnesium is an essential component of the chlorophyll molecule and also plays an important role in plant metabolism. When magnesium is deficient, leaf chlorosis is visible, the older leaves developing a pale area along the margin of the leaf. Yield responses to the application of magnesium (as magnesium sulphate) are rare.

Micronutrients

There are many elements that are considered to be important in tea production, although only a few produce deficiency symptoms when they are in short supply. They include so-called micronutrients, for which the quantities required are very small, such as zinc, copper, iron, boron, manganese and molybdenum. Their functions are well described by Bonheure and Willson (1992) and by Hajra (2001). Only zinc and copper are briefly considered here.

Zinc

Zinc is an important micronutrient (or trace element) in tea. It was in the 1960s that its importance was recognised in Sri Lanka by Tolhurst (1961), and soon afterwards in South India and afterwards in East Africa (Box 12.4) and North-East India. There are distinctive deficiency symptoms, of which perhaps the most obvious are small sickle-shaped leaves, together with shortened internodes, so that shoots developing from axillary buds appear in the form of a rosette.

The deficiency is corrected by foliar applications of zinc oxide (3.0–4.5 kg ha^{-1} as a 1% foliar spray) with two or three split applications annually, or zinc sulphate 1–2% (w/v), with no more than 12.5 kg ha^{-1} in 4–6 split applications. Soil applications are ineffective. Yield increases of 10–20% have been reported (but it is not clear on what

---

> **Box 12.4** Zinc – a Surprise Discovery
>
> Tolhurst (1969) noticed that tea growing under bamboo laths resting on wire in an artificial shade experiment in Kericho, Kenya (alt. 2200 m), did not show symptoms of zinc deficiency, whereas in unshaded tea the symptoms were marked. Pandal-shaded tea also out-yielded tea that was unshaded. Measurements showed that an appreciable amount of zinc dissolved by rain and dew from the galvanised wire netting was dripping onto the foliage below. The crop was responding to zinc, not to the pandal shade. By chance (and keen observation by Tolhurst), an experiment designed to answer questions about shade had instead provided an answer to a completely different problem.

base yield level these occurred). It is assumed that removing zinc as a limiting factor allows the crop to respond to other inputs. (Based on a review by Mouli *et al.*, 2012, and Anon., 2012.)

Copper

Copper is an essential component of the group of enzymes (oxidases) that control the oxidation process during the manufacture of black tea. If copper is deficient oxidation occurs very slowly, and even when the time specified for oxidation is increased, green infusions and light liquors are still produced. A foliar spray of copper sulphate (0.5–1.0%) is recommended when the leaf copper content is 5 ppm or less.

NPK: Recommended Commercial Application Rates

Whatever the uncertainties associated with the interpretation of the results of fertiliser experiments, and their variability from place to place and year to year, someone has to take a decision based on the best available evidence (Box 12.5). This is normally the job of the tea research institutes, and the advice being given to their members is summarised below.

North-East India

In Assam, the recommended annual NPK application rates for mature tea (Anon., 2012) are based on the current yields (in bands of 500 kg made tea ha^{-1}, averaged over a pruning cycle). The recommendations for low- and high-yielding fields are given below as examples:

- Low-yielding: for yields not exceeding 1500 kg ha^{-1}, up to 90 kg N, 20 kg P_2O_5 and 90 kg K_2O ha^{-1} (depending on the K status of the soil) are recommended.

Box 12.5 Commentary: Intelligent Guesswork

The most profitable level of fertiliser is not easy to determine, as has been indicated. It gets worse. There is usually a drive for more crop, to 'spread overheads', i.e. fixed costs. The value of an extra kilogram of tea is therefore its sale value less the variable costs of producing it. Establishing exactly what these costs are is by no means straightforward. There are the obvious ones – the extra fertiliser needed to produce it, harvesting, manufacturing variables, packing and transport. But some fixed costs contain variable elements. Take depreciation. Machinery may be depreciated at standard rates, often determined by tax legislation. But if a piece of machinery has to process twice as much tea it will wear out, and need to be replaced, in half the time. So if you express the life of a machine in tonnes of tea rather than in years, depreciation becomes a variable. And so on. Nasty. One reputable research establishment once said that as long as the sale value of the additional crop exceeded the cost of the additional fertiliser, it was profitable to apply it. We now know that this may be far from the case.

In 1973 I inherited a fertiliser regime in which levels of nitrogen were determined by estimates of future crop. This meant that high-yielding estates got high levels of nitrogen, and low ones, low. At that time the tea was recovering from long years of starvation, so it would have been reasonable to assume that low-yielding estates were more starved than high-yielding ones. There was therefore a respectable case for reversing that policy and giving extra nitrogen to the low-yielding areas. For political reasons (my predecessor was now my boss), the best I could do was apply a blanket dose to all estates. It worked.

I am not a great believer in symptoms of potash deficiency. I invariably found that the condition in which my boots were visible under the bush could be cured by the application of more nitrogen. This is not to minimise the importance of potash. In Mufindi we introduced irrigation estate by estate over a number of years. This greatly increased yields and therefore the offtake of potash, which, as a cost-saving measure, we were not applying. (After all, there were no visible deficiency symptoms.) After a number of years, a nasty variety of *Phytophthora* began to appear, estate by estate, year by year, in exactly the same order in which irrigation was introduced . . . Hidden hunger?

T. C. E. Congdon

- High-yielding: for yields between 2500 and 3000 kg ha^{-1}, these values are increased to 140–165 kg N, 50 kg P_2O_5 and up to 165 kg K_2O ha^{-1}.

And in Darjeeling, where it is cooler:

- Low-yielding: for yields not exceeding 600 kg ha^{-1}, up to 60 kg N, 20 kg P_2O_5 and up to 60 kg K_2O ha^{-1}.
- High-yielding: for yields between 1000 and 1400 kg ha^{-1}, 90–120 kg N, 20 kg P_2O_5 and up to 120 kg K_2O ha^{-1}.

These recommendations are based on the results of a number of trials conducted at Tocklai and on commercial tea estates in North-East India, supported by survey data from a large number of tea estates (Anon., 2012). The general rule is to apply N and K fertilisers in two equal splits if the application rate exceeds 100 kg ha^{-1}, with the first application in the spring, during the early rains when the soil is moist, and the second in August/September.

Kenya

The 1986 edition of the *Tea Growers' Handbook* (TRFK, 1986) states that about 230 kg N ha^{-1} should be applied annually to tea plants three and more years from planting. Subsequently, this should be increased to 300 kg N ha^{-1} for 'high-yielding' clonal tea. Single applications are recommended in areas with only one rainy season, and two applications where the rainfall is bimodal, if only to improve the overall uniformity of distribution. A well-known tea company undertakes its own research on selected estates representing low and high altitudes. Whole-field plots are used for all treatments except those with low nitrogen applications. The optimum economic application is believed to be between 150 and 200 kg N ha^{-1}, but extra N (up to 275 kg N ha^{-1}) is applied to fields that are expected to yield in excess of 2500 kg ha^{-1}. The N:P:K compound fertiliser recommended for use in Kenya is formulated in the following ratios: NPK 25:5:5 or 25:5:10 or NPKS 25:5:5:5.

Tanzania

The advice given to producers in Tanzania on fertiliser application is based on the results of research at the Ngwazi Tea Research Station since 1986, mainly with irrigated tea. These recommendations are summarised in Box 12.6.

The advice given to prevent potassium deficiency in Tanzania was:

- At the highest N level (450 kg N ha^{-1}), with the tea well irrigated and yielding *c*. 6000 kg ha^{-1}, an annual uniform application of 225 kg K$_2$O ha^{-1}, split 150 kg K$_2$O ha^{-1} at the start of the rains followed by 75 kg K$_2$O ha^{-1} at the start of the warm dry season.
- At a low N input, with the tea partially droughted and yielding 3000 kg ha^{-1}, an annual application of 150 kg K$_2$O ha^{-1} at the start of the rains.

But advice such as this always needs to be tempered by the economic context and practical experience (Box 12.7).

South India

Detailed monitoring of the nutrient contents of tea plants in South India by Venkatesan (2006) at a number of sites showed that, for a selection of seedlings and clones, on average, a consistent 60% of the nutrient applied (e.g. N) was taken up by the tea plants, yielding the equivalent of 2000 kg ha^{-1} of made tea. The absolute quantity of nitrogen

Box 12.6 Fertiliser Applications to Tea in Southern Tanzania

In southern Tanzania, following the results of experiments at the Ngwazi Tea Research Station, the 'normal' practice is to apply annually 250–300 kg N ha^{-1}, in the form NPK (acronym as traditionally used) 25:5:5 or 25:5:10 or as separate elements, possibly with additional sulphur (S) depending on cost. In this location with a single rainy season, the applications are split, with half applied as urea (46% N) in July/August to the irrigated tea (corresponding to the start of spring), perhaps dissolved in the irrigation water, except in the year of prune. A second application of urea is then made (spread by hand) after the start of the rains. This should be completed by the end of December. In the year of prune about 160–180 kg N ha^{-1} can be applied as sulphate of ammonia (21% N, 24% S) in the December following pruning.

For a crop yielding 6000–8000 kg made tea ha^{-1} (the target yields for new plantings) about 120–160 kg K ha^{-1} and 30–40 kg P ha^{-1} are removed annually in the crop. Potassium can be applied as muriate of potash (KCl, at 60% K$_2$O or 50% K) and phosphorus as single superphosphate (20% P$_2$O$_5$ or 9% P). Measurements made at Ngwazi suggest that annual applications of 125 kg K ha^{-1} should be enough to maintain adequate soil potassium reserves for a crop yielding 5000 kg ha^{-1}. As the soils are known to be deficient in phosphorus, soil and leaf analysis on a systematic basis is advised to check if and where deficiencies exist, and to identify any remedial action needed.

In view of the low levels of sulphur in the soil, additional sulphur may be needed (as sulphate of potash and/or sulphate of ammonia). For a crop yielding 6000–8000 kg ha^{-1}, the annual crop removal of sulphur is about 120–160 kg ha^{-1}. Copper and zinc are also deficient.

Box 12.7 Commentary: Volatility

A few practical points for managers, often neglected:

- Before ordering fertilisers, obtain samples from suppliers. Have them analysed. When you receive the bulk fertiliser, take samples and have them analysed. Compare. I have done this and had some hair-raising results. Hire a good lawyer.
- Buying compound fertiliser should be cheaper than making up your own from 'straights'. But get quotes for straights and compare. At least if you mix your own you will get the proportions right, and mixing is relatively inexpensive to do.
- Fertiliser is expensive stuff, and can be applied to crops other than tea. In some circumstances it can therefore become 'volatile'. So you may not be putting on as much as you think – or have paid for. The more you split the annual dose, the more opportunities there are for volatility, and the more likely it is that supervision will become lax.

T. C. E. Congdon

removed in the crop in a year averaged about 90 kg ha^{-1}. This meant that it was necessary to apply an extra 160 kg N ha^{-1} to reach the total recommended annual nitrogen application (250 kg N ha^{-1}) in order to maintain a crop yielding at least 2.0 t ha^{-1} growing in a soil with 'low' or 'medium' organic matter content. Provided that the organic matter status of the soil was 'medium', this was equivalent to a return of 100 kg of processed tea produced for every 4–5 kg N applied. Venkatesan (2006) claims that this can result in an annual saving of 40 kg N ha^{-1}, although the argument is difficult to follow.

The recommended rates of nitrogen application in South India during the 1980s varied depending on the actual and expected yield (Ranganathan and Natesan, 1987). Low-yielding crops were given extra N (30–40 kg ha^{-1}) because a lack of N was one reason (an important limiting factor) why the crop was low-yielding, which may or may not have been true. For yields above 3000 kg ha^{-1}, extra N is applied (above the base rates of 290, 250 and 220 kg N ha^{-1} for crops classified as low-, medium- or high-yielding) at the rate of 10 kg for every 100 kg (1:10) of additional crop expected. It is like a reward paid in advance of delivery.

In South India, the recommended amount of potassium to apply is based on the quantity of nitrogen fertiliser to be applied. It is also dependent on the potassium status of the soil. When yields are below 3000 kg ha^{-1}, and sulphate of ammonia is the N source, the amount of potash to apply is calculated as a fixed ratio (N:K$_2$O) of the quantity of N applied. For soils with a low/medium N status this ratio is 10:4, and for soils with a high N status it is 10:5. For yields above 3000 kg ha^{-1}, the corresponding ratios are 10:8 for both levels of soil fertility.

For soils with a medium N and low K status (e.g. a kaolin), the recommendation to tea growers in South India is to apply N:K$_2$O in the ratio 2:1 when using sulphate of ammonia. For urea, the ratio is 1:1. For high-yielding (> 3000 kg ha^{-1}) tea fields the ratio is constant (1:1) regardless of the N source (Ranganathan and Natesan, 1987). It is not clear on what experimental evidence this level of precision in the advice is based.

Timing: because of the risk of leaching and surface run-off removing the nutrients, particularly potash and nitrogen, growers are advised to divide the annual NK application into four splits (for fields yielding below 3000 kg ha^{-1}), with the timing dependent on the rainfall distribution and amounts, or at least five applications for fields yielding more than 3000 kg ha^{-1}.

Subsequently, Verma and Palani (1997) revised these recommendations to take into account the year of prune and the height of pruning. If pruning is below 0.45 m, they recommend a N:K$_2$O ratio of 1:2; from 0.45 to 0.60 m a ratio of 2:3; and if more than 0.60 m a ratio of 1:1. A system of NK auditing was also introduced, based on the crop harvested in the preceding 2–3 months.

As phosphate (P$_2$O$_5$) is to a great extent immobile in the soil, it can be applied in two or more splits at intervals of one or two years. To minimise the risk of its removal in run-off, and to increase its availability, it is considered best to bury the P-fertiliser at a depth of 150–250 mm (Verma and Palani, 1997).

Sri Lanka

A different approach is recommended in Sri Lanka (Wickremasinghe and Krishnapillai, 1986). The quantity of N to apply is decided by considering the highest yield obtained in the corresponding year of the pruning cycle over the last three cycles. This is considered to represent the potential yield of that field (but I'm not sure why – it represents the highest yield achieved in recent years, which is not the same as the potential yield). For each 200 kg ha^{-1} yield increment between 800 and 2000 kg ha^{-1}, an extra 20 kg N ha^{-1} is applied (above a base level of 80 kg N ha^{-1} for yields below 800 kg ha^{-1}). From 2000 to 4000 kg ha^{-1} an extra 40 kg N ha^{-1} is applied for each 500 kg ha^{-1} yield increment. Thus 360 kg N ha^{-1} is applied if the yield target is 4000 kg ha^{-1}. Split applications (three) are recommended in all except the year of prune, when the application level is reduced. It is not evident on what basis these recommendations are made. Is this what is known as a pragmatic approach?

Summary: Crop Nutrition

There is a diversity of advice. In the view of Grice (1979), who had experience in Asia and Africa, the way to decide how much nitrogen to apply is to use:

intelligent guesswork based on sound experimental evidence

After numerous field and laboratory experiments, thousands of soil and leaf samples analysed, the development of computer models and years of experience, this is still probably the best way forward, but with the emphasis on the word *intelligent*. Practical considerations are important: for example see Box 12.7.

- Tea is grown on a wide range of soil types: alluvial, volcanic, sedimentary, organic . . . in origin.
- Tea soils have been formed, or they exist, under high-rainfall conditions, with intense leaching of the products of weathering.
- This has resulted in acid soils, a prerequisite for successful tea production.
- The optimum pH for tea is between 5.0 and 5.6.
- The predominant clay mineral is kaolinite; deep well-drained soils are preferred, but tea can grow well where the water table is maintained through drainage at or below 1.0 m.
- The first statistically designed fertiliser experiments with tea began in the 1930s in Sri Lanka.
- The value of these experiments was constrained by our inability to think beyond the limits of our experience.
- It is not possible to give specific recommendations on how much nitrogen to apply.
- Yield/nitrogen response curves can be interpreted in many different ways: on the basis of the law of diminishing returns (asymptotic), or yields declining once a threshold has been exceeded (quadratic), or linear segments.

- The shape and position of the response curves vary between sites and from year to year, making it difficult to plan ahead.
- The 'optimum' amount of nitrogen to apply will vary depending on the costs of application and the value of the tea (green leaf or made tea)
- Beware of graphs that only show computer-generated curves with no real data, suggesting unjustified precision.
- Check to see if another type of response curve fits the data.
- Tea quality declines with the amount of nitrogen applied, and quality varies with the geographical location.
- Little has been reported on how yields respond to reductions in nitrogen applications.
- Potassium is the second most important element in the nutrition of the tea crop, but surprisingly little is known about the yield responses. It is particularly important in the year of prune.
- Little is known about the phosphorus requirements of tea. Its availability is limited in acid soils, except in the surface mulch layer where the pH is slightly higher and tea roots can proliferate.
- Sulphur deficiency symptoms are commonly known as sulphur yellows. Following the replacement of sulphate of ammonia by urea as a nitrogen source, sulphur deficiency is now more prevalent than it was.
- Zinc and copper are important micronutrients. There are distinctive deficiency symptoms for zinc. Foliar sprays are effective.
- Lack of copper inhibits the oxidation process in the manufacture of black tea. A foliar spray can correct a deficiency.
- Each major tea-producing country has developed its own methodologies for advising growers on the selection, quantities and application of fertilisers.

The uncertainties associated with the interpretation of leaf and soil analyses as a guide to sustainable nutrient management of tea are considered in Chapter 13.

Notes

1. Ferrasols have a highly weathered subsurface horizon dominated by iron and aluminium hydrous oxides with some kaolinite-type clay minerals, and sometimes quartz, but with virtually no weatherable materials. The soils have a low cation exchange capacity with a low level of exchangeable bases; they are usually deep and well drained, reddish or yellow coloured with a stable granular structure, resistant to erosion, and although high in clay they behave like loam soils. They are acid, have low fertility and are deficient in phosphorus. Acrisols are less strongly leached than ferrasols, and lack some of the granular structure, but otherwise they are very similar.
2. Regretfully the quadratic equations are cited to a level of precision that cannot be justified, e.g. yield = $3478.1 + 13.717x - 0.0171x^2$; $r^2 = 0.8664$; yield maximum occurs at 401 kg N ha^{-1}.
3. The abbreviation NPK fertiliser is strictly incorrect. It should read as N P_2O_5 K_2O. It would be better if we always referred to the elements: phosphorus (P) and potassium (K) rather than to

their oxides (e.g. phosphate P_2O_5 and potash K_2O), as we do with nitrogen (N). The phosphorus (P) content of phosphate is about 40%, and the potassium (K) content of potash is about 80%.

4. Karirana (1°6′S 36°39′E, alt. 2260 m); Timbilil (0°22′S 35°21′E, alt. 2180 m); Changoi (0°29′S 35°14′E, alt. 1860 m); Sotik Highlands (0°35′S 35°5′E, alt. 1800 m); and Kipkebe (0°41′S 35°5′E, alt. 1800 m).

13 Pores for Thought

Gaseous Exchange

Wight (1938) described the structure and function of the leaves of a tea plant 80 years ago when working in Assam. In one square inch (645 mm^2) of the leaf surface he found, for example, that there were between 10,000 and 20,000 stomata, small pores which in tea are present only on the lower surface of the leaf (Figures 13.1 and 13.2). In fact, these numbers are a factor of 10 out (Box 13.1). There are 100,000–200,000 in a square inch (equivalent to 155–310 stomata mm^{-2}). It's a lot, anyway!

Wight emphasised the role of stomata in controlling the loss (egress) of water vapour from the leaf through the pores into the surrounding air (a process we know as transpiration), and the entry (ingress) of carbon dioxide (CO_2, a gas) into the leaf. On entering the leaf, the CO_2 dissolves in the water saturating the cell walls surrounding the intercellular spaces adjacent to the pore. The CO_2 then passes in solution from cell to cell until it reaches the palisade cells, where, in the presence of chlorophyll and under the influence of sunlight, the process of photosynthesis occurs. The CO_2 combines with the elements of water (hydrogen and oxygen) to form a sugar. For every unit volume of CO_2 taken in by the leaf, an equal volume of oxygen is released. This process, which as Wight (1932) stated allows plants to 'feed themselves', also serves to purify the atmosphere for animals, since we breathe in oxygen and expire carbon dioxide.

Although they may seem far removed from practical issues arising on a tea estate, fundamental studies of stomatal behaviour, photosynthesis, transpiration and plant water status are important for helping us to understand the mechanisms responsible for the observed responses of tea to its environment. For example, such studies help in explaining where shade trees are likely to be beneficial, in identifying drought-resistant clones, in understanding that responses to irrigation can be limited by dry air, and in explaining the complex responses of tea to shelter belts (Chapter 16).

Stomatal Behaviour

A range of techniques has been used to monitor the behaviour of tea stomata. These include the infiltration score technique, the pressure-drop porometer, the steady-state constant-flow porometer, the non-ventilated transient (or dynamic) porometer, and direct observation. Unfortunately, they have not always given results that are consistent or directly comparable. The Scholander pressure chamber is also a useful way of

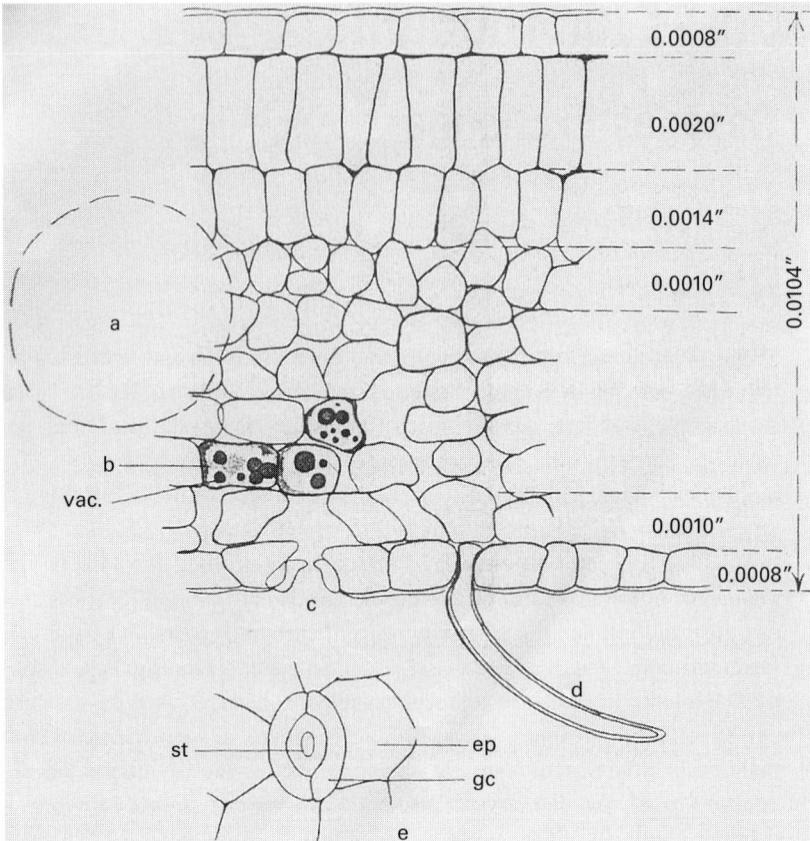

Figure 13.1 Cross-section of a tea leaf, showing a stomatal pore (c) on the lower surface (from Wight, 1938). ep, epidermis; gc, guard cell; st, stomatal pore; vac, vacuole.

measuring plant water status (or, strictly, the xylem water potential) in the field (Box 13.2, Figure 13.3).

In early work in Tanzania and Kenya, the infiltration technique, based on isopropyl alcohol, was used to study stomatal behaviour (i.e. to measure the degree of opening of the widest stomatal pores) in relation to environmental variables, and specifically to water stress. Preliminary measurements showed that the most consistent estimates were obtained from healthy, fully grown, yet still supple leaves that were fully exposed to sunlight (Carr, 1971a). In Tanzania, partial stomatal closure was observed during the middle of the day on nearly all occasions when diurnal assessments were made (Carr, 1968). Similar diurnal patterns of stomatal opening were observed during dry weather for a selection of clones in Kenya. Stomata were always wider open in the rains than during the dry season in both (previously) irrigated and unirrigated plants (Figure 13.4). Differences in the dryness of the air (saturation deficit) were believed to be largely responsible for both the diurnal and the seasonal differences in the degrees of stomatal opening observed (Carr, 1977a; Othieno, 1978b).

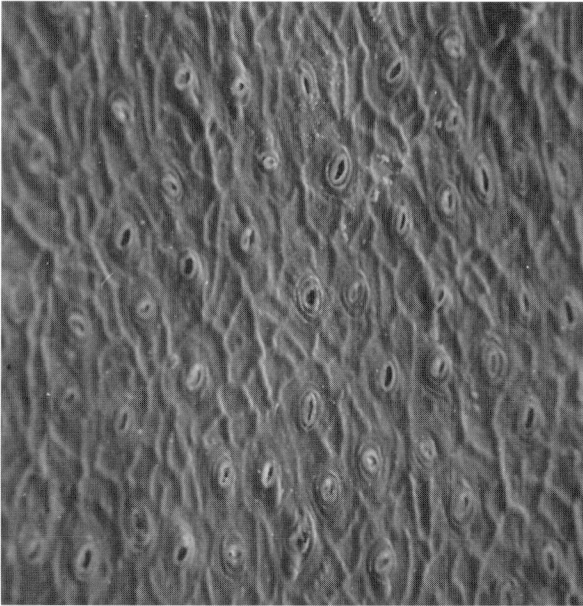

Figure 13.2 Stomata are present only on the lower surface of a tea leaf (density *c.* 200 mm^{-2}).

Box 13.1 Stomatal Densities on the Lower Surface of Tea Leaves

A number of scientists have calculated the density of stomata, which occur only on the lower (abaxial) surface of the leaf of a tea plant. Values have varied, from about 190 mm^{-2} (Fordham, 1971) to 130 mm^{-2} (Squire, 1976) to 150–200 mm^{-2} (Samson *et al.*, 2000; Olyslaegers *et al.*, 2002). Ng'etich and Wachira (2003), in a comparison of common cultivars grown in Kenya (all diploid), reported densities in the range 240–312 mm^{-2} with significant differences between individual clones (triploid and tetraploid cultivars had smaller densities). It is not known whether these differences in density are important.

In Tanzania, differential stomatal closure between well-irrigated plants (BBT-1) and unirrigated plants was observed when the potential soil water deficit (SWD) exceeded about 200–300 mm (Carr, 1968, 1969a). There was then a progressive decline in relative opening until the potential SWD reached about 500 mm, followed by all-day closure (Figure 13.5). BBT-1, a China-type cultivar, is unusual in many respects, being very drought-tolerant on an annual yield basis, but responsive to irrigation in the short term. Because of these unusual properties, it is worthy of study.

The slope of the linear relationship between the xylem water potential (Ψx) and the infiltration score is a possible indicator of drought tolerance. For example, compared with heterogeneous mature seedling tea, the stomata of BBT-1 are relatively insensitive to large changes in the shoot water potential (range –0.1 to –1.5 MPa). Thus, there was

---

Box 13.2 Measuring Water Stress in a Tea Plant Using a Pressure Chamber

The Scholander pressure chamber has proved to be a useful way of measuring the water status of tea plants in the field. Figure 13.3 shows the cut end of a tea shoot protruding from the pressure chamber. In one image the stem end is still dry; in the other, sap (with bubbles) has appeared at the stem surface. The pressure in the chamber at which sap is first visible corresponds to the tension of the water (negative pressure) in the xylem vessels, which is a measure of the plant water status (or stress level) when the stem was severed.

The technique was first used on tea by Carr (1971b) in Tanzania and subsequently by other scientists, mainly in Africa. It has been used to help to:

- identify the critical water potential (–0.7 to –0.8 MPa) beyond which the growth of shoots (and yield) is limited
- identify the corresponding potential *soil water deficit* (SWD) and *saturation deficit* (SD) of the air
- explain the seasonality of yield
- understand the effects of shelter from wind on crop water use
- determine the relationship to stomatal behaviour
- compare clones for drought tolerance
- assess the relative yield potential of clones
- evaluate the effects of methods of bringing tea into bearing on crop water status
- compare the responses of seedling tea bushes and their clones to water stress.

It deserves to be used more widely.

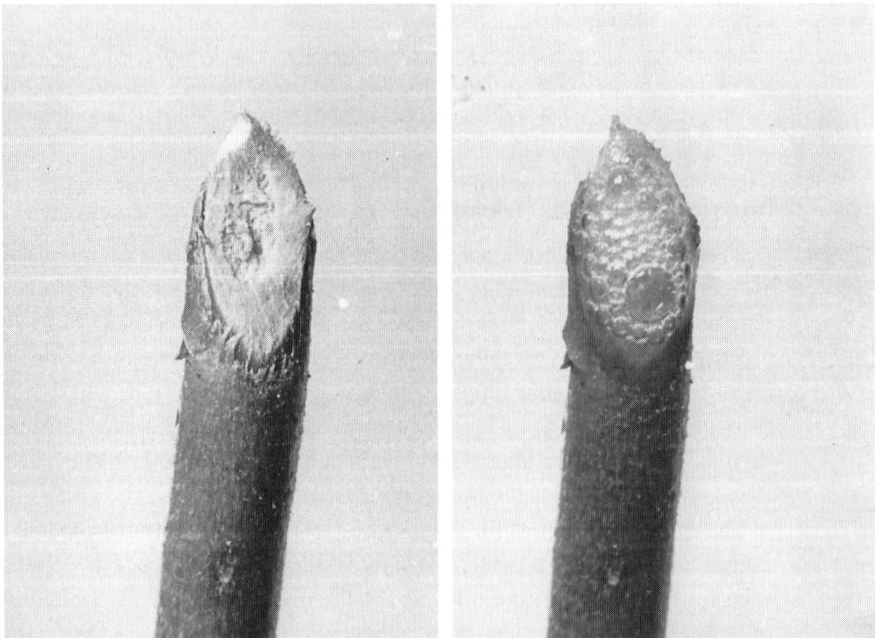

Figure 13.3 Cut end of a tea shoot protruding from a pressure chamber.

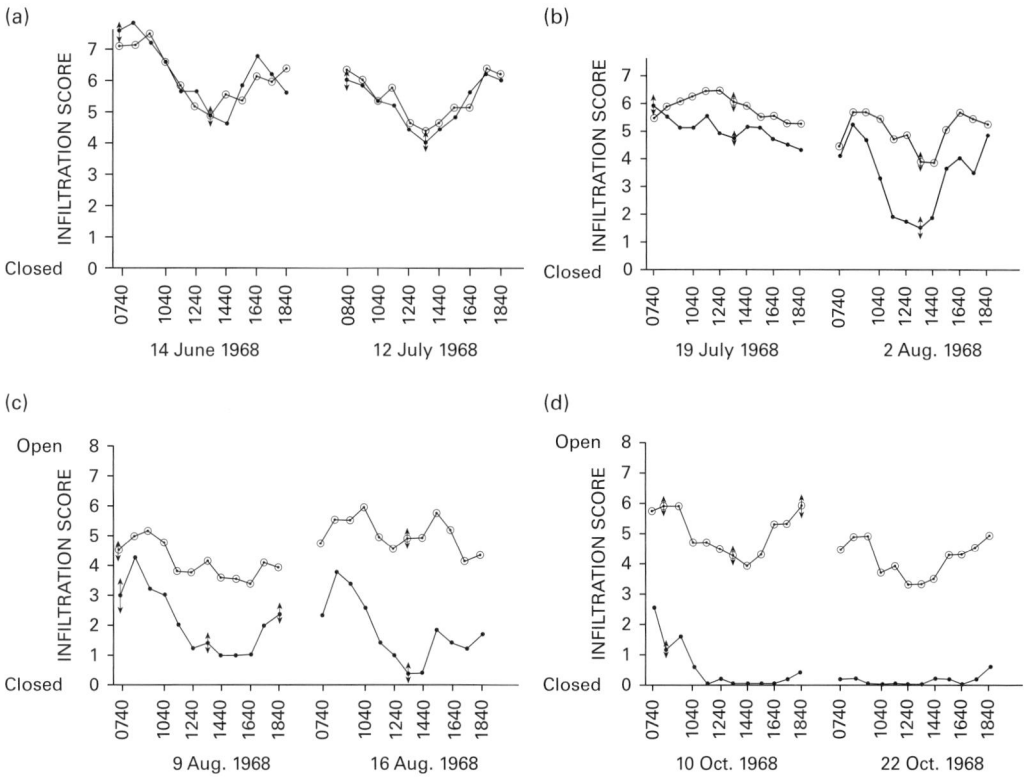

Figure 13.4 Examples of the diurnal changes in the degree of stomatal opening, based on the infiltration technique for well-watered field-grown (solid circles) tea (clone BBTI) and progressive soil drying (open circles) over a 14-week period from mid-June until october (the dry season). (a) Phase 1: typical diurnal pattern; partial midday closure (sunny day). (b) Phase 2: first sign of differential closure (cool, cloudy). (c) Phase 3: progressive closure from early morning, recovery late afternoon (getting warmer). (d) Phase 4: complete stomatal closure throughout the day, no sign of recovery in evening (hot).

an eightfold difference in the slopes of the regression line for seedling tea (−0.71) and BBT-1 (−0.09) (Carr, 1971a). A follow-up study during dry weather in Kenya confirmed that there were differences between clones in the slope of this relationship, with a seedling population being classified as drought-sensitive and clone TRIEA-6/127 as relatively drought-tolerant, but with none as extremely tolerant as BBT-1 (Carr, 1977b). As the soil dried (to a maximum potential SWD of 300 mm), the daily minimum shoot water potentials declined (to −2.0 MPa), more in some clones than others. There was some evidence of a genotype × environment interaction for both variables. The infiltration score and the xylem water potential were both negatively correlated with air temperature (range 13–28 °C) and saturation deficit (0.06–2.5 kPa).

In Malawi, Fordham (1971) used a pressure-drop porometer as well as the infiltration method to measure both diurnal and seasonal changes in stomatal opening of mature Assam-type seedling plants. Progressive closure of the stomata was observed from midday onwards in both irrigated and unirrigated tea, and there were marked seasonal changes

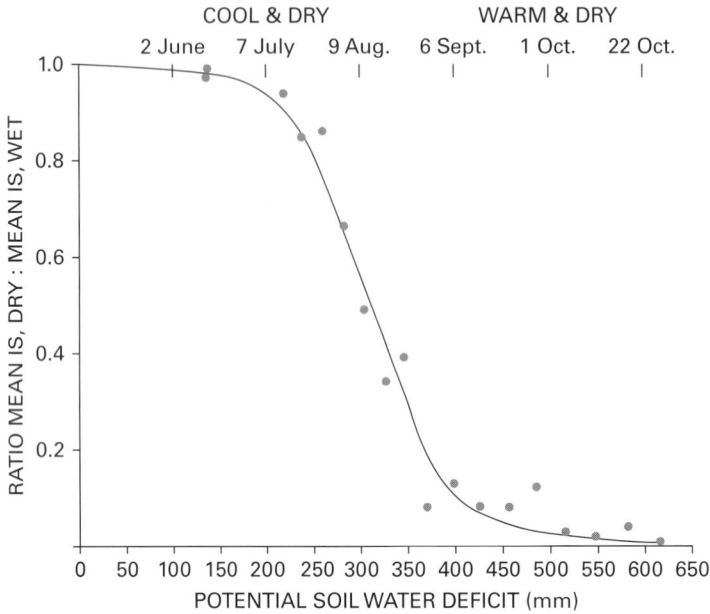

Figure 13.5 Relation between relative stomatal opening (ratio of infiltration score of unirrigated to irrigated tea) and the potential soil water deficit.

associated with the dry season. Also in Malawi, Squire (1976), using a diffusion porometer in the cool dry season, showed leaf conductances (measured on the second leaf of an actively growing shoot of several clones) increasing from low values in the morning, peaking during the middle of the day, before declining throughout the afternoon. Later, Squire (1977) used silicone rubber impressions of leaves to estimate stomatal opening in the wet season. Again, in contrast to the results obtained with infiltration liquids, the stomata were wider open at midday (3.5–4 μm) than at dawn or dusk (2 μm).

In contrast to these observations by Squire (1976) in Malawi, diurnal changes in stomatal conductances recorded in Sri Lanka (6°55'N, alt. 1382 m) were similar to those summarised above using the infiltration technique. In the early morning, conductances (clone TRISL-2025, measured with a portable infrared gas analyser on recently mature leaves) were large, decreasing towards midday and increasing again in the afternoon (Mohotti and Lawlor, 2002). The pattern was the same regardless of the degree of shade or nitrogen level. There were negative linear relations between conductance and leaf temperature (range 16–33 °C), saturation deficit of the air (range 0.5–3.8 kPa) and illuminance (range 500–2000 μmol m^{-2} s^{-1}).

Burgess (1992) attempted to make direct comparisons of diurnal patterns of stomatal behaviour using steady-state and transient porometers and the infiltration technique. The stomatal conductances of six irrigated clones were also compared with the corresponding infiltration score. Despite taking all the precautions possible (for example, in the calibration of the porometers), it was difficult to reconcile the results obtained. Relationships between conductance (transient porometer) and the infiltration score were linear but varied with the clone. As a result of this experience, Smith *et al.* (1993a) urged caution when using the results of porometry to identify drought-tolerant clones, but there are also doubts about the reliability of the infiltration technique.

In an interesting short-term pot experiment in Colombia, Hernandez *et al.* (1989) demonstrated convincingly the relative sensitivity of tea stomata (and also those of coffee and cacao) to the dryness of the air. Conductance in all three species (all were shaded) declined rapidly as the saturation deficit increased from 0.5 to 4.0 kPa, while transpiration rates were reduced when the saturation deficit exceeded 1.0–1.5 kPa. By comparison, sunflower stomata (not shaded) were less sensitive to the dryness of the air, and transpiration continued to increase over the range 0–4.0 kPa.

Photosynthesis

Fundamental studies on photosynthesis in Assam by Hadfield (1968) helped to explain the function of shade trees in tea. In summary, in locations where the ambient air temperature regularly can exceed 30 °C, leaf temperatures can be several degrees warmer than this and exceed 30–35 °C, which is above the optimum for photosynthesis in tea. This research revolutionised our attitude to shade trees, and led to rational, science-based decisions about where shade trees were likely to be beneficial. This topic is developed further in Chapter 16.

Subsequently, using a portable gas exchange system, photosynthetic rates (A) of individual mature leaves (clone TRIEA-6/8) at the surface of the canopy were monitored during the warm dry season in southern Tanzania (Smith *et al.*, 1993b). Rates increased up to an illuminance (photon flux density) of about 1000 μmol m^{-2} s^{-1} but above this value they remained relatively constant. There was an asymptotic relationship between photosynthetic rates and stomatal conductance, with photosynthesis increasing with conductance over the range 8–100 mmol m^{-2} s^{-1}, but with minimal changes in photosynthesis when conductance exceeded 100 mmol m^{-2} s^{-1}. Irrigation and fertiliser increased photosynthetic rates both by increasing A per unit leaf area and by reducing damage to the canopy, thus increasing the proportion of healthy leaves available to intercept light. Although there was a broad temperature (air and leaf) optimum for photosynthesis (range 20–36 °C), irrigation-induced increases in photosynthesis rates could be accounted for by increases in conductance and associated reductions in leaf temperature (Figure 13.6).

The effects of fertiliser were more complicated and are too difficult to summarise here.

By comparison, Barman *et al.* (2008) at Tocklai in North-East India (26°47′N, alt. 97 m) identified for a selection of tea clones an optimum illuminance for photosynthesis of about 1200 μmol m^{-2} s^{-1} (possible range 1000–1400), while the optimum leaf temperature was considered to be about 26 °C (range 25–30), values comparable with those reported by Smith *et al.* (1993b). Similarly, Barman *et al.* (1993) found that, at the same low-altitude site in Assam, leaf temperatures were highest at midday in both shaded (65% of incident light) and unshaded plants. They were 2 °C (shaded) and 5 °C (unshaded) above the ambient air temperature. Stomatal conductance and transpiration rose to maximum values at midday, but both declined

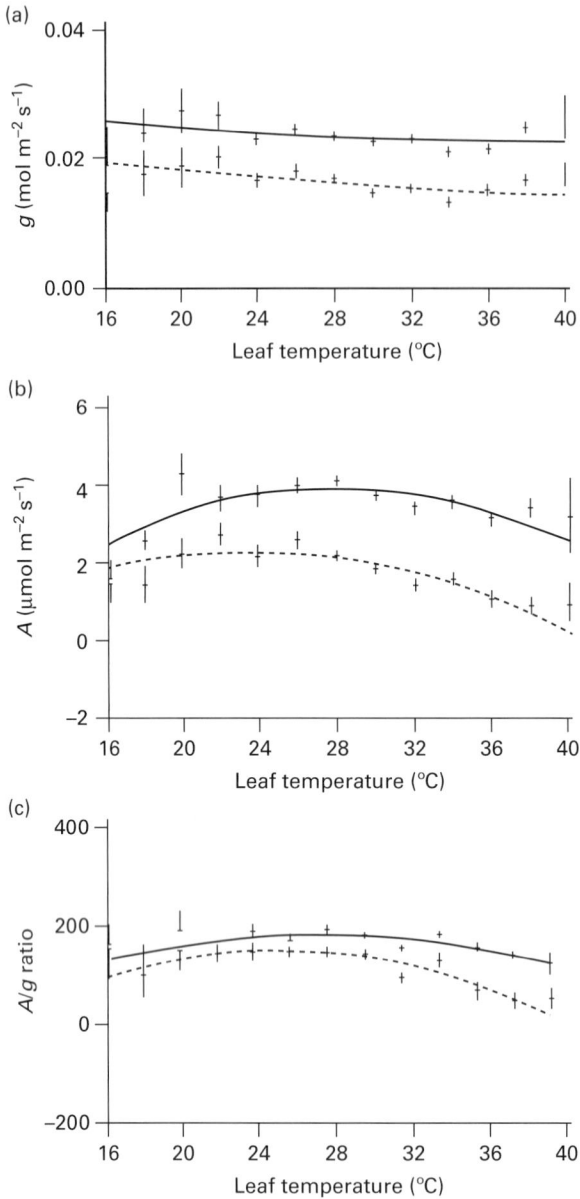

Figure 13.6 Mean values of (a) stomatal conductance (g), (b) photosynthesis (A), and (c) the A/g ratio as recorded at increasing leaf temperatures in well-irrigated (upper line) and partially-irrigated tea (from Smith *et al.*, 1994).

sharply in the afternoon, especially in full sun. Photosynthesis rates were highest in the morning, declining as the day advanced under both shaded and unshaded conditions. They were 33% higher under shade than in full sunlight.

Similarly, in Sri Lanka, Mohotti and Lawlor (2002) found, at a high-altitude site (1400 m), that photosynthetic rates increased rapidly from zero at dawn to a maximum between 0800 h and 0900 h followed by a progressive decrease during the remainder of the (bright and sunny) day, even when the environmental conditions were less severe during the late afternoon (irradiance, temperature and saturation deficit declined). Photoinhibition was implicated in the (complicated) explanation put forward to explain this diurnal pattern in photosynthetic rates, linked to the corresponding diurnal pattern in stomatal conductances, as summarised above.

In Assam, Barman *et al.* (2008) showed how net photosynthesis gradually increased as the photosynthetic photon flux density increased from 200 to 1200 μmol m^{-2} s^{-1}, before stabilising and then declining. At a light intensity of 2200 μmol m^{-2} s^{-1} net photosynthesis (of the canopy) was half that recorded at 1200 μmol m^{-2} s^{-1} (7 compared with 14 μmol m^{-2} s^{-1}).

Previously, Squire (1977) had monitored representative diurnal changes in photosynthesis in the wet, cool and dry seasons in southern Malawi using the radioactive ^{14}C technique. Photosynthesis of leaves on the bush surface was light-saturated when irradiance reached 350–400 W m^{-2} (equivalent to a light intensity of about 800–900 μmol m^{-2} s^{-1}). Declines in photosynthesis observed in the afternoon appeared, in large part, to be closely related to falls in the xylem water potential (range −0.4 to −1.5 MPa, measured with a pressure chamber; Box 13.2). Photosynthesis did not decrease at the start of the cool season when the yield of tea declined, while irrigation in the dry season had little immediate effect on shoot growth, but increased photosynthesis. As Tanton (1982b) demonstrated, when the saturation deficit of the air exceeds 2.3 kPa the rate of shoot extension is restricted (Figures 13.7 and 13.8), even for an irrigated crop. These observations are important to our understanding of the allocation of assimilates within the tea plant and seasonal yield distribution, and in explaining responses to irrigation (see Chapters 7 and 15).

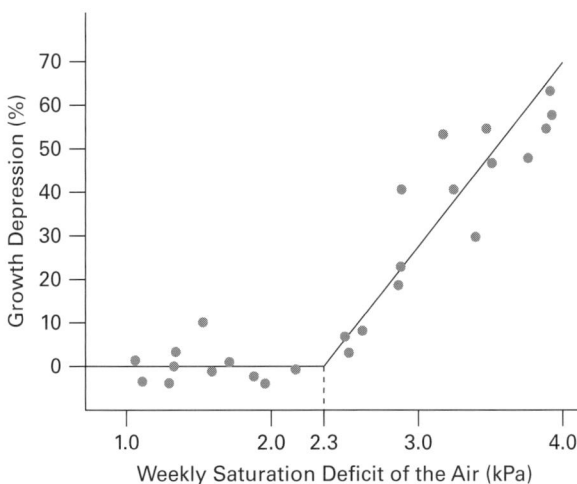

Figure 13.7 Relation between depression in shoot extension rate and the mean weekly saturation deficit of the air. The critical value above which shoot extension rates decline (relative to shoots growing in a humid atmosphere) is *c.* 2.3 kPa (from Tanton, 1982b).

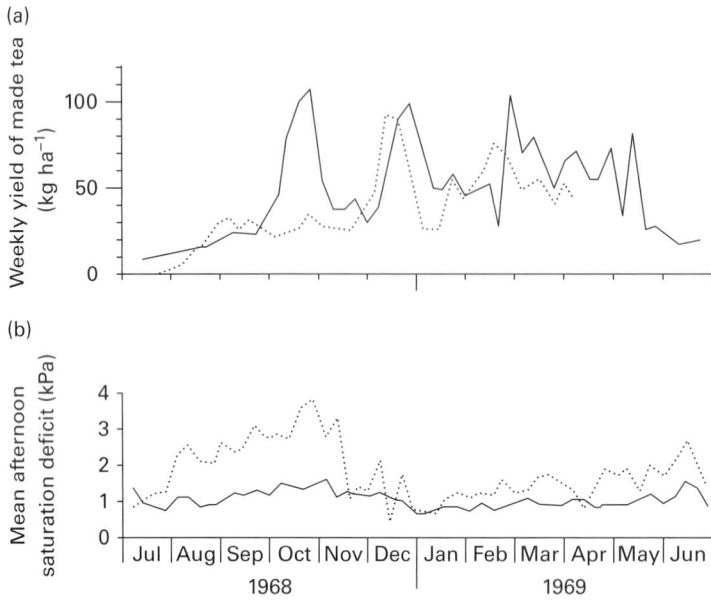

Figure 13.8 Comparison of (a) weekly yield distribution and (b) mean afternoon saturation deficit for Mulanje, Malawi (dashed line) and Mufindi, Tanzania (solid line).

In Darjeeling, Kumar *et al.* (2011) found that, unlike shoot extension, photosynthesis in China-type clones continued throughout the winter months when minimum air temperatures fell below 12 °C, sometimes as low as 5 °C. Previously Kabir (2002) had monitored seasonal changes in photosynthesis at Darjeeling Tea Research Centre, Kurseong (25°55′N, alt. 1240 m) and several other indicators of stress in three clones (Bannockburn 157, Phoobshering 312 and Tukdah 78). Photosynthesis continued through the winter (when the ambient air temperature = 17 °C) until drought became the dominant factor in April. Unfortunately, it is not possible to draw many transferable lessons from this study, except that it confirms that photosynthesis continues under conditions where shoot growth has stopped.

As De Costa *et al.* (2007) concluded in their detailed review of the photosynthetic process:

the photosynthetic apparatus and partial processes of tea show specific adaptations to shade. Maximum light saturated photosynthetic rates are below the average for C3 plants and photoinhibition occurs at high light intensities. These processes restrict the source capacity of tea.

Transpiration

Although portable gas analysers have been used to monitor instantaneous rates of transpiration (Smith *et al.*, 1994), they do not allow whole-plant water use to be determined. More accurate estimates of transpiration are needed when scheduling irrigation for immature tea or, for example, comparing the water use of different clones. Kigalu (2007) has described the successful use of sap flow meters, based on the stem heat flow method, for measuring transpiration rates of individual plants in a plant density experiment (three years after field planting) in Tanzania. There were differences in water use (on a per unit leaf area basis) between the two well-watered clones, with AHP-S15/10 (with a spreading habit) transpiring faster for most of the day than BBK-35 (more upright). Similarly, there were differences in diurnal patterns of water use between plants (both clones) grown at low density (8333 ha^{-1}) or at very high density (83,333 ha^{-1}). The same method was used by Samson et al. (2000) to measure water stress in tea in South Africa, with some success. As a result of these initial studies, the sap-flow technique was considered to be a possible discriminator for identifying clones sensitive to drought stress. Dry air conditions are considered to be an important limiting factor to tea production in South Africa. In Sri Lanka, Anandacoomaraswamy et al. (2000) used a similar method (the heat pulse technique) to measure transpiration of tea in the field with and without *Grevillea* shade trees.

Drought Tolerance

Compensatory leaf growth following the relief of water stress is common in many leafy crops. It also occurs in tea (Box 13.3). Rates of expansion of leaves of plants (China-type, clone BBT-1) that had not been irrigated during the dry season were 75% greater, over the four-week period following the start of the rains, than those of adjacent plants that had either been partially or fully irrigated (Figure 13.9). In addition, the final area of individual leaves was 74% greater than that of the corresponding leaves of plants that had been well watered at all times (Carr, 1969a).

In a study similar to one in Kenya described earlier, the responses to drought of seedling plants were compared with those of their clones during an extended dry season (Carr, 1977b). There was no evidence to suggest that the response of a clone to drought,

---

Box 13.3 Compensatory Leaf Expansion After the Relief of Water Stress

Increases in the areas of leaves (China-type, clone BBT-1) on plants that had been either partially irrigated or fully irrigated during the previous dry season, over the four weeks following the start of the rains, are shown in Figure 13.9. The corresponding rates of expansion, over the same four-week period, of leaves from adjacent plants that had not been irrigated are also shown. Not only is the rate of leaf expansion 75% greater in the previously droughted leaf, but the final size is 74% larger than leaves that were well watered at all times (Carr, 1969a). This compensatory growth following the relief of water stress is commonly seen in other leafy crops, for example sugar beet leaves.

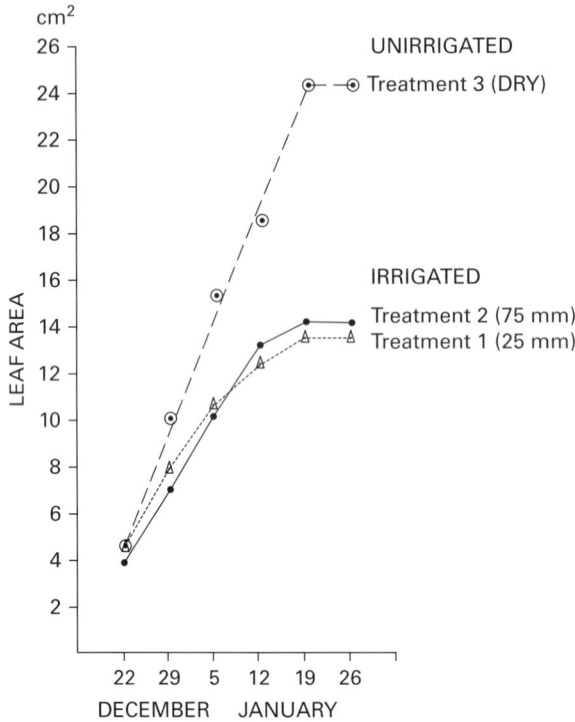

Figure 13.9 Compensatory leaf expansion after release of water stress on previously droughted crop.

in terms of the behaviour of the stomata and changes in its plant water status, could have been predicted from that of its seedling 'parent' (ortet). It was postulated that, since the seedlings were originally identified in a field on account of their comparative vigour (but probably not observed under drought conditions) they must have been able to compete effectively with their neighbours for water. This advantage would not be maintained when clonal plants were competing with neighbouring plants that were genetically identical. Using the same criteria as described above (the Ψx / infiltration score relationship), clone TRIEA-7/4 was considered to be relatively drought-tolerant. The stomata of this clone remained open during the dry season, and the shoot water potential remained high, suggesting a very efficient root and water transport system throughout the plant (or possibly changes in osmotic potential, as suggested by Karunaratne et al., 1999). By contrast, the stomata of clone TRIEA-7/14 began to close early in the dry season and were slow to reopen when the rains began (an example of *drought avoidance* as opposed to *drought tolerance*).

In Malawi, Squire (1976) found that the xylem water potential (Ψx), measured in the cool and rainy seasons, was least in clones that gave the largest yields (recorded over a full year), and postulated that measurements of Ψx with a pressure chamber might be used to screen new clones for high productivity at an early stage of growth, although the physiological link between Ψx and yield was not obvious. In a comparison of six irrigated clones in Tanzania, Smith *et al.* (1993a) were not able to confirm this relationship.

In a detailed comparison of six young (< 3 years) contrasting clones in southern Tanzania, Smith *et al.* (1994), using a portable gas analysis system, found that stomatal conductances (*g*) of clones AHP-S15/10 and TRIEA-6/8 were consistently at least 10% higher than those of clones BBT-1 and BBT-207 (both China-type). The rates of photosynthesis (*A*) in clones BBT-1, TRFCA-SFS150, AHP-S15/10 and TRIEA-6/8 were always greater than those in BBT-207. Irrigation increased *g*, *A* and the *A/g* ratio (a measure of instantaneous water use efficiency) in all six clones. Clones also differed in the relationship between leaf temperature and *A*. By contrast to the results summarised above (under *Photosynthesis*) for 'mature' clone TRIEA-6/8 plants, irrigation also increased the temperature optimum for photosynthesis, and reduced photoinhibition at high illuminance (Smith *et al.*, 1993b).

In the same experiment, clones BBT-1 and TRFCA-SFS150 were classified as 'drought-resistant' (based on the relative annual yield loss), although different mechanisms were involved (Carr, 2010a, 2010b). Drought-resistance indicators for both these clones, and also for the high-yielding AHP-S15/10, were strongly related to high Ψx values (less negative) in the dry season. Smith *et al.* (1993a) were of the view that

there is now sufficient accumulated evidence for the relationship between drought resistance and Ψx (measured with a pressure bomb) to be used to help to identify drought resistant clones during a dry season.

Smith *et al.* also believed that further investigations were justified into establishing the relationships between *A* and *A/g* and an appropriate measure of drought resistance. Subsequently in Kericho, Kenya, Tuwei *et al.* (2008b) evaluated the effects of grafting on drought tolerance. This was based on an index derived from the ratio of yields in a drought year with those in the previous or subsequent year, or equivalent period, with only a mild drought. They considered that xylem water potential measurements (least negative) of a clone under well-watered conditions may be related to its drought tolerance as a rootstock, and could be helpful in a selection programme. By contrast, the performance of clones as drought-resistant scions, based on visual symptoms, was correlated with stomatal conductance values.

Among the morphological leaf traits studied by Olyslaegers *et al.* (2002) in South Africa, stomatal density, pore diameter and pore depth were not linked consistently to stress tolerance. Cuticle thickness was also not a good indicator. In contrast, leaf conductances were greater and leaf water potentials lower in two clones considered to be sensitive to very dry air (TRFCA-PC113 and TRFCA-SFS204) compared with two clones thought to be tolerant (TRFCA-PC114 and TRFCA-SFS150), but this observation was site-dependent. Previously, Samson *et al.* (2000), in a comparison of the same four clones under controlled conditions, considered leaf-related sap flow measurements (which are related to transpiration rate) to be a promising discriminator for identifying clones sensitive to drought stress induced by dry air conditions (considered to be an important limiting factor to tea production in South Africa: see Figures 13.7 and 13.8). Similarly, Nijs *et al.* (2000) compared TRFCA-PC113 and

TRFCA-PC114 using a canopy-level energy balance approach together with measurements of stomatal conductance and leaf water potential of individual leaves. Neither approach could distinguish clearly between the two clones, which differed (visually) considerably in their response to water stress, although clone TRFCA-PC114 (tolerant) did exhibit greater stomatal control in young leaves, and associated higher (less negative) water potentials, than (susceptible) clone TRFCA-PC113.

Measurements made by Damayanthi *et al.* (2010) during an extended dry season in Sri Lanka highlighted differences between eight clones in their responses in terms of a number of parameters. Drought-tolerant cultivars (e.g. TRISL-2025, CY9 and DG7) maintained a higher xylem water potential under dry soil conditions than drought-susceptible clones (e.g. TRISL-2023, TRISL-2026 and TRISL-2024) as a result of osmotic adjustment. Photosynthetic and transpiration rates of the first three clones also declined less fast as water stress increased.

Summary: Gaseous Exchange

- Stomata in tea only occur on the abaxial (lower) surface of the leaf.
- A range of techniques has been used to monitor stomatal behaviour, but results are not always directly comparable.
- In particular, evidence for diurnal changes in stomatal opening is inconsistent, partly depending on the technique used.
- Nevertheless, it appears that stomata are sensitive to temperature and/or dry air.
- Rates of photosynthesis increase up to an illuminance of about 1000 μmol m^{-2} s^{-1}, and then remain relatively constant.
- The optimum leaf temperature for photosynthesis is in the range 25–30 °C.
- Rates of photosynthesis can vary between clones but are not linked directly with yield.
- Sap flow meters have been used with apparent success to monitor actual transpiration by young clones.
- Measurements of xylem water potential with a pressure chamber have proved to be a very useful way of quantifying plant water status, and possibly drought resistance, in the field.
- The sensitivity of stomata to changes in xylem water potential varies between clones, and this may also offer a drought tolerance/avoidance selection procedure.
- There is no apparent direct relationship between the responses of a clone to water stress and that of its ortet.
- Caution is urged when using porometry for identifying drought-resistant clones.
- Several other possible indicators of drought tolerance have been identified, and novel ways of reducing transpiration have been tested, with limited success.
- Compensatory leaf growth occurs after the relief of prolonged water stress.

14 More Pores for Thought

The Answer Lies in the Soil

If that is the answer, what is the question? Is soil fertility sustainable? The soil system is a living entity composed of an assemblage of organisms, all of which are fuelled by carbon derived from plant residues. This process creates conditions favourable for plant growth. A field of tea represents a dynamic system in which soil conditions change as nutrients are lost and imperfectly replaced. The capacity of the plants to absorb nutrients from the soil solution also alters as the soil becomes more acid and loses its structure. To facilitate the *sustainable intensification* (as current jargon describes it) of tea production, it is necessary to develop soil management procedures that will, at the least, maintain yields and, by using resources efficiently, reduce or limit any adverse environmental impact that may occur.

Central to the recycling of nutrients are the soil microorganisms. Kibblewhite (2011) highlighted the important role that organic matter plays in sustaining microbial life in a tea soil. The amount of organic carbon stored within the soil (and the litter layer) in tea plantations is substantial (due to the decomposition and incorporation of fallen leaves and prunings) and increases with the age of the tea. For example, in China, organic carbon in the soil continued to increase for as long as 50 years after planting. From field surveys and the literature, Li *et al*. (2011) estimated the total amount of carbon stored in tea plantations in China to be 316 Tg,[1] of which 83.3 Tg were stored in above-ground biomass, 8.0 Tg in the litter layer, and 225 Tg in the soil. When spread over 1.6 million ha of tea (the area of tea in China at that time) this was equivalent to 197.5 t C ha^{-1}. Although this is less than in the indigenous forest, it is still an important component of the carbon balance in a tea-growing region.

The tea industry was one of the early leaders among commodity crops in looking to research to provide solutions to the field (and processing) problems, by employing its own scientists and paying for its own research. But, as seen in Chapter 12, research findings from statistically designed, replicated field trials can vary considerably from year to year and from one country to another and can be contradictory even within one region.

This chapter focuses on considering ways in which the fertility of the soil can be sustained. The uncertainties associated with interpreting the results of leaf and soil analyses as a guide to sustainable nutrient management are reviewed, while attempts are made to explain the reasons for yield decline over time. Ways of maintaining biodiversity in the soil to ensure its long-term sustainability are summarised, as are aspects of organic production methods.

Leaf and Soil Analysis

There are still many uncertainties concerning leaf analysis as a diagnostic tool for determining the nutrient status of tea and hence the crop's fertiliser requirements. Specifying what part of the plant to sample and when is one issue. The other is how best to interpret the results of the analyses. The answer lies in the soil, but where is it in the soil? Nor has it yet been found possible to specify the critical concentrations that signal nutrient deficiencies (or excesses). Interpreting the results of soil analysis can be difficult. The problem again, in part, is one of sampling, including the number and size of samples needed to obtain representative and realistic values (Box 14.1).

In 1970, as a result of an extensive review of the literature, Willson (1970) concluded that:

it is impossible to discern from the tea literature [published during the previous 70 years] . . . a clear lead to a method [of foliar analysis] of wide application.

Twenty years later, Barua (1989) came to a similar conclusion in his comprehensive text on tea:

despite some progress much remains to be done towards standardizing the procedure for sampling leaves for nutrient analysis. Unless this is done, leaf analysis data cannot be relied upon as a supplement to fertilizer trials.

Box 14.1 Soil Sampling

When sampling soils for physical, chemical or biological analysis, it is necessary to agree the intensity of sampling (soil and leaf) needed in order to get representative samples that justify the cost of analysis The size of the soil (or leaf) sample is by necessity small relative to the volume of soil (or the number of leaves) in a field. Consider for example a 1 ha area of tea:

$$\text{Mass of dry soil in 0.15m depth of soil in 1 ha } (0.15 \text{ m} \times 10{,}000 \text{ m}^2)$$
$$= 1500 \text{ m}^3 \times 1300 \text{ kg m}^{-3}$$
$$= 1950 \text{ tonnes}$$

where the dry bulk density of the soil is 1.3 g cm^{-3}

Sample size = 10 sampling locations per ha × 500 g site^{-1} ha^{-1} = 5 kg moist soil, much of which will be discarded = 4 kg dry soil @ 25% water content (depending on the purpose of the survey, each sample may be analysed separately or they can all be bulked and sub-sampled for one analysis).

$$\text{Ratio of sample size to mass of bulk soil} = 4 : 1950 \times 1000 \text{ kg}$$
$$\approx 1 : 500{,}000$$

The message: *the soil sample size relative to the volume of soil being sampled is very small.*

Table 14.1 Critical nutrient concentrations in (1) the uppermost mature leaf of tea as determined by the Tea Research Foundation of Kenya (Owuor and Wanyoko, 1983) and (2) the third leaf on a tea shoot (Bonheure and Willson, 1992)

Nutrient	Deficient Mature	Deficient 3rd leaf	Borderline Mature	Sub-normal 3rd leaf	Adequate Mature	Normal 3rd leaf
N (%)	< 3.0	3.0	3.0–3.5	4.0	> 3.5	5.0
P (%)	< 0.15	0.35	0.15–0.17	0.40	> 0.17	0.50
K (%)	< 1.20	1.60	1.20–1.50	2.0	> 1.50	3.0
Mg (%)	< 0.10	0.05		0.10		0.30
Zn (ppm)	< 10	20		25		50
Cu (ppm)		10		15		30

Over the years different leaves have been proposed for routine analysis (Table 14.1). The advisory service introduced in East Africa in 1969 was based on the analysis of (1) the first leaf plus terminal bud, and (2) the third leaf. This was later modified to analysis of the third leaf only. In Sri Lanka, by contrast, the mother leaf (the leaf that subtends an actively growing shoot) is considered to be the most appropriate leaf to analyse. Elsewhere, it is the whole shoot (three leaves plus a bud) that is analysed (Bonheure and Willson, 1992).

Results also vary with the season (Tolhurst, 1971), and between clones, as well as with the position and age of the leaf. The availability of one mineral element can also influence the uptake of another. Where this is the case, the N:K and N:P ratios are then used, for example, as extra guides. The fact that there is not a consensus on norms has not stopped guideline concentrations being proposed to indicate when there *may* or *may not* be a nutrient deficiency (Figure 14.1).

Hilton (1975), in a review of the experience of leaf analysis as a diagnostic tool in central Africa, identified the third leaf from an apical shoot sampled during the second crop peak as giving the best reflection of the nutrient status of tea. He also considered that the N:P and N:K ratios could be used as possible indicators of yield-limiting deficiencies (when values exceed 20 and 3 respectively).

The problems of interpretation are highlighted by measurements made in the fertiliser experiment described in Chapter 12, which was sited in the Kericho District of Kenya and continued for 18 years (Owuor *et al.*, 2012). Changes in the nutrient status of a high-yielding (up to 10,000 kg made tea ha^{-1}) clone (AHP-S15/10) were monitored through regular soil and leaf analysis every three months. These time intervals were chosen to represent the four different seasons experienced in the region. Different nitrogen rates (0–600 kg N ha^{-1}) were provided by two fertiliser formulations (NPKS 25:5:5:5 and NPK 20:10:10). These were broadcast once a year as a single application. The bushes, planted in 1970 at a density of 13,400 plants ha^{-1}, were pruned at four-year intervals.

Both fertiliser formulations reduced the pH of the soil by similar amounts, indicating that this acidification was due to the nitrogen content of the fertiliser, not to the sulphur content. The greater the quantity of nitrogen applied, the larger the reduction in the pH, falling from > 5 to < 4 at all depths (down to 300 mm) by the end of the experiment.

Plant Analysis Report

Broad Spectrum Leaf Analysis

Crop Nutrition & Environmental Laboratory Services

Customer:	Mufindi Tea Company	Crop Type:	Tea	Date Received:	17-Oct-05
Farm Name:	Stone Valley Estate	Crop Stage:	Mature leaves, actively flushing plants	Report Date:	18-Nov-05
Contact Person:	Paul Bebingoton	Comments:		Sample ID:	2259PA0041

Field Name: KINGA 3 (UP3)

History (last 3 analysis)

Parameter	Unit	Result	Optimum	Very Low	Low	Optimum	High	Very High	Symbol	Current			
Nitrogen	%	2.55	3.50		�				N	2.55			
Phosphorus	%	0.20	0.30			▩			P	0.20			
Potassium	%	2.98	1.60				▩		K	2.98			
Calcium	%	1.10	0.40					▩	Ca	1.10			
Magnesium	%	0.44	0.18					▩	Mg	0.44			
Sulphur	%	0.18	0.10			▩			S	0.18			
Manganese	ppm	93	100						Mn	93			
Boron	ppm	44	11			▩			B	44			
Zinc	ppm	19	25			▩			Zn	19			
Iron	ppm	58	120		▩				Fe	58			
Copper	ppm	4.90	12.00	▩					Cu	4.90			
Molybdenum	ppm	1.49	0.10					▩	Mo	1.47			

COMMENTS

Very Low Copper:	Low levels of copper can cause stunted growth with distortion of young leaves and growing points. Necrosis of apical meristems occurs. Low copper causes poor fermentation which reduces final product quality.
Low Nitrogen:	Low nitrogen level results in stunted growth, yellowing of young leaves and premature falling of old leaves.
Low Iron:	Low levels of iron causes interveinal chlorosis, which turn white and dry up. Amino acids and nitrates accumulate when Fe is deficient.
High Potassium:	Plants with excess K will be deficient in Mg and Ca due to imbalance.
High Magnesium:	Excess magnesium rarely occurs under field conditions.
Very High Calcium:	Excess calcium can result in short internodes while leaves do not reach full size before yellowing and curling back. Leaf edge blacken and leaves distort, crack and fall off.
Very High Molybdenum:	Molybdenum toxicity rarely occurs.

RECOMMENDATIONS

Check nitrogen units in fertiliser program. Apply high N foliar to boost N levels in addition to granular fertiliser. Calcium and magnesium unusually high for tea and could indicate that soil pH is too high. Check soil pH and acidity if necessary or poor yields noted - use ammonium sulphate or elemental sulphur as fertiliser sources. Uncommon for iron deficiency in tropical acid soils but if there is presence of chlorosis in younger leaves, add iron as a foliar feed. Apply a high copper foliar feed (i.e. Coptrel) to boost copper levels and improve leaf quality.

Figure 14.1 The fact that there is no consensus on norms for soil and leaf analysis has not stopped guideline concentrations being proposed, as, for example, by this commercial service offered in Kenya.

Increasing the quantity of fertiliser applied raised the nitrogen (N) content of the mature leaf (the uppermost leaf left behind at the last harvest, or the third leaf on a shoot), but reduced the potassium (K) content. The available phosphorus (P) in the soil increased, especially in the 0–100 mm layer, confirming the low mobility of this element. There was no change in the concentration of leaf sulphur (S).

In contrast, the leaf P content declined with the increase in N applied. This was despite an increase in the available P in the soil. As the P content in the soil did not change much over time, the result suggests that with the decline in soil pH, more P was fixed and rendered unavailable to the plant. Similarly, there was a reduction in the soil available K, despite the additional K being applied. This was thought to be due to the displacement of potassium ions by ammonium ions, which resulted in more K being leached. The decline in the uptake in P and K with increasing N applications had no effect on yields. Application of NPKS fertiliser, but not NPK, increased the soil available S. The issues of importance to our understanding of the nutrition of tea from these results are summarised in Box 14.2.

Box 14.2 Nutrient Analysis of Tea in Kenya

The detailed study by Owuor *et al.* (2012) described in the text identified the following issues, which are important for our understanding of the nutrition of the tea plant:

- The soil pH declined with increasing applications of N fertiliser.
- Sulphur was not responsible for the decline in the soil pH.
- The acidification of the soil may have long-term implications to agriculture, by making the soil less productive in the future.
- No additional N uptake occurred at application rates above 300 kg N ha^{-1}. This is similar to the yield responses recorded in the same experiment (see Owuor *et al.*, 2008 and below).
- The efficiency of N uptake was similar from both compound fertilisers (NPKS 25:5:5:5 and NPK 20:10:10).
- Both soil and mature leaf P contents declined with increased applications of nitrogenous fertiliser. This was despite the additional P being applied, which resulted in a build-up of P in the surface layers of the soil.
- Similarly, both soil and leaf K contents declined with increased levels of fertiliser applied.
- Applying N and K fertilisers at the same time did not improve K uptake, but probably resulted in the displacement of K ions by ammonium ions, which were subsequently leached.
- High rates of N fertilisers will therefore reduce the soil K levels, especially when the soil pH is low.
- This will result in K deficiency, low yields and eventually the death of the tea bush (responses similar to those seen where tea becomes 'moribund').
- It may be necessary to review the need to apply NPK fertlisers as compound mixtures.
- In Kenya, mature tea is considered to be deficient when tea leaf K levels fall below 1.2%; at 1.2–1.5% they are borderline, and adequate when beyond 1.5%. Leaf K levels were above 1.2% at all times in this study.
- Similarly, P levels in a mature leaf should not fall below 0.15% (considered to be deficient). They are borderline at levels between 0.15% and 0.17%, and adequate at levels above 0.17%.
- Generally East African soils are well supplied with P. This, together with prunings (every four years) and leaf fall, both left *in situ*, helps to sustain production through the recycling of nutrients.

Declining Yields

When established plants are removed prior to replanting, the supply of carbon to the soil is cut off and soil organic carbon levels fall rapidly. The priority for management during this interim stage is to try to ensure that as much surplus nitrogen (crudely, the amount of N applied as fertiliser less the amount removed in the harvested crop) as possible is assimilated into the soil organic matter before the soil microbes consume the last of the readily metabolised carbon. This restricts the amount of nitrate-N leached below the limits of the root zone. In an established crop, the more total dry matter that is produced, the greater the amount of carbon that can be returned to the soil and used to sustain the microbial population.[2]

Bacteria and fungi synthesise and secrete enzymes such as phosphatases, proteases, ureases and pectinases. Together these enzymes form an important part of the soil matrix, contributing to nutrient availability and soil fertility. In other words, they are an indicator of soil health. In a response to the 'general feeling among planters and scientists that the fertility of old forested soils is being exploited by tea cultivation', the activities of enzymes in naturally forested soils (deep latosols) in South India were compared with those found in similar soils in an estate that was under tea cultivation (Venkatesan and Senthurpandian, 2006). This study was located in the humid regions of the Western Ghats (8–$13°N$, altitude range 300–2500 m). The soils were sampled to depths of 2 m. In general, enzyme activity declined with depth. Phosphorus availability was high down to a depth of 1.25 m (this seems to be deep for P), while exchangeable K and organic matter declined rapidly below a depth of 0.25 m. The pH was similar in both forested and cultivated soils. The soils under tea cultivation were apparently richer in urease activity at all depths than the soils under indigenous forest (Box 14.3). Protease activity was almost nil below a depth of 0.50 m in both types of soil. Hydrolytic enzymes in general were more active in the surface soil layers where the effective feeder roots of tea are most prolific. There were positive correlations between the activities of all enzymes and soil organic matter content in both soils. So, in this case, tea cannot be said to be more exploitative of soil fertility than the indigenous forest.

Box 14.3 Urease Activity and the Minimum Time Interval Between Applications of Urea-Based Fertiliser

Urease is an enzyme that plays a role in breaking down urea and releasing ammonium ions. By measuring urease activity in a soil, the rate of hydrolysis of urea can be assessed. This in turn can give an indication of the time interval between any two applications of fertiliser. In a comparison of six tea soils in South India, 27 days was found to be the minimum interval between two applications of urea-based fertiliser at three sites, 33–36 days at two others, and only 18 days at Munnar (Venkatesan and Senthurpandian, 2006).

Box 14.4 Changes in Soil Properties Under Tea Cultivation Over Time (up to 76 Years) in Kericho, Kenya

Under good management the tea bush can remain productive for more than 100 years, but peak yields, or so we are told, usually occur between 20 and 40 years after planting. How this is determined when so many other factors are changing is unclear.

There then follows a reduction in yield. To determine the reasons for this decline, a study was undertaken in the Kericho District of Kenya ($0°22'S$ $35°21'$ E, alt. 2178 m) (Kamau *et al.*, 2012). The properties of the topsoil (0–0.20 m) in fields that had been cleared of trees and planted with tea 14, 29, 43 and 76 years earlier were compared with those of the soil in the neighbouring indigenous Mau Forest Reserve.

The soils were all classified as nitisols with 46–59% clay and 33–38 g kg^{-1} of organic carbon. Although there were small differences between sites, no clear trends emerged over the time sequence in any of the chemical properties of the soils.

The main difference was in the soil pH and associated variables (e.g. extractable aluminium and manganese), but the trend was not consistent. The forest soil had a pH of 4.7, compared with an average of 3.9 in the soils under tea (range 3.5–4.5). The C:N ratio was consistent with values between 9.4 and 10.9 in the tea soils. Similarly, the microbial C:N ratio remained constant over time (range 4.1–4.7), while total microbial activity (measured spectrophotometrically by hydrolysis of fluorescein diacetate) ranged between 0.62 (± 0.13, $n = 15$) (at 29 years after planting tea) to 0.90 (± 0.15, $n = 15$) (after 76 years). These values compare with 0.87(± 0.06, $n = 3$) for undisturbed forest soil.

None of the differences recorded was large enough or consistent enough to explain yield decline over time. Indeed, when looked at another way, the soil properties under the forest had been maintained despite (or because of) many years of intensive tea cultivation.

Kamau *et al.* (2012) undertook a similar study in Kenya, the details of which are given in Box 14.4. Not one of the differences in soil properties recorded was large enough, or consistent enough, to explain the decline in tea yield over time. Indeed, when looked at another way, the soil properties under forest had been maintained despite (or because of) many years of intensive tea cultivation.

Since the 1960s, when paraquat was introduced to the tea industry as a herbicide, the great majority of tea can now be grown in weed-free, undisturbed soil. As no hoes are needed to control weeds, damage to the tea roots is avoided and the time taken for the tea canopy to cover the ground is reduced. Under these conditions, a surface layer of decomposing organic matter builds up from leaf-fall and prunings, which the tea roots can exploit for nutrients. This mulch layer also means fewer weeds (Hainsworth, 1969; Willson *et al.*, 1975).

In a detailed study in Assam (covering 22 estates), Phukan *et al.* (2011) were unable to identify differences in the physical, chemical or microbiological properties between soils that had been rehabilitated before replanting (4–5 years before sampling), those which had not been rehabilitated, old tea (more than 50 years) due to be replanted, and virgin forest. The only possible exception was the organic carbon status, which was higher in the majority of tea estates that had been rehabilitated than in those that had not. One difficulty faced in interpreting studies of this kind is the small size of the samples, and knowing how representative they are of the bulk soil (Box 14.1).

The export of nutrients in processed tea is referred to in Chapter 12. Recently, a detailed analysis of the nutrient balance in the smallholder subsector of the Kenyan tea industry has been published (Sitienei and Kamau, 2013). Using data supplied by the Kenya Tea Development Agency, it was estimated that annual imports of the major nutrients (NPK) for the smallholder totalled 14,560 t of nitrogen, 2800 t of phosphate and 2800 t of potash. This was mainly formulated as compound NPK 25:5:5. But it is the quantity that is mined from farmers' fields and exported in black tea that is of particular interest: 9600 t nitrogen, 2700 t phosphate and 3100 t potash. These figures were based on an assumption that 95% of smallholder tea was being exported. This represents a large drain on resources. Marketing more tea as a high-value tea extract rather than as black leaf, and encouraging more internal consumption, are possible ways to restrict this loss of nutrients from the country (or recycling all the used tea bags!). Care is needed to avoid nitrate pollution of water bodies such as Lake Victoria from neighbouring tea estates in the Nandi Hills District of Kenya (Maghanga *et al.*, 2013) (Box 14.5).

Organics

There is a strong, but very variable, demand in some countries for 'organic' produce, grown without the use of inorganic fertilisers.

Before the advent of inorganic fertilisers, known colloquially by growers as 'artificials', the only nitrogenous manures available were organic substances of animal or

---

Box 14.5 Nitrate Pollution

In the Nandi Hills of Kenya there are about 5000 ha of mature tea, yielding on average about 3500 kg ha^{-1}. Fertiliser is applied at an annual rate of about 140 kg N ha^{-1}, usually as NPK 25:5:5 compound and usually in two split applications, some of it from the air (Figure 14.2). In order to assess the impact of this fertilisation on the water quality, Maghanga *et al.* (2013) sampled the water in 10 rivers passing through one group of estates over a three-year period. The soils are well drained nitisols. Samples were taken before and after fertiliser application and analysed for nitrate-nitrogen (NO_3^--N) ions. At all times and places the nitrate-N levels were well below the maximum contaminant level (MCL) for drinking water of 10 mg L^{-1}. Relatively high levels were recorded in only one of the 10 rivers (range 4.9–8.2 mg L^{-1}) and that was only in one year when fertiliser was first being applied, indicating possible contamination by surface run-off. There was evidence of water contamination upstream of the tea estate, which was contributing to NO_3^--N pollution of Lake Victoria.

Figure 14.2 On some large estates fertiliser is applied from the air. This photograph shows the aerial application of zinc in Malawi – aeroplane vintage unknown (WS). *A black and white version of this figure will appear in some formats. For the colour version, please refer to the plate section.*

vegetable origin, in which nitrogen is combined with carbon and oxygen mainly in the form of a protein. Examples of organic fertilisers include cattle manure, formed by rotting down a mixture of urine-soaked straw with dung; the residual cakes of various seeds after the oil has been expressed; refuse from slaughter houses; and chicken manure.[3] All these substances have to be broken down, by soil organisms, into simpler compounds such as ammonia that, in the presence of oxygen, form nitrates. Nitrates are soluble in water and so can be used by plants. The bulky manures also supply humus to the soil.

Cooper (1946) reported the results of a series of experiments in Assam, beginning in 1931, in which a selection of organic (e.g. dried blood and horn meal, both mixed with bone meal) and inorganic materials were compared with each other and with inorganic compounds (including sulphate of ammonia), plus an unfertilised control. All these treatments supplied the equivalent of 45 kg N ha^{-1}, later increased to 67 kg N ha^{-1}. Over the 15 years that the experiment ran there were no consistent differences in yields between the various sources of nitrogen, and all of them, in particular sulphate of ammonia, out-yielded the control. The least effective source was oilcake; dried blood did quite well; horn meal performed better than expected; nitrate of soda was erratic, but as good as any other manure (but it can only be used in alternation with sulphate of ammonia); sulphate of ammonia was easily the best form in which to apply nitrogen to tea. Other organic sources mentioned by Cooper (1946), together with detailed instructions on how to prepare and store them, include composted straw and leaf mould. March was the recommended time to

apply fertiliser/manure in North-East India. There was no benefit from split applications. In round figures, an annual application of 45 kg N ha^{-1} increased yields in
North-East India by about 4 maunds per acre (about 400 kg ha^{-1}), except in year 1
when it was 1 maund.[4]

More recently, Kekana et al. (2012) reported the preliminary results (two years'
data) of a long-term soil fertility trial using organic fertiliser at Kangaita (alt.
2020 m) situated to the east of the Great Rift Valley in Kenya. Cattle manures
'enriched' with different rates of NPKS inorganic fertilisers were compared with
straight organic manures, and an NPKS control. The inputs were applied at rates
equivalent to 0, 75, 150 and 225 kg N ha^{-1}. Treatment effects on the nutrient status of
the soil (at 100 mm increments down to 0.60 m) and the plant (mature leaf) were
recorded. Enriched manures (up to 150 kg N ha^{-1}) increased the soil pH, increased
the P content of the leaf, and reduced the K content of the leaf and the soil when
compared with NPKS fertiliser. Yields were increased in one of the years by enrich-
ing the organic fertiliser. It is not possible from the way the data were presented to
quantify these differences. This approach to crop nutrition, using an inorganic
fertiliser, is known as *integrated soil fertility management*. This can include planting
leguminous cover crops such as boga medeloa (*Tephrosia candida*), pigeon pea
(*Cajanus indicus*) and cow pea (*Vigna vexillata*). Although the development of the
young tea plants may be depressed, there can be benefits when the intercrop, or its
loppings, is buried within the tea rows.

In Assam, at Tocklai, Baruah et al. (2011) compared the soil and leaf nutrient status of
tea that had been converted to organic management in 2009, and monitored in the
following two years during the transition period, with 'conventional' tea (clone T3E3,
planted in 1968). Several organic treatments were compared, including farmyard manure, vermicompost, mulch, oilcakes (neem or castor or kusum), P-enriched vermicompost, and wood ash, in various combinations The inorganic treatment received annual
applications of 140 kg N, 50 kg P$_2$O$_5$ and 140 kg K$_2$O ha^{-1}. Pests and weeds in the
organic treatments were controlled with neem extract and by mechanical means, and in
the inorganic treatments by chemicals. The organic carbon status was maintained at
satisfactory levels in all the organic treatments (range 0.9–1.09%) and in the 'conventional' control (0.93%), supplemented in all the plots by leaf litter and prunings from the
tea and leaf litter and loppings from the shade trees. Soil pH was similar in all the
treatment combinations (about 4.8). There were small increases in available phosphate
in some organic treatment combinations, and in the cation exchange capacity, compared
with the conventional treatment. In contrast, soil available potash levels declined from
the first year after conversion and remained low in all the organic treatments for the
duration of the trial (Baruah et al., 2011).

Although Baruah et al. (2011) did not specify which leaf/shoot was sampled for foliar
analysis, the shoot phosphate contents were similar in all treatments in both years of
conversion (possibly due to the universal application of rock phosphate), including
the inorganic control. The foliar application of vermiwash, produced from high-potash-
content biowaste – from Japani habi (*Mikania micrantha*) and kolmou (*Ipomea carnea*) –
may have prevented a decline in the available soil potash levels.

During the first year of conversion, yields were uniformly low in all treatment combinations, at 1300–1400 kg made tea ha^{-1}. During the second year the inorganic treatment out-yielded the organics by 200–300 kg ha^{-1}.

In Sri Lanka, Wekumbura *et al.* (2010) compared the partitioning of nutrients and dry matter between and within roots, stems and leaves in 12-year-old plants which had been converted to organic production methods (as specified by the International Federation of Organic Agriculture Movements, IFOAM), using either neem oilcake or tea-waste compost, with a so-called 'conventional' inorganic treatment. Partitioning of assimilates and nutrients was similar in all the treatments. Unfortunately, the relative amounts of nutrients applied in each method were unclear. Previously, Kathiravetpillai (1993) described early field-scale attempts in Sri Lanka at producing tea using organic manures. These included coconut poonac, groundnut cake, rubber seed cake, spent tea leaf and compost. The process of compost making was described.

In a related study at the Singampatti Group of estates in Tamil Nadu (8°N, alt. 1300 m), the prunings from tea plants in the organic fields were chopped and buried in trenches (0.5 m wide and 0.3 m deep) in every alternate row. Before the trenches were closed, compost (10 t ha^{-1}) was added together with castor or neem cake (1.25 t ha^{-1}). A further 10 t ha^{-1} compost (comprising 20% cow dung slurry plus cattleshed waste, together with Guatemala grass and boragina) were broadcast. The tea was grown under high shade (*Grevillea robusta*, spaced 10 m × 10 m), low shade (*Erythrina lithosperma*) and 'in-between shade' (*Gliricidia sepium*). Despite the care with which the conversion process was implemented, and organic methods of cultivation put into practice, the initial optimism that organic cultivation would succeed was not realised. Much of the tea has since been returned to conventional management (V. S. Sharma, personal communication, 2014). Organic production was not sustainable.

This is not really surprising, as there is little experimental evidence to support these 'organic' practices. One has to be a 'believer' to take the risk of conversion, since the financial benefits (in terms of higher prices for the processed tea) are not guaranteed while the additional costs of production continue to rise. Prices go down as well as up.

Having said that, in the context of sustainability, the global scene with respect to the availability and cost of inorganic fertilisers is not good. The cost of fixing nitrogen from the atmosphere is dominated by the price of energy, in particular the cost of oil (Greenwood, 1982). In January 2016 the price of oil fell to below US\$30 a barrel, which was good news for agriculture, but the price only a few months previously had been in excess of US\$100, which was bad news, and the price continues to fluctuate.[5] Phosphate rock is a non-renewable resource, and current global reserves may be depleted in the next 50–100 years as global demand for P increases. These reserves are under the control of only a few countries, mainly Morocco, USA and China, and thus, like oil, subject to international political influence (Cordell *et al.*, 2009). Similarly, there is a finite amount of potash in the world, which we must learn to use sensibly.

Summary

- Foliar analysis is only useful as a diagnostic technique when remedial applications of potassium are needed, not to help guide routine fertiliser applications.
- Lack of agreement on what to sample and when has inhibited the development of foliar analysis as a diagnostic tool for determining the nutrient status of tea.
- There is no agreement on the critical concentrations of the principal elements signifying deficiencies.
- The availability of one mineral element can influence the uptake of another.
- Notwithstanding all these uncertainties, guidelines are used commercially to provide advice to tea growers.
- The displacement of potassium ions by ammonium ions can result in the leaching of potassium.
- Organic matter plays an important role in sustaining microbial life in a tea soil.
- A management priority is to ensure that as much surplus nitrogen as possible is assimilated into the soil organic matter.
- Enzymes, secreted by bacteria and fungi, are a good indicator of soil health.
- The market for organic tea is inconsistent (i.e. volatile). Field evaluation of organic materials, which are not always readily available in the quantities needed, is difficult.

Notes

1. 1 Tg is a teragram $= 10^{12}$ g.
2. Data presented in Chapter 7 indicate that typical total annual dry weight gains of tea in Africa are about 21 t ha^{-1} (including roots), of which 5 t ha^{-1} are shoots (= yield) and 7 t ha^{-1} are roots. It goes without saying that, for the crop to be sustainable, prunings should not be removed from the field for firewood.
3. Chicken manure (certified organic) is imported from South Africa to Tanzania by road, a distance of over 1000 km, for use on an organic tea estate in the Southern Highlands. One wonders how the late President Julius Nyrere would have reacted if he had known that the country's precious foreign exchange was being used to import chicken manure, using scarce fuel, from his erstwhile enemy (the apartheid-run Republic of South Africa) for a luxury consumer market in the West. Unfortunately, it is not possible to certify that the droppings from indigenous chickens are organic, although that undoubtedly is the case. Some things do not have a rational explanation!
4. 1 maund = 37.32 kg.
5. 1 kg of oil produces about 45 MJ energy. About 90 MJ are needed to synthesise 1 kg of fertiliser N (Greenwood, 1982).

15 Water Productivity

More Crop per Drop

Tea is traditionally grown in areas of high rainfall. Annual average totals of 1200–1500 mm are often quoted. But tea also grows in places where total rainfall can exceed 7000 mm, for example in parts of Assam, while totals of 4000 mm are experienced in South India and parts of Sri Lanka. In such places, the water management issues are associated with the safe disposal of excess water and subsurface drainage. At the other extreme, tea is now grown in areas where the annual rainfall total may be as little as 900 mm, such as in parts of Tanzania and Zimbabwe. Here irrigation is necessary.

It is, of course, not the total rainfall that is important, but rather it is the distribution of the rain month by month relative to the potential rates of evapotranspiration. In many tea areas, evapotranspiration from a full canopy of tea (ETc) during the summer months is between 3 and 6 mm d^{-1}, equivalent to 90–180 mm month^{-1}, depending on the season.

Tea producers, whether they irrigate or not, want answers to straightforward questions. These include: Which soil and water conservation techniques are most appropriate and effective? Which plant spacings and plant population densities are best suited to irrigated/rain-fed conditions? How much water (or rather how little) is needed for irrigation? What are the effects of irrigation on annual yield and on crop distribution? How is water best allocated between immature and mature plants, or between cultivars? How much extra fertiliser is needed, particularly nitrogen, for irrigated crops? When and how is the best time to apply it?

In addition, planners of new projects want to know the suitability of a site, and its yield potential with and without irrigation. Those planters who are no longer expanding the area of tea (or replanting) want to know how best to manage the crop when there is a drought. Everyone wants to know how to maximise water productivity. In other words, how do you get 'more crop per drop'?

This chapter is divided into two parts. Part 1 reviews the methods used to estimate the water requirements of tea, Part 2 covers the results of tea irrigation experiments and how they can be interpreted. The vast majority of the published research on these topics has been done in eastern Africa, in particular at what is now known as the Ngwazi Tea Research Station (see Box 3.7 for a description of the climate and soils at NTRS).

Part 1   Water Requirements

Everyone knows how to measure rainfall. Or at least they think they do. Measuring or estimating evapotranspiration is not quite so straightforward.

One approach to estimating the water use of tea (ET) is by considering each component of the hydrological water balance in turn and determining ET by difference:

$$ET = P - R - D \pm \Delta W$$

where P is precipitation

   R is run-off

   D is deep drainage

   ΔW is the change in soil water storage.

This estimate of ET is only as good as the accuracy of the weakest component.

Precipitation

Let us begin with rainfall – or precipitation in general, since in Georgia some of the precipitation comes in the form of snow, and elsewhere (e.g. western Kenya) it sometimes/often falls as hail.

Rainfall depends on the latitude, as modified by altitude and topographic features such as proximity to the sea, an inland lake, an escarpment or a mountain. Rainfall is not as easy to measure as one might assume. Much depends on the siting of the rain gauge relative to these features, since rainfall can vary a great deal over short distances (Box 15.1).

Box 15.1 Rainfall Measurement

Rain gauges have to meet strict specifications for the data to be reliable and comparable with other places. The cheap plastic gauges available in shops are not good enough. The siting of a gauge is also important. Too often rain gauges are to be found in the most inappropriate places, such as next to the office roof (sometimes collecting splash and run-off) or under a tree (it was only small when the rain gauge was first positioned there!). The recommended height of the rim is usually 300 mm above the ground (to minimise the distortion caused by wind eddies). The sample size is also very small. One rain gauge of 127 mm diameter (a recognised size) recording rainfall representative of an area of 1 ha has a sample size ratio of 1 in 1.2 million, while for 100 ha it comes to 1 in 120 million, and for 1000 ha 1 in 1.2 billion. Only nationally recognised designs of rain gauge should be used, and guidelines on siting should be followed. We are back to stating that the role of the weather recorder is a high-status job that should be recognised as such, and appropriate training provided. Sometimes it never rains at weekends!

Run-off

Studies at a full-scale catchment level in the Mau forest in Kenya showed that under a full crop cover of tea run-off only represented 1% of the annual rainfall. This was similar to the run-off recorded from an adjacent forested catchment (Blackie, 1979).

By contrast, in young tea there is always a risk of run-off and soil erosion unless precautions are taken to slow the flow of water across the soil surface (Box 15.2; Figure 15.1). Mulches restrict evaporation from the soil surface, as well as

---

Box 15.2 Run-off and Soil Erosion

Run-off and soil erosion were monitored for three years on a 10% slope in Kericho, Kenya, during the establishment of young tea (Figure 15.1). In the first year after planting, soil loss totalled 160 t ha^{-1} from both the manually weeded and the herbicide-weeded plots. This was reduced to 40 t ha^{-1} when oats were used as an intercrop, and to virtually nothing when there was a grass mulch (plus herbicide). In the second year, the greatest soil loss was from the herbicide plots (*c.* 70 t ha^{-1}); on the hand-weeded plots (soil loss *c.* 50 t ha^{-1}), the weeds were left in place on the soil surface. Once the ground cover had reached 60% there was virtually no run-off or erosion from any of the plots. This, then, is the target crop cover to attain as soon as possible after planting (Othieno, 1975).

Figure 15.1 Soil and water conservation experiment. Once the crop cover exceeds 60% the amount of run-off becomes negligible (Kericho, Kenya, MKVC).

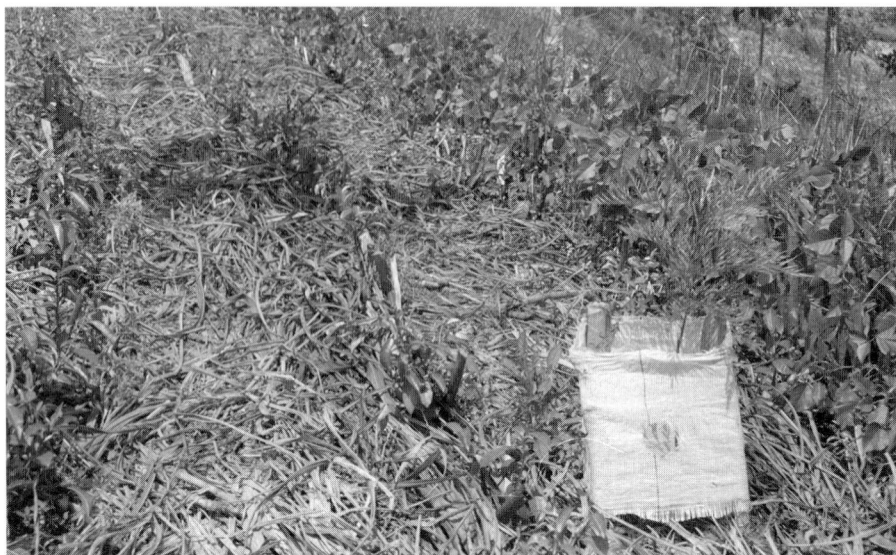

Figure 15.2 Mulching is a very effective way of reducing run-off (Sri Lanka, MKVC).

minimising run-off (Figure 15.2), but beware of fire. Contour drains should direct any run-off that does occur into down-drains and for disposal away from the field (Hudson, 1992).

When installing run-off control measures, such as cover crops like Guatemala grass, always begin at the top of the field (Figure 15.3). This reduces the risk of the measures being installed lower down the field from being destroyed by run-off from above. Think too of the risk of nitrate pollution downstream and the control methods needed to minimise that risk. Figure 15.4 illustrates the massive input of human labour needed to construct by hand the terraces on steep eroded land due to be replanted with tea in Sri Lanka. Save Our Soil, the title of a project to conserve soil in Sri Lanka, is a good working slogan (Figure 15.5). Soil takes hundreds of years to form but can be lost from the catchment in hours.

Deep Drainage

When the soil profile is at field capacity (see Box 15.3), excess rainfall will either run off or percolate through the soil from the surface before appearing as outcrops of the water table in streams and rivers. How much of the total rainfall this represents will depend very much on local conditions, but, as an example, Blackie (1979) found on a mature tea estate in Kenya that deep drainage represented 40% of the 2021 mm mean total annual rainfall. In parts of Assam, Bangladesh and Rwanda high water tables are a feature of the low-lying areas where some tea is grown, and open ditches and sometimes field drains are needed to dispose of excess soil water (Figures 15.6 and 15.7).

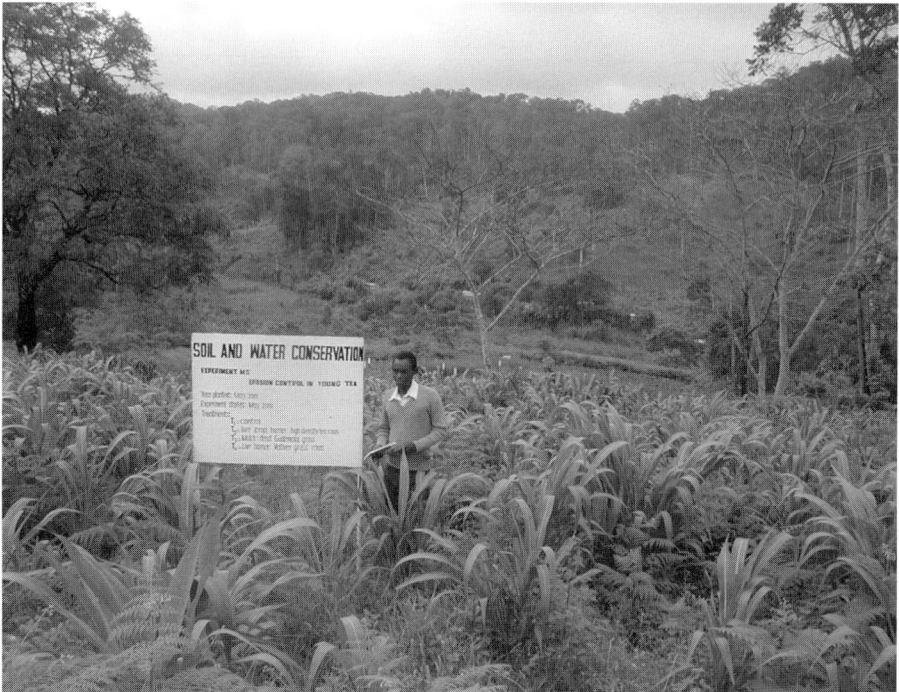

Figure 15.3 There is limited justification for using cover crops such as Guatemala grass to rehabilitate the soil before replanting (Marikitanda, MKVC).

Figure 15.4 (a, b) Rehabilitation of eroded tea land in Sri Lanka: construction of soil conservation terraces (MKVC). *A black and white version of this figure will appear in some formats. For the colour version, please refer to the plate section.*

Water Storage

When considering the amount of water available in a soil there are several fundamental questions that need to be answered. These include: (1) the effective soil depth from which water can be extracted, and hence (2) the total amount of available water in the root zone, (3) what proportion is easily available, and (4) at what point as the soil

Figure 15.5 Soil and water conservation has become an important topic in Sri Lanka because of the degradation of soil over many years (Save Our Soil project). The sign lists the cover crops being tested (MKVC). *A black and white version of this figure will appear in some formats. For the colour version, please refer to the plate section.*

Figure 15.6 Disposing of excess water in areas where annual rainfall can exceed 4000 mm is an important aspect of estate management. Note the down-drains (South India, MKVC). *A black and white version of this figure will appear in some formats. For the colour version, please refer to the plate section.*

Figure 15.7 Stone drains assist in the safe disposal of surface water run-off in areas of high rainfall (South India, MKVC).

dries (known as the critical soil water deficit) does the actual rate of water loss (transpiration) fall below the potential rate?

For example, at the end of the six-month long dry season in southern Tanzania, rainfed seedling tea plants, eight years from planting, were rooting to depths of 4.3 m, and drying the soil to permanent wilting point down to 3.6 m (Carr, 1974). This corresponded to an actual soil water deficit (SWD) of about 330 mm (or 80% of the estimated total available water in the root zone). The corresponding potential SWD, which assumes no restriction on water availability in the soil, was about 1000 mm ($ETc = 0.85\ ETo$). Similar results were obtained in Malawi, where, by the end of the dry season, the actual SWD under unirrigated mature seedling tea reached 380 mm with water extraction at a soil depth of 3–3.5 m (Willatt, 1973).

Subsequently in Tanzania, roots of mature clonal tea plants (clone TRIEA-6/8), were observed at depths below 5 m and water was being extracted from depths greater than 4 m. Based on a water balance model derived from neutron probe data, the estimated annual *actual* water use (ETa) ranged from 800 mm (unirrigated tea) to 1200 mm (fully irrigated). It was also possible to demonstrate that when SWD exceeded 60 mm, ETa declined linearly until, by the end of the dry season, the estimated *actual* SWD for unirrigated tea was about 330–350 mm. This represented 95% depletion of the extractable or available water (Stephens and Carr, 1991b). These results confirmed the earlier findings obtained with seedling tea in Tanzania and Malawi.

Box 15.3 Soil Water Retention

Water is held in the soil against gravity by capillary forces within the soil pores and by adsorption onto soil particles, particularly clays. The maximum amount of water that can be retained in this way is known as *field capacity* or *the upper limit of available water*. The lower limit is called the *permanent wilting point*. The water retained between these two limits is known as *available water*. The pore size distribution determines the ease of availability of this water. It is more difficult for roots to extract water from small pores than from large ones.

   A useful practical exercise with which to demonstrate these principles of water availability in soils involves the use of a domestic sponge.

1.   Place the sponge in a bucket of water until the sponge is saturated with water. Remove the sponge from the bucket. Water will immediately begin to pour out of the sponge, rapidly at first and progressively more slowly.
2.   Eventually water is only dripping out of the sponge slowly. At this point, gravity is balanced by the capillary (surface tension) forces retaining water within the pores. The water content of the soil (sponge) in this equilibrium state is the *field capacity*.
3.   Now begin squeezing the sponge. More water will exit the sponge, initially from the large pores and progressively from the smaller pores as you squeeze harder.
4.   Eventually no more water emerges, although the sponge is still moist, however hard you squeeze. This equates to the *permanent wilting point*.
5.   The volume of water that came out of the sponge between *field capacity* and *permanent wilting point* is the *available water*.

Evaporation

The rate of water loss by tea is a function of both the plant (crop cover and height) and environmental factors (i.e. air temperature, humidity, solar radiation (sunshine) and wind speed). The Penman equation, as modified by McCulloch *et al.* (1965) for use at high altitudes, has been widely used in East Africa to estimate evaporation from an open water surface (Eo) or evapotranspiration from a reference crop (ETo), usually short grass or alfalfa (lucerne). This estimate is then multiplied by a crop-specific factor (Kc), which for a healthy, well-watered tea crop covering the ground has a generally accepted value of 0.85 (Squire and Callander, 1981)

$$ETc = Kc \times Eo$$

There are two principal components within ETc: transpiration by the crop (T) and evaporation from the soil surface, from the crop residue surface or from water intercepted by the foliage (E). Ideally, they can/should be estimated separately.

$$ETc = E + T$$

On days when there is frequent rainfall and the soil surface remains wet, the value of Kc will be close to unity, even in young tea. Once the soil surface dries, evaporation will be restricted and transpiration will be the dominant component of ETc with a value proportional to the crop cover (%). As the soil dries, the stomata begin to close and transpiration is restricted, and the value of Kc declines.

Based on these principles, Dagg (1970) developed the following relationship for estimating water use by young tea, using data obtained from a large weighing lysimeter in Kericho (Kenya) planted with young tea:

$$ETc = Eo[0.96a + (1 - a)0.9n]$$

where ETc is the monthly evapotranspiration total
Eo is the corresponding open water evaporation total
a is the fraction of the soil covered by the crop at noon
n is the fractional number of rain days per month.

The crop factor can be divided into two parts, one for the evaporation from the soil surface and one for transpiration from the crop. This allows the frequency of rainfall events (or wetting of the soil) to be taken into account in a semi-rational way. In addition, Monteith (1972) modified the Penman equation to allow water use by a crop to be calculated without using an *empirical* crop factor, Kc (Box 15.4):

$$ETc = Kc \times ETo$$

This was achieved by including appropriate values to represent the resistance to water vapour diffusion (1) from within the crop canopy to the crop surface (the canopy resistance, r_c) and (2) from the surface of the crop canopy into the surrounding air (the aerodynamic resistance, r_a). The Penman equation, now known as the Penman–Monteith

---

Box 15.4 The Penman–Monteith Equation

Howard Penman first described the Penman equation (which bears his name) in 1948 when working at Rothamsted Experimental Station in the UK. Based on the first principles that govern the evaporation process, it uses standard weather data (daily hours of bright sunshine, mean daily air temperature, average daily saturation deficit and 24-hour wind speed) to estimate evaporation from an open water surface (Eo), or evapotranspiration from a reference crop (ETo), which is usually short grass or alfalfa. A crop factor (Kc) is then used to convert these estimates to the potential water use by a specific crop ($ETc = Kc \times ETo$). Subsequently, the equation was developed further by John Monteith, after which it became known as the Penman–Monteith equation. This is now the standard method used internationally for estimating crop water requirements, providing there are reliable weather data available for the location.

equation, is the standard method used internationally for estimating crop water requirements (Allen *et al.*, 1998).

Because tea has a dense, flat canopy, the mature crop behaves more like short grass than a tall, aerodynamically rough crop like maize. The thick boundary layer above the crop canopy effectively decouples the foliage from the surrounding atmosphere. This means that transpiration in tea is mainly determined by the net radiation (energy balance) term in the Penman equation while the aerodynamic term (a function of the dryness of the air and the prevailing wind speed) plays a relatively small role.

Evaporation Pans

Where weather data are not easily available, the evaporation pan offers a practical alternative way of estimating transpiration by the crop. The principal disadvantage is that the siting, exposure, colour and size of the pan can all influence the results obtained (see Figure 3.1, Chapter 3, and Box 15.5).

Box 15.5 Evaporation Pans: a Comparison

What a difference colour makes! (from Burgess, 1994).

Evaporation from two types of tank was recorded at NTRS. Both tanks were constructed locally with available materials, although paint of the right colour was not easy to obtain!

The 'British Meteorological Office' Pan

This square pan had galvanised iron sides (1.85 m long and 0.60 m deep). It was sunk into the ground with the rim 0.10 m above ground level. The water level in the tank was kept at ground level on a daily basis. It was screened (with a 50 × 50 mm square mesh, 3.5 mm diameter wire). The inside and outside of the tank were both (eventually) painted aluminium.

The 'Kenyan' Pan

This circular pan (diameter 1.21 cm, depth 0.25 m), which is similar to the United States Weather Bureau (USWB) pan, was made of 6 mm thick iron sheet. It sat on a wooden platform with the base 150 mm above ground level. The water level was maintained at a level 50 mm below the rim on a daily basis. There was no screen. This tank was also painted with aluminium paint (eventually).

Evaporation rates were 17% greater from the USWB pan when compared over similar periods, and when both tanks were painted aluminium.

$$E(\text{US pan}) = 1.17(\pm 0.014)E(\text{UK pan})$$

$$R^2 = 0.95; \ n = 17$$

Pan evaporation was very sensitive to the colour of the pan (± 8%).

Box 15.6 Tea Water Requirements

In summary:

1. The generally accepted value for the crop factor (Kc) for tea with complete ground cover is 0.85 ($ETc = 0.85\ ETo$).
2. Mature tea can extract water from depths greater than 4 m; the corresponding total depth of water extracted from the soil can reach 330–350 mm.
3. The total annual water use (ETa) by rain-fed tea at NTRS was about 800 mm, and by well-irrigated tea 1200 mm. ETa declined linearly when the SWD exceeded 60 mm.
4. The Penman–Monteith equation can be used to provide realistic estimates of ETc.
5. A sunken and screened 'British' evaporation pan also provides a good estimate of ETc in the dry season.
6. For young tea, ETc can be estimated with a model that includes crop cover and the number of rain days.

In a comparative study at NTRS, Burgess (1994) found that evaporation rates from a 'British Meteorological Office' screened, sunken pan (Euk) were virtually identical to the calculated Penman–Monteith (Epm) value during the dry seasons.

$$Euk = 0.98 \pm 0.009Epm$$

$$R^2 = 94\%;\ n = 350$$

Previously, Stephens and Carr (1991a) had shown that evaporation from a screened, sunken British pan matched closely the value calculated using the McCulloch–Penman method for estimating Eo (open water evaporation):

$$ETc = 0.99Eo(\text{McC–P})$$

$$R^2 = 93\%;\ n = 20$$

For a summary of tea water requirements see Box 15.6.

Part 2   Irrigation

Tea producers considering whether to irrigate, or not, want an answer to a very straight-forward question. Will it pay to irrigate tea? The answer, as ever, is that familiar response, 'It all depends!' as no two situations will be the same. You then have to go back to first principles, using your knowledge of the soils (depth, water-holding capacity, root depth and water-table depth) and the climate (the frequency, extent and duration of dry spells), and undertake your own site-specific evaluation.

Figure 15.8 Line-source experiment at Ngwazi Tea Research Station. The crop closest to the line-source is fully irrigated, but the amount of water applied declines with distance in both directions so that at the extremities the crop receives only rainfall. To be effective, a line-source irrigation system should only be used when there is no wind. This usually occurs very early in the morning or late in the evening (Tanzania, MKVC). *A black and white version of this figure will appear in some formats. For the colour version, please refer to the plate section.*

Evaluating Drought Resistance/Tolerance

Although the experiments described in this chapter focus on clones that are popular or of scientific interest in eastern Africa, the approach adopted in attempting to quantify the relative sensitivity to drought/responsiveness to irrigation should be of wide international interest. Rarely has a critical evaluation of how clones respond to periods of dry weather been attempted. More often it is visual appearance that provides the evidence, but this is only one of many ways in which clones can be assessed.

Although it is not an entirely arbitrary distinction, the experimental results considered below are differentiated between immature (< 7 years from field planting) and mature (> 7 years) tea plants.

Much of the research reported in detail below was undertaken at the Ngwazi Tea Research Station (NTRS) in southern Tanzania. This site was specifically chosen because of the long dry season experienced every year (Figure 15.8).

Immature Tea

There is a general view that irrigating young tea is worthwhile (Figures 15.9 and 15.10). It can increase plant survival and improve early growth, leading to cumulative yield benefits over time (Dale, 1971). In Malawi, Willatt (1970) showed that irrigation increased survival of 18-month-old rooted cuttings, and improved rooting depth and lateral spread. Similarly, in Kericho, Kenya, Othieno (1978a) compared the responses of five young contrasting clones to supplementary irrigation during the dry season. As a result of better survival and early growth, cumulative benefits were recorded over

Figure 15.9 Micro-catchments serve a valuable function when planting young tea under dry conditions. These can either collect rain or hold water carried to the field (MKVC).

the first two years of yield recording. There was, however, little improvement in yield distribution, with low yields from both irrigated and unirrigated plants during the dry season when other factors were probably limiting shoot growth, such as dry air, or low night temperatures (Othieno, 1978b).

Scheduling Irrigation in Immature Tea

Based on measurements with a neutron probe made under young tea at NTRS, Burgess (1992) developed a simple model for estimating the cumulative water loss. Given the degree of ground cover, soil available water capacity, maximum rooting depth and ETc, the model partitions crop water use into transpiration, soil evaporation and drainage. Using this approach, Burgess (1993a) described a simple method for scheduling irrigation of immature tea. For each of three pan evaporation rates (3, 4 and 5 mm d^{-1}) graphs

Figure 15.10 In Africa away from the equator (e.g. in southern Tanzania, Malawi and countries to the south) there is a single dry season that can last from four to six months. In these areas, tea is often irrigated during the dry season.

were presented from which the potential SWD could be predicted for different degrees of ground cover, given the number of days since the soil profile was last at field capacity.

Mature Tea

In the 1950s and 1960s, a number of enterprising individuals in Africa and India experimented with sprinkler irrigation of mature tea (tea not being suited to flood irrigation). This interest was stimulated by the availability of portable aluminium pipes and robust sprinklers.

Results were disappointing (mainly because not enough water was applied, and because there were other more important limiting factors to yield, e.g. couch grass). There was also a reluctance amongst some managers to create extra work during the dry season when there were important leisure activities to enjoy! It was for good reason they were known as 'irritation' experiments. In addition, the sequence and duration of the seasons made interpretation of the results difficult. It was time for some scientific input.

One of the first tea irrigation experiments reported from Africa was undertaken at NTRS over the period 1967–1970 (Carr, 1974). In this high-altitude seasonal location, there is an extended dry season alternating with one rainy season. Frequent irrigation doubled the annual yields (from about 1000 to 2000 kg ha^{-1} processed tea; these are low yields by today's expectations) obtained from mature (planted 1959), heterogeneous seedling tea, previously heavily shaded, and improved crop distribution to some extent. Irrigation was applied at potential soil water deficits (SWDs) ranging from 25 to 150 mm, with the soil profile wetted to field capacity at each irrigation. The limiting SWD for both *annual* and *dry-season* yields for this deep-rooting tea (4.3 m) was in the range 100–150 mm, equivalent to about 25% depletion of the available water. There was some evidence of 'compensatory' shoot growth following the relief of water stress in the less frequently irrigated plots, and especially in the 'dry' unirrigated plots after the start of the rains. In each of the irrigated treatments, an average of about 700 mm of water was applied during the dry season. This corresponded to a *yield response to irrigation* totalled over the three-year period of about 1.4 kg ha^{-1} mm^{-1}. For one treatment, in which only half the calculated quantity of water was applied at each irrigation occasion (75 mm at a potential SWD of 150 mm), the corresponding yield response was about 1.8 kg ha^{-1} mm^{-1}.

The principal objective of another experiment at that time was to determine whether it is necessary to adjust the frequency of irrigation during the dry season, and particularly in the 'winter' months when shoot extension rates are slow (Carr, 1971a; Nixon and Carr, 1995). The results clearly showed that there were no adverse effects on total *annual* yields. However, crop distribution was dependent on the timing of irrigation. For clone BBT-1 (which has a high base temperature for shoot extension, 14 °C, and consequently low winter yields) there were no adverse effects from irrigating less frequently at this time (full replacement at a potential SWD = 150 mm) compared with more frequent irrigation (SWD = 50 mm). Indeed, when the treatment combinations were reversed in the warm dry season, yields were greater in the previous SWD = 150 mm treatment than in the control (irrigated at a SWD = 50 mm throughout the dry season). Similarly, compensatory growth during the rains made up for loss in yield during the warm dry season in the treatment that was irrigated less frequently at that time (SWD = 150 mm). The timing of nitrogen applications modified these responses. The large peaks in production that occurred after the relief of stress (induced by low temperature or dry soil) were the result of synchronisation of shoot growth, which is more extreme in this clone than in most others.

At about the same time, similar experiments were being undertaken independently in southern Malawi (16°05′S, alt. 650 m), where there is an equivalent seasonal climate to that found in southern Tanzania (Dale, 1971). Although most conditions were similar at both sites, there were marked differences in the dry-season yield responses to irrigation. In particular, there was variability between years in Malawi despite similar quantities of water being applied. In some years, yields in the dry season were limited by the hot (air temperatures > 30 °C), dry air (saturation deficits > 2.0 kPa) conditions experienced at that time in this relatively low-altitude site. In such a year, *yield responses to irrigation (plus rainfall) in the dry season* were only 0.3 kg ha^{-1} mm^{-1}, about one-third of those obtained in less extreme years (Carr *et al.*, 1987). Later, an analysis of commercial yields in Malawi suggested a *dry-season yield loss* corresponding to about 1.2 kg ha^{-1} mm^{-1} increase in *potential SWD* over the range 100–650 mm. This compared to an equivalent value of 1.4 kg ha^{-1} mm^{-1} for Kericho, Kenya (Carr and Stephens, 1992) (Box 15.7).

What to Plant?

This question goes way beyond the scope of this chapter. Here the focus is on how best to identify clones that do relatively well under dry conditions – i.e. those that exhibit

---

Box 15.7 Water Productivity

Unless it is clearly defined, there can be confusion over the use of this and related terms.

Rain-Fed Crop
- Annual yield 3000 kg ha^{-1}
- *ET* crop 950 mm
  - transpiration 800 mm
  - evaporation 150 mm
- Rainfall 1300 mm
- *Transpiration efficiency* = 3000/800 = 3.8 kg ha^{-1} mm^{-1}
- *Water-use efficiency* = 3000/950 = 3.2 kg ha^{-1} mm^{-1}
- *Water-use efficiency (rainfall)* = 3000/1300 = 2.3 kg ha^{-1} mm^{-1}

Irrigated Crop
- Annual yield 4500 kg ha^{-1}
- *ET* crop 1450 mm
  - transpiration 1250 mm
  - evaporation 200 mm
- Irrigation 500 mm

Water-use efficiency (irrigation) = (4500 − 3000)/500 = 3.0 kg ha^{-1} mm^{-1}

---

drought-tolerant or drought-resistant characteristics – and also clones that are responsive to irrigation. The results of several long-term experiments at NTRS, designed to provide answers to questions like these, are summarised below. Many were based on the line-source experimental design for applying differential quantities of water to a crop, as illustrated in Figures 15.11 and 15.12. This put a limit on the number of clones that could be evaluated. The first two experiments described here each included six clones, sourced from Kenya, Malawi and Tanzania. The relative capacity of tea clones to tolerate drought, or conversely to respond to irrigation, was assessed by comparing total annual yields, dry-season yields only, cumulative yields over a number of years post-planting, yield losses due to drought, and visual observation of the effects of drought. Interpretation of the results was not always easy.

Experiment 1

The yield responses of six contrasting clones, all of which were commercially and/or scientifically important in eastern Africa, to differential drought treatments were recorded from planting in 1988 until 2003. To establish the plants, the experimental area was uniformly irrigated for the first two dry seasons post-planting. Differential drought treatments were first imposed towards the end of the dry seasons in 1990 and 1991, for 16 and 13 weeks respectively.

Figure 15.11 A line-source irrigation experiment for evaluating responses of different clones to water stress – see text for further details (Ngwazi Tea Research Station, Tanzania, MKVC).

Figure 15.12 The plants in the foreground are fully irrigated, and those in the distance are rain-fed only, with those between being partially irrigated. Crop yield/water-use production functions can be developed by plotting yield against water applied or used. (Tanzania, MKVC).

Annual Yields

Annual yields obtained in these two years decreased as the *maximum SWD* (simulated using a water balance model) increased. Clones AHP-S15/10 and BBT-207 were identified as drought-sensitive (probably because of their sensitivity to a stem canker, *Phomopsis theae*), and TRFCA-SFS150 and BBT-1 as drought-tolerant; clones BBK-35 and TRIEA-6/8 were intermediate in response.

Reporting the same experiment, Nixon (1996c) demonstrated how responses to drought changed as the clones matured. Figure 15.13 shows this, and also how the shape of the response curves relating annual yields to the maximum potential SWD varied considerably between clones, with BBT-1 again being particularly tolerant of drought, and TRIEA-6/8 and AHP-S15/10 both relatively sensitive.

In 1993/94, 5–6 years after planting, annual yields declined rapidly once the SWD exceeded about 250–300 mm. This was the year before the first formative prune.

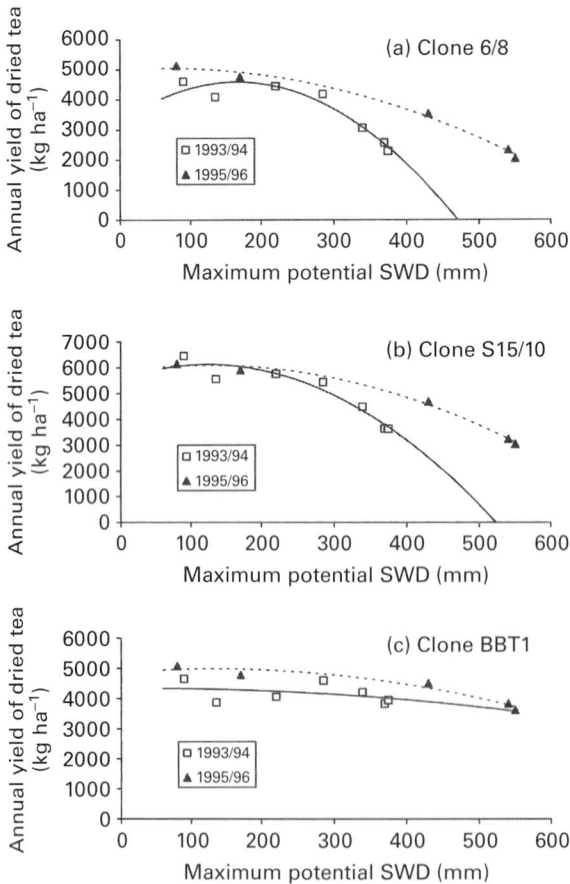

Figure 15.13 An example of the relationships between yield and the potential soil water deficit (i.e. the dryness of the soil) for three contrasting clones over two years. As the crop matures, its sensitivity to water stress declines.

The decline was less rapid in 1995/96, 7–8 years after planting (the year after pruning). There was no change in the shape of the response curve for BBT-1 (an extreme China-type clone), which remained flat. Clones TRFCA-SFS150, BBK-35 and BBT-207 were similar in response to TRIEA-6/8.

The annual yields of dry tea under well-irrigated conditions in 1995/96 were all in the range 5300–6300 kg ha^{-1}, with AHP-S15/10 yielding the most, and BBT-1 the least. There were considerable differences between the extremes amongst the six clones in their responses to drought. Assuming a linear yield decline once the SWD exceeded 300 mm, the annual yield losses (or yield gain from irrigation) were least with BBT-1 (only 1 kg ha^{-1} mm^{-1} in 1993/94 and still only 2 kg ha^{-1} mm^{-1} in 1995/96), and most with clone BBT-207 (as much as 15 kg ha^{-1} mm^{-1} in 1993/94, reducing to 7 kg ha^{-1} mm^{-1} in 1995/96) (Nixon, 1996c).

By 2002/03, there were further changes in the annual yield responses to drought (potential SWD). For clones TRIEA-6/8 and BBT-207 it appeared to be linear, with an average yield loss for both clones of 6.6 kg ha^{-1} mm^{-1}; for AHP-S15/10, BBK-35 and TRFCA-SFS150, it was two-step linear with yields declining by 4.4 kg ha^{-1} mm^{-1} once the potential SWD exceeded about 370 mm; for clone BBT-1 the relationship had not changed (Mizambwa, 2004).

Dry-Season Yields

When a similar analysis was conducted for dry-season yields, clone AHP-S15/10 was again identified as drought-sensitive in both years, but not BBT-207. Overall, AHP-S15/10 was the highest-yielding clone, reaching 5600 kg ha^{-1} in the fourth year after planting in the field when well irrigated (Burgess and Carr, 1996a).

Cumulative Yields

Ranking clones on the basis of cumulative yields over the first two pruning cycles (1990–1994 and 1994–1999) identified clone AHP-S15/10 as the highest-yielding under well-watered conditions, with 5000–8000 kg ha^{-1} (12–17% more tea than each of the other five clones). These were in the order TRFCA-SFS150, BBT-207, BBK-35, TRIEA-6/8 and BBT-1 (TRIT, 2000). Under droughted conditions BBT-1 yielded 3000 kg ha^{-1} (11%) more than TRFCA-SFS150, followed by BBK-35, TRIEA-6/8 and BBT-207.

Yield Loss Due to Drought

In terms of yield loss due to drought, BBT-207 suffered most, with a reduction in cumulative yield of 24,500 kg ha^{-1} (–58%), followed by TRIEA-6/8 (22,000; –54%), BBK-35 (20,000; –48%), AHP-S15/10 (20,000; –41%), TRFCA-SFS150 (15,000; –34%) and lastly BBT-1 (9000; –24%). Based on absolute yields under dry conditions, as well as by both relative and absolute yield loss, BBT-1, AHP-S15/10 and TRFCA-SFS150 could be identified as the three most drought-tolerant, with BBT-207 and TRIEA-6/8 the most drought-susceptible.

Visual Assessment

Based on visual assessments of the severity of drought symptoms in 2001, the clones were ranked in the order TRIEA-6/8 (worst), BBT-207, AHP-S15/10 = BBK-35 (but variable), TRFCA-SFS150, and BBT-1 (best).

Experiment 2

In a similar experiment with six other clones, AHP-12/28 (and also TN-14/3, TRFCA-PC113 and especially TRFCA-PC105) based on relative yield loss could be described as drought-sensitive, but in absolute terms it was the third highest-yielding under dry conditions and second highest under wet conditions. TRFCA-PC81 was the 'best' clone using any of the three criteria, and TRFCA-PC105 the 'worst'.

Cumulative Yields

Under well-watered conditions cumulative yields were similar for clones TRFCA-PC 81, AHP-12/28 and TRIEA-31/8, totalling 26,000 kg ha^{-1} over the five years 1998/99 to 2002/03. These were followed by TRFCA-PC113 (22,000), TN-14/3 (19,000) and TRFCA-PC105 (18,000). Under drought conditions, the clones were ranked in a similar order, with yield losses ranging from 9000 kg ha^{-1} (−35%) for AHP-12/28 to 7000 kg ha^{-1} for TRFCA-PC81 (−24%) and TN-14/3 (−37%).

Conclusions from Experiments 1 and 2

Of the 12 clones tested, the following four performed best under *well-watered conditions:*

AHP-S15/10, TRFCA-PC81, AHP-12/28, TRIEA-31/8

And, *under dry conditions:*

TRFCA-PC81, BBT-1,[1] AHP-S15/10 and TRFCA-SFS150.

What Plant Density?

This experiment is described in detail in Chapter 6, Box 6.3.

How Much Nitrogen?

The yield responses of mature clonal tea (TRIEA-6/8) to irrigation and fertiliser were studied at NTRS in a long-term (1986–2004) field experiment again based on the line-source technique. Stephens and Carr (1991a) reported the results for the first three years of the experiment in terms of *yield loss for each millimetre increase in the potential SWD*. In all three years there were significant linear or slightly curvilinear reductions in yields during the *dry season* with increasing potential SWD (up to *c.* 700 mm).

At relatively high fertiliser levels (300 kg N ha^{-1}), the rate of yield loss (dry tea) in the dry season was equivalent to 1.9–2.7 kg ha^{-1} (mm SWD)$^{-1}$, and for unfertilised tea 0.9–1.4 kg ha^{-1} (mm SWD)$^{-1}$. In the first two years, the benefits from irrigation continued into the rains, but in the third year there was evidence of some compensation, with yields from the unirrigated tea exceeding those from irrigated tea during the early part of the rainy season.

Annual Yields

Annual yields from unirrigated tea in well-fertilised plots increased from 2000 to 3200 kg ha^{-1} over the three years and, for the equivalent irrigated tea, from 3400 to 4900 kg ha^{-1}, reflecting in part cumulative benefits. Irrigation increased the proportion of the annual crop harvested in the dry season by up to 45%.

Relative Yield Loss

Following Doorenbos and Kassam (1979), relative yield loss during the dry season was plotted against the corresponding relative reduction in *actual water use* for each of the first four years of the experiment (1986–1989). The slopes of the response curves (Ky) were similar (except in the year of prune), and independent of fertiliser level, with a pooled value of 1.3, meaning that the reductions in yield *during the dry seasons* were proportionally more than the reductions in crop water use (Stephens and Carr, 1991b).

After pruning in November 1990, annual yields continued to increase in the following three years, reaching 6000 kg ha^{-1} in well-irrigated and fertilised plots by 1991/92 (Burgess, 1993b). *Annual yield responses to increases in the maximum potential SWD* were now consistently curvilinear at fertiliser inputs corresponding to 300 kg N ha^{-1} and above but, apart from in the year of prune (1990/91), linear at nitrogen levels less than these. It is not clear what caused these differences in the shape of the response curves. For example, the yield loss for each millimetre increase in the potential SWD at 225 kg N ha^{-1} was between 1.7 and 2.5 kg ha^{-1}. By contrast, at 300 kg N ha^{-1}, it increased from 0.75–1.8 kg ha^{-1} mm^{-1} at SWD = 200 mm to 2.3–3.0 kg ha^{-1} mm^{-1} at a SWD = 400 mm, and 4.9–11.8 kg ha^{-1} mm^{-1} at SWD = 600 mm.

Afterwards, Stephens *et al.* (1994) analysed the *annual* and *seasonal yields* and the corresponding *actual water-use* (*ET*a) totals for the first eight years of the experiment (1986–1993). For plots receiving annual applications of 225 kg N ha^{-1} (the commercial N level applied at that time), yields increased over this period from 1800 to 3700 kg ha^{-1} in the unirrigated tea and from 3200 to 5700 kg ha^{-1} in a well-irrigated crop. *ET*a did not vary greatly between years, averaging 700 mm and 1200 mm respectively. There was therefore an upward trend in water productivity over time from 2.3 to 5.3 kg ha^{-1} mm^{-1} for unirrigated tea, and from 2.7 to 4.6 kg ha^{-1} mm^{-1} for irrigated tea. In each year, *annual yields* declined linearly with increases in the maximum estimated *actual soil water deficit* (SWDa) at rates of up to 6.3 kg ha^{-1} for each millimetre increase in SWDa. The slope and shape of the relationships between yields and water use were strongly dependent on fertiliser application levels (range 0 – 450 kg N ha^{-1}).

In the cropping year 1995/96, irrigation was withheld for seven weeks prior to the start of the rains to allow a potential SWD of 230 mm to accumulate in the normally well-irrigated treatments. Although some crop may have been lost at this time, there was no obvious compensation in yield following the start of the rains as anticipated. At the 300 kg N ha^{-1} input level, the average *annual yield response to water (irrigation plus rainfall)* received during the year (1700 mm) was 3.4 kg ha^{-1} mm^{-1}, close to the values obtained in the previous three years (3.1–3.4 kg ha^{-1} mm^{-1}). Similarly, the yield response to drought over these four years showed a consistent curvilinear relationship between the *annual yield* (range 6000–4000 kg ha^{-1}) and the *potential maximum SWD* (range 50–800 mm). As the SWD increased, yields declined at an increasing rate, from 1 to > 6 kg ha^{-1} mm^{-1} SWD. For comparison, the corresponding linear mean was 3.2 kg ha^{-1} mm^{-1} SWD (Nixon, 1995, 1996a).

The weight of prunings recorded in June 1996 (all foliage above 0.40 m removed) reflected the cumulative effect of treatments on dry matter production in the five and a half years since the last prune (in November 1990). At the 300 kg N ha^{-1} level, dry weights increased from about 20 to 27 t ha^{-1} with increase in water applied, and from 7 to 30 t ha^{-1} as nitrogen fertiliser (as NPK) inputs increased from zero to 450 kg N ha^{-1} in the well-irrigated treatment (Mizambwa, 1997).

Frost

After pruning in June 1996, the experiment was severely damaged by frost. The tea was collar-pruned to aid recovery, and at the same time plastic barriers were installed to 1.0 m depth between the fertiliser treatments in order to restrict movement of nutrients between adjacent plots. The fertiliser treatments were also modified, with nitrogen alone becoming the principal variable, not NPK. For the next two years, the whole experiment was irrigated uniformly during the dry season until complete ground cover (> 95%) was again achieved. Afterwards, for the five years from 1999/2000 to 2003/04, differential irrigation treatments were reinstated. At the 300 kg N ha^{-1} level (which is close to the optimum for well irrigated tea), *annual* yields increased linearly ($R^2 = 64\%$; $n = 25$) with the *depth of irrigation water* applied (range 10–820 mm) during the dry season, from about 4300 kg ha^{-1} (rain-fed) to 6900 kg ha^{-1} (fully irrigated) at an average rate of about 3.1 ($\pm$ 0.49) kg ha^{-1} mm^{-1} (Kimambo, 2005).

Conclusion

For planning purposes, an annual yield response of 3–4 kg ha^{-1} for each millimetre of effective irrigation for well-managed clonal tea, after allowing for a residual potential SWD at the end of the dry season of 200 mm, is probably realistic, providing there are no other limiting factors. For low-input/low-yielding farming systems, it is proportionally less.

Mature versus Immature Comparisons

In an attempt to understand better the processes leading to reduced drought sensitivity with age, Nixon *et al.* (2001) compared the responses of mature and young clone

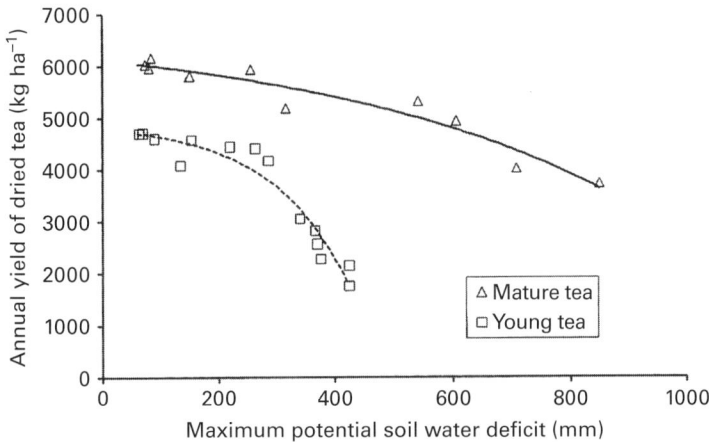

Figure 15.14 As the tea bush matures so its sensitivity to drought declines. Tea is most susceptible to drought in the third and fourth years after planting in the field, when the root:shoot ratio is out of balance.

TRIEA-6/8 plants in the two experiments about 22 and five years respectively after field planting (Figure 15.14). For the mature crop, annual yields (1993/94) did not decline until the maximum potential SWD exceeded about 300–400 mm, and 200–250 mm for the young tea. At deficits greater than these, yields declined rapidly in young tea (up to 22 kg ha^{-1} mm^{-1}) but relatively slowly in mature tea (up to 6.5 kg ha^{-1} mm^{-1}). The apparent insensitivity of the mature crop to drought was principally due to compensation during the rains for yield lost during the dry season. Differences in dry matter distribution and shoot:root ratios also contributed to these contrasting responses. Thus the total above-ground dry mass of well-irrigated mature plants was about twice that for the corresponding young plants, but the dry mass of structural roots was four times greater, and for fine roots eight times greater. In addition, each unit area of leaf in the canopy had six times (by weight) more fine roots to extract and supply water than did young plants.

Conclusions from All Experiments Described Above

The line-source irrigation system provided a successful way for evaluating a selection of contrasting clones to drought/irrigation and for assessing water productivity. It was also an excellent visual demonstration site for visitors.

With strictly controlled harvesting, the annual yield potential of some clones (e.g. AHP-S15/10, TRFCA-PC81, AHP-12/28 and TRIEA-31/8) exceeded 7000 kg ha^{-1} (dry weight) within six years of planting in this part of southern Tanzania.

Water productivity increased as yields rose.

Response Curves

- The shape of the annual yield/water-use or irrigation response functions changed with time from planting, and varied between the clones tested.

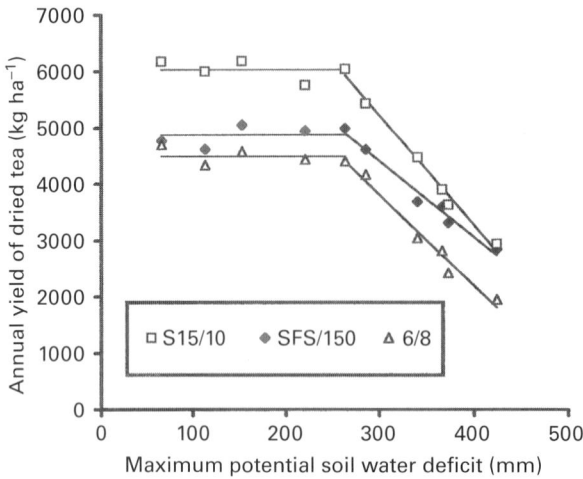

Figure 15.15 Example of two-step linear crop-yield water-deficit production functions for three clones.

- For some clones the response function was sometimes linear (e.g. AHP-S15/10), for others it was curvilinear (e.g. BBK-35), and for others it could sometimes be two-step linear (Figure 15.15). But it was not always consistent for individual clones between years.

Yield Responses
- Critical potential SWDs for *annual* yields varied between *c.* 50 mm and 200–300 mm, depending mainly on plant age. During the dry season, it is < 50 mm for immature tea.
- For 5–7-year-old plants (the year before the first prune), *annual* yield losses due to drought/gains from irrigation varied between 5–6 and 10–15 kg ha^{-1} mm^{-1} once the critical SWD was exceeded.
- Clone BBT-1 was an outlier; its (lack of) *annual* yield response to irrigation/ drought was totally different to any other clone. This was partly the result of yield compensation, more than in other clones, when drought stress was relieved by rain or irrigation.
- In general, the sensitivity to drought declined after six or seven years from planting (the year after the first prune) when it tended to stabilise at around 3–6 kg ha^{-1} mm^{-1}.
- Clones TRIEA-6/8 and BBT-207were both drought-sensitive when young, but clone TRIEA-6/8 in particular became more drought-tolerant with age. Care has to be taken when defining what is meant by drought tolerance.
- Increases in drought tolerance as plants mature are associated with reductions in the shoot:root ratios with time from planting.

Figure 15.16 Uprooting a bush infected with *Phomopsis theae* (RHVC).

- Based on cumulative yields, clones BBT-1, TRFCA-SFS150 and AHP-S15/10 can be considered to be drought-tolerant. TRFCA-SFS150 also has a low base temperature for shoot growth, while AHP-S15/10 recovers slowly from pruning. TRFCA-PC81, AHP-12/28 and TRIEA-31/8 also performed relatively well.
- Part of the benefit from irrigation with some clones was due to the control of the disease *Phomopsis theae* (Figure 15.16).

Composite Plants
- There is some evidence for worthwhile (*c.* 10%) yield benefits from selected scion–rootstock combinations (e.g. TRIEA-31/8 on TN-14/3), including improved drought tolerance.

Plant Density
- Large cumulative yield benefits (up to about 8000 kg ha^{-1}) are possible from high-density planting (20,000–40,000 ha^{-1}) in the first six years after planting, and also in the year following the first prune.

> **Box 15.8** Consensus Views of Commercial Irrigators on Irrigation of Tea in Southern Tanzania (Nixon,1995)
>
> 1. Profitability of irrigation of mature tea is very dependent on the shape of the yield/water-applied response curve.
> 2. Indications are that it is not profitable to apply full water requirements on well-fertilised tea.
> 3. A soil water deficit of 200–250 mm can be allowed to accumulate prior to the start of the rains without a loss in the annual yield.
> 4. But this will have an impact on crop distribution, with more crop in the rains and less in the dry season.
> 5. Young tea is very sensitive to water stress, with yields declining rapidly when soil water deficit exceeds 200–250 mm.
> 6. No one mentioned quality!

- There was no clear evidence that the optimum density depended on the crop water status. Well-watered and droughted plants responded to density in similar ways (from Chapter 6).

These complex findings have implications in terms of clone selection for improved water productivity under rain-fed or irrigated conditions.

Irrigation Systems

Conventional semi-portable sprinkler irrigation systems are generally used to irrigate tea in those areas of Africa where tea is irrigated. Poor design, excessively wide sprinkler spacings, and the adverse effects of wind have led to uneven water distribution, and water wastage. This has led to a commercial interest in drip irrigation as an alternative way of applying water to the crop.

Möller and Weatherhead (2007) evaluated in detail the technical and financial performance of the first commercial drip irrigation scheme (55 ha) in Njombe, southern Tanzania (Kibena Tea Ltd; 9°12′S, alt. 1860 m; deep clay loam soil, xanthic ferrasol). They found that the uniformity of irrigation water distribution and the efficiency (related to water losses) were both considerably better than those previously reported for adjacent sprinkler-irrigated areas. Scheduling irrigation (on three- or four-day intervals) using tensiometers offered potential water savings of 26% compared with the water balance approach currently used. Root systems adapted to the spacing of the dripper lines (1.2 m, single row; or 2.4 m apart, alternate rows). Gross margins were sensitive to the price and yield of processed tea, the amount and cost of fertiliser applied (fertigation was practised) and electricity costs.

There were considerable savings in labour compared to sprinklers. However, damage by rodents to the lateral pipes and to the emitters was a problem here, and

Figure 15.17 A major problem of drip irrigation in tea is susceptibility to damage of the pipes by pluckers and by rodents (Tanzania, MKVC).

especially on a nearby 200 ha drip-irrigated scheme (Figure 15.17). On this larger project, the saving in labour was offset by the extra people needed to identify and repair leaks in the pipes. Overall, however, drip irrigation was considered to be a technical success (Box 15.8).

In a supporting large-scale (9 ha) drip-irrigation field experiment at the same site, Kigalu *et al.* (2008) compared different levels of water application (0.25, 0.50, 0.75 and 1.00 times the cumulative SWD, one dripper line per row; together with 0.25 and 1.00 times SWD, dripper line alternate rows). Water was applied at three- and four-day intervals, over four successive dry seasons (2003/04 to 2006/07). Unfortunately, shortages of water in three of these years meant that irrigation was curtailed before the end of the dry weather. As presented, the results are not easy to explain or interpret. Unless confirmed, the results need to be treated with caution.

Annex to Chapter 15 Research and Recommendations Reported from Elsewhere

USSR

More than 50 years ago, Lebedev (1961, 1962) and Petinov (1962) reported the results of some fundamental research on the water relations of tea undertaken in Azerbaijan and Georgia (then both part of the USSR). Both locations had a Mediterranean-type climate with a hot, dry summer. By applying water daily between 1100 h and 1600 h (for three-minute spells four times every hour) yields were increased by 50% compared with tea that was irrigated conventionally at 10-day intervals. They believed that the change in microclimate afforded by daily intermittent irrigation, for example by reducing the air temperature and raising the humidity, was the reason for its success. Several indicators of plant water status, such as leaf temperature and stomatal opening, supported this view. Subsequent research in Africa has confirmed the validity of this interpretation (e.g. Tanton, 1982b; Carr *et al.*, 1987).

India

Little formal research on tea irrigation appears to have been reported from India, although there is a prolonged dry season in the north-east during the winter months. This is particularly severe in Terai, Dooars, Assam, North Bank, Nowgong, Cachar and, to some extent, parts of Darjeeling. The general advice given to growers in the north of India is as follows (Tea Research Association, 2012). It is not clear on what evidence these recommendations are made:

Maximum response of irrigation is generally obtained in the best sections of existing mature tea areas. For this it will be also essential to identify and remove other limiting factors. The best results are expected to come from irrigating unpruned or early light skiffed teas. In general, depending upon rainfall received in October, irrigation should commence from November and continue till March/April. The first application in November can be a little more than the estimated field irrigation requirement followed by five more applications, each at an interval of three weeks. In severely drought prone years, irrigation in April may be necessary but after the rainfall in April exceeds 75–125 mm, irrigation can be discontinued. As in the case of young tea, the irrigation schedule should not be interrupted except after heavy rainfall exceeding 38 mm and then for not longer than 2–3 days. The quality of the water used for irrigation should be checked prior to use.

Some situations are particularly prone to drought, for example the exposed south and south-western slope in Cachar, Darjeeling and elsewhere, where there is very sandy soil especially when this is underlaid by stony layers and where the soil is shallow. This is the case along the 'Jhora' edges in the Dooars and Terai.

In South India Varadan (1996) made a brave attempt to develop an irrigation scheduling/planning model for tea for growing in seven different regions. Several assumptions were made, some of which were probably erroneous. The effective rooting depth for mature tea was assumed to be only 0.3 m and the allowable soil water depletion level as an arbitrary 0.5. The available water capacity of the soils was assumed to vary from 80–100 mm m^{-1} in some regions up to 120–140 mm m^{-1} elsewhere. Reference crop evapotranspiration was calculated using the rarely reported Hargreaves temperature method for each of the seven regions. ETo peaked in March with values ranging between 4.3 mm d^{-1} (Nilgiris) and 6.0 mm d^{-1} (Wayanad). The crop factor was taken to be a realistic 1.0 in all cases. The corresponding predicted irrigation intervals for these two areas were five and seven days respectively, with net water applications of 26 and 30 mm. Rainfall appears to have been ignored. Unfortunately no attempt was made to relate this analysis to the results of tea irrigation research done elsewhere, when different assumptions may have been made and conclusions reached, particularly with respect to the frequency of irrigation. Unfortunately, Varadan (1996) made no attempt to quantify the yield benefits to be expected from irrigation, using the international literature; only the frequency and peak irrigation water requirements in each region were specified. The analysis is open to criticism.

Note

1. BBT-1 does not make good tea, but could make a good rootstock (see Chapter 5).

16   A Shady Business

Teas Need Trees

One generation plants a tree, the next generation enjoys the shade – old African saying

Trees are an important component of several tea production processes. First, they are associated with the long-established but controversial tradition of growing tea under shade, and second, on many estates, wood burning provides the energy needed to process the tea in the factory. Trees are also used to provide shelter from wind, either in clumps placed in strategic wind-exposed places, or as lines planted within the tea across the direction of the prevailing wind. Leguminous woody plants, planted on the contour, can also play a role in the control of run-off and soil erosion on sloping land, while tea itself can be an important component of some agroforestry-based crop production systems. Each of these uses will be considered briefly in this chapter.

Shade Trees

Shade trees were introduced into the tea plantations in North-East India in the second half of the nineteenth century in an attempt to reproduce the shady conditions of the forest, the natural home of the large-leaved Assam-type tea plant. One of the first trees to be used was a legume, *Albizia chinensis* (sau tree), and Buckingham (1885) described it as being 'highly beneficial'. Other leguminous trees were also found to be 'useful', and by the first quarter of the twentieth century most planters in North-East India accepted shade as a regular feature on tea estates (Box 16.1). A tradition developed whereby a good stand of shade trees came to be associated with good tea culture. This was the tradition exported into other tea-growing areas such as Sri Lanka and South India, with little objective evidence to support the practice. Planters moving to East Africa to establish the tea industry in the 1930s chose a non-legume species (*Grevillea robusta*) to provide shade. But, as Barua (1989) points out in a comprehensive review of the role of shade trees in tea, the term 'shade' means different things depending on location. For example in Assam, shade trees were, and still are, expected to provide more or less continuous cover for the tea (Figure 16.1). By contrast, in Sri Lanka, South India and previously East Africa, it was rare to see continuous cover, as the trees were widely spaced (Figure 16.2).

At about the same time, in the early twentieth century, sulphate of ammonia was being introduced to estates as an alternative (to organic matter) source of nitrogen. This led

---

Box 16.1 Shade Tree Species Planted in India in 1940

The Indian Tea Association published a pamphlet in 1940 describing, in a series of excellent drawings, 12 of the 20 or so leguminous 'jungle trees' that had been or were being evaluated as possible shade trees (Winter and Bora, 1940). These included *Erythrina ovalifolia* (large leaves, deciduous, rapid grower), *Albizia lucida* (deciduous, leafless for short time in early rains), *A. procera* (upright, deciduous), *A. lebbeck* (deciduous, slow growing), *A. odoratissima* (small and slow-growing), *A. moluccana* (evergreen, shallow rooting, easily blown over), *A. stipulata* (deciduous, wide spreading, easily established), *A. myriophylla* (a woody creeper), *Derris robusta* (hardy, easy to establish, deciduous), *Dalbergia assamica* (resistant to most common diseases), *Cassia siamea* (deciduous) and *Leucaena glauca* (common in Java).

In the introduction to the pamphlet, it was stated that '*shade in moderation is considered by most practical means to have a beneficial effect on the growth of tea. It is however an accepted fact that excessive shade has a detrimental effect both on crop and quality. What constitutes excessive shade has never been determined and is still liable to much controversy.*' It went on to say '*it is fairly certain … that the manurial effect of the fallen leaves will prove to be much more important economically than the shading effect of the canopy.*' Already, the confounding effects on tea productivity of shade per se and the addition of nitrogen from the leguminous shade trees were recognised.

Figure 16.1 In the tea plantations in North-East India, shade trees were introduced in the nineteenth century to reproduce the shady conditions of the forest. This image shows low-level shade provided by a legume tree (*Albizia chinensis*) (Bangladesh, MKVC).

Figure 16.2 Shade trees used to be common in East Africa, but are now rarely seen. This photograph shows *Grevillia robusta* shade trees in Kericho, Kenya, before their removal (MKVC).

scientists to look at the value of shade by comparison with responses to nitrogen fertiliser. The large-scale shade trials conducted by Cooper (1926) and Wight (1959b) in Assam soon showed that there was a strong interaction between the response to shade and the response to nitrogen. Namely, the response to nitrogen was greater in unshaded tea than in shaded tea. Nevertheless, legume-shaded tea even without nitrogen yielded more than unshaded tea with nitrogen.

In an ambitious attempt to understand the fundamental mechanisms influencing the response of tea to shade, Hadfield (1974a) described in great detail the light climate, including spectral analysis, at the surface of the canopy in shaded tea areas in North-East India, and the effects of various species of shade trees on this climate. The aim was to identify mixtures of species and spacing of shade trees to provide uniform, reproducible light conditions in order, for example, to improve the precision of field experiments under both clear sky and overcast conditions.

Results of a long-term (15-year) shade experiment in Malawi reported by Shaxon (1968) showed that all species of shade tree tested caused a marked reduction in yield during the rainy season (November to April). The degree of depression in yield was apparently related to the amount of shade cast, while significant responses to nitrogen fertiliser were recorded only under full sunlight. In Kenya, McCulloch *et al.* (1965) adopted a different experimental approach. Yields of green leaf were recorded for a period of three years from rows of tea successively further away from the rows of non-leguminous *Grevillea robusta* shade trees at two sites (Limuru and Kericho). Yields were greatest where the tea was least shaded. In Kericho, Kenya, artificial shade was compared with *G. robusta* shade trees and an unshaded control over a period of 10 years. The largest yields were obtained from bushes beneath the artificial shade, provided by bamboo laths resting on galvanised wire netting. One explanation for this unexpected response, proposed by Tolhurst (1969), was that the crop was responding to zinc being leached out of the galvanised wire. In other words, it was nothing to do with shade as such, purely an artefact (see Chapter 12). At this cool, high-altitude site (2178 m), yields from the tree-shaded treatment were similar to those from the unshaded control treatment.

In order to understand the results more fully, Ripley (1967) reported the results of detailed measurements of the microclimate at this Kericho site in the rainy season and again during dry weather. The predominant effect of the 5–6 m tall shade trees, spaced at 9 × 9 m intervals, was to reduce the radiation at canopy level, but only by 20% overall. There was a slight reduction in wind speed and in the minimum air temperature. Artificial shade reduced radiation by 40% and reduced the daily range of air temperature. The data were averaged over one-month periods. It is possible that the low temperatures were the dominant limiting factor, as a result of which the tea crop was unable to respond to full light intensity.

Subsequent measurements made 20 years later, over a four-year period on the same experiment, showed that the heavy shade cast by the now very large *G. robusta* trees reduced yields compared with the unshaded control treatment, by up to 50%, even after the shade trees had been lopped. This loss in yield was the result of a reduction in the number of tea shoots per unit area (Othieno and Ng'etich, 1992).

Because of the many interacting factors involved, unravelling this complex story about shade trees, a puzzle that occupied many hours of debate amongst planters as well as scientists up to the 1970s, has not been easy. Factors considered at different times include increased circulation of nutrients from depth through leaf fall, the supply of mulch and organic matter from fallen leaves, protection from hail damage and shelter from wind. But science eventually came to the rescue through the fundamental studies carried out by Hadfield (1968, 1974a, 1974b) in Assam. In summary, in locations where the ambient air temperature regularly exceeds 30 °C, tea leaf temperatures can be several degrees warmer than this and exceed 30–35 °C, which is above the optimum for photosynthesis in tea. Large horizontal leaves (Assam-type) heat up more than small vertically inclined leaves (China-type). The small upright leaves also allow light to penetrate deeper into the leaf canopy. Shading, by cooling the leaf, limits how often 35 °C is exceeded. Shade is therefore likely to be beneficial in hot, windless areas where large-leaved Assam-type clones or jats are grown, such as in Assam. Shade is unlikely to be of value in cool, high-altitude areas. As a result, shade trees are now rarely seen in eastern and central Africa, but are prevalent in North-East India and Bangladesh. They can still be seen in South India and parts of Sri Lanka, but there they are usually widely spaced (Figure 16.3).

In weighing up the advantages of removing or keeping shade trees, it should not be forgotten that they provide social and environmental benefits, including increased biodiversity, firewood for smallholders and shade for pluckers. Perhaps shade trees in tea should be treated as a form of intercropping within an agroforestry project and

Figure 16.3 The definition of what is effective shade varies from country to country. Does this example – taken in South India – qualify as shade, or is it serving an ecological function (MKVC)?

evaluated accordingly (see below). Leguminous trees used as shade in mature tea include the Ceylon sau (*Falcataria moluccana*), an evergreen, fast-growing tree, which is favoured in Sri Lanka but 'not generally liked' in North-East India. Dadap (*Erythrina indica*) is common in Sri Lanka, where it is kept lopped, but it is only seen occasionally in North-East India.

To note:

Nothing is new! An announcement in Bulletin No. 57, *Proceedings of Workshop on Tea Research Past, Present and Future* (UPASI Tea Research Foundation, September 2013, p.13) states:

A research project has been sanctioned by the Tea Board under the 12th Five Year Plan to carry out collection and evaluation of suitable alternate shade/fuel trees for South Indian tea plantations.

I hope they read Winter and Bora's (1940) pamphlet first!

Fuelwood – Hot Stuff

When the cost of oil became prohibitively expensive in the 1970s an alternative source of energy was needed to provide the energy to operate the tea factory. Fuelwood was seen to be one answer. The trees now grown for this purpose are predominantly *Eucalyptus* species (Figure 16.4) As pressure on land (and water) increases so it is becoming necessary to treat the production of fuelwood in similar ways to those one might use to produce the manufactured tea. In other words, treat trees as a commercial

Figure 16.4 *Eucalyptus grandis* is probably the most widely planted of all the eucalypts. It is a popular tree, grown for its fuelwood to provide the energy needed to run a tea factory (Tanzania, MKVC).

crop and pay attention to the details of silviculture (Schönau, 1983). This section summarises the main agronomic issues that should be addressed when growing fuel-trees as a crop. Reference is made to the definitive text on eucalypts by Eldridge *et al.* (1994) and to the reports prepared by Paul Jacovelli when working as a consultant to CDC East African teas in the 1990s.

Not everyone is enthusiastic about the value of eucalypt plantations, and there is much opposition to them in some countries, including India. As Eldridge *et al.* (1994) stated:

the fault is not with the trees but with the people whose combination of enthusiasm and ignorance led to the planting of eucalypts in unsuitable places. It is not appropriate to have large areas of close spaced trees where water is a more important commodity than wood.

Limited information on eucalypts as crop plants means that proponents and opponents are not sufficiently well informed to debate their value, but their contribution to the tea industry is not questioned, providing responsible land use planning is practised.

Species

There are literally hundreds of species (more than 500) of eucalypts (commonly known as gum trees), the vast majority of which are indigenous to Australia. Only a few are cultivated commercially. They come in a great range of shapes and sizes, from tall trees to small shrubs. As a primary source of *Eucalyptus* oil, the blue gum (*Eucalyptus globulus*) is one of the most commonly cultivated species around the world. Similarly, *E. grandis* is probably the most widely planted of all the eucalypts. Grown for hardwood production, including fuelwood, there are in excess of 2 million ha of these trees planted worldwide.

The natural distribution of *E. grandis* in Australia is between latitudes 33°S in New South Wales and 16°S in northern Queensland, where the dry season lasts about three months, and from sea level to an altitude of 1100 m. *E. grandis* will generally grow faster than any other tree species on good sites in subtropical and warm temperate climates. Under favourable conditions, it can yield between 30 and 50 m^3 ha^{-1} per year.

By describing the conditions in Australia where *E. grandis* is indigenous, an indication of where it can be successfully cultivated can be given. This information is then collated with climatic data collected from sites elsewhere in the world where *E. grandis* is under evaluation. Using this approach, Booth and Pryor (1991) specified the broad geographic limits within which *E. grandis* can be grown. The mean annual rainfall should be within the range 700–2500 mm; the dry season should not last longer than seven months; the mean maximum temperature in the hottest months should not exceed 34 °C nor be below 25 °C; similarly the mean minimum temperature in the coldest month should be between 3 °C and 18 °C, and the mean annual temperature between 14 and 25 °C, while the absolute minimum temperature should not be less than –8 °C. This approach does not specify the optimum conditions for *E. grandis*.

E. grandis has the following attributes:

- excellent natural stem form
- excellent branching characteristics
- adaptable to a wide range of sites
- responds to good management, including weed control
- relatively easy to grow from seed
- regrowth after bush fires
- vigorous regrowth after coppicing for fuel.

On the negative side:

- low wood density, commonly around 400–500 kg m^{-3} from young trees
- intolerant of weed competition, especially grasses
- susceptible to various diseases (particularly to fungal stem and leaf pathogens)
- intolerant of flooding and frost.

Other species include *Eucalyptus urophylla*, which occurs naturally on Timor and adjacent islands north of Australia, and *E. camaldulensis*. *E. urophylla* is well suited to wet/tropical regions with dry seasons lasting from two to six months at altitudes up to 3000 m. Growth rates can exceed 30 m^3 ha^{-1} per year. *E. camaldulensis* is the most widely spread of all eucalypts, being indigenous throughout Australia with the exception of Tasmania. It is mainly a riverine species growing on silts and sands. It is widely planted on infertile and dry sites, particularly in the Mediterranean region and North Africa, but its popularity is increasing in South-East Asia and Brazil.

Silviculture

There is considerable variability between eucalypt seed provenances (or seed sources) in a number of attributes. It is advisable only to use improved seed. The best in terms of growth and form are often from the Coff's Harbour and Buladelah areas of New South Wales. Breeding programmes are under way in a number of countries. *E. grandis* seed is available from South Africa, Zimbabwe, France and Australia. Standard propagation procedures are followed in a well-sited nursery. The preferred spacing is 3.0 × 1.5 m (density = 2222 trees/stems ha^{-1}). One month after planting, 100 g single superphosphate (10.5% P) can be applied to each tree. Within three weeks after planting out any vacancies should be infilled to ensure uniformity. It is essential to control weeds during crop establishment and to take appropriate precautions against fire. In favourable sites canopy closure occurs after about 12 months, after which the lower branches begin to be shed.

Estimating Production

The basic tree measurements used to estimate the stocking levels (i.e. the number of trees/unit area) and individual tree (and plantation) volumes are described here.

1. Stem diameter: this is measured using a special diameter tape placed around the circumference of the tree at breast height, 1.3 m above the ground.
2. Tree height: this is the vertical distance between the base of the tree and its uppermost point (the tip). This is best measured using a clinometer.
3. Tree volume (m^3) = tree height (m) × basal area (m^2) × a 'form factor'.

The 'form factor' is a correction to allow for the taper of the whole tree (= 0.33 for *E. grandis*). See Philip (1994) for more detail. An appropriate, statistically sound, tree selection sampling procedure is needed when estimating production.

Yield Prediction

When developing or expanding a tea project, it is first necessary to undertake an inventory of any existing fuel stocks and then to predict forward the amount of fuelwood that is likely to be needed (which is dependent on the amount of tea to be processed). In addition, the expected rate of development of the trees over the anticipated duration of the rotation needs to be estimated. This estimate is based on knowing the shape and position of the curve relating the mean annual increment in tree volume (MAI) to the age of the trees. These are known as MAI curves. Ideally they should be developed under the local conditions but, as this can take several years to complete, examples of curves obtained from trees grown under similar conditions can be used to make the initial estimates. In the examples given by Paul Jacovelli in his consultancy report, the MAI (for *E. grandis*) reached its peak (about 50 m^3 ha^{-1} a year) at a tree age of about seven years in equatorial Uganda (stem density = 2222 ha^{-1}), but in South Africa, with its marked seasonal climate, it took 10 years from planting for the MAI to reach its peak (about 30 m^3 ha^{-1} a year, with a stem density of only 1000 ha^{-1}) (P. Jacovelli, personal communication). The year during which MAI reaches a maximum is regarded as the ideal time to harvest. Beyond this time interval, competition between trees within the stand increases, and this eventually leads to the death of some trees.

In Uganda, the maximum MAI values reached in year 5 were assumed, for planning purposes, to be 55 m^3 ha^{-1} a year (from a possible range of 50–70). These estimates should have been ratified by measurement. In climatically suitable areas in Africa, Papua New Guinea and Brazil, annual growth rates of 70–90 m^3 ha^{-1} have been recorded. Assuming a density of 400 kg m^{-3}, this equates to annual dry matter yields of up to 36 t ha^{-1} (Eldridge *et al.*, 1994).

Fuelwood Efficiency

This can be measured either as kg (made tea) m^{-3} (fuelwood), or the inverse mass or volume of fuelwood needed to produce one kilogram of tea. In Uganda one company achieved 350 kg (made tea) m^{-3} of fuelwood, whereas a sister company in Malawi could only achieve 100 kg (made tea) m^{-3}.

The most obvious causes of this large difference are poor storage of the wood and inadequate drying (Figure 16.5). Good practice involves the following actions:

- The wood should be cut into relatively small slices (0.3 m long × 0.15 m diameter) and allowed to dry for at least six months. The wood is ready to be used when it is air dry.
- The grate should be stacked with logs positioned uniformly to ensure even burning. The logs need to be burnt in strict rotation according to their drying sequence.
- Dry wood has an energy equivalent of about 20,000 kJ kg^{-1}. Wet wood (say 50% moisture content) has an energy equivalent of nearly half this value. Planting, growing, harvesting, transporting and handling the additional trees needed to make up the difference if wet wood is used have a cost.

Figure 16.5 Firewood should be stacked under cover to allow it to dry out. Here is an example of good practice (Tanzania, MKVC).

Water Use and Productivity

Despite popular views to the contrary, several detailed studies on the water use of eucalypts indicate that they use no more water than comparable tree crops in terms of water productivity, that is the amount of dry matter produced per unit of water transpired by the trees. In particular, concern has been expressed in many areas of the world over the effects of large-scale planting of eucalypts on the hydrology of small watersheds. For example, because of the likely adverse impact on low water flows, Sikka *et al.* (2003) advised caution when planning large-scale conversion of natural grasslands to blue gum plantations in the catchments of hydroelectric reservoirs in the Nilgiris mountains of South India. Both low and peak flows were reduced as a consequence of the change from grass to trees (*E. globulus*). The changes in flow were more pronounced during the second rotation (after coppicing) than during the first.

Transpiration

A range of techniques has been used to measure eucalypt transpiration, which is largely a function of the leaf area index (L) and the saturation deficit (SD) of the air. Drought avoidance is facilitated through low L values and large seasonal changes, the near-vertical arrangement of the leaves, the sensitivity of stomata to the SD, deep rooting, osmotic adjustment to maintain the turgor pressure in the leaves, and homeostatic adjustment of the hydraulic properties of the conducting tissues. In the dry zone of South India, water availability is the principal limiting factor to growth, but, when water is available, nutrient limitations become important. Removal of both water and nutrient stress can result in a fivefold increase in volume growth in the first year (Calder *et al.*, 1993).

Micrometeorological methods were used by Kallarackal and Somen (1997a) to monitor transpiration in a four-year-old coppiced plantation (*E. grandis*) at Muthanga in South India (11°39′N 76°21′E, alt. 750 m). Water loss ranged between 2.5 and 6.5 mm d^{-1} depending on the season, which is equivalent to an annual total of 1200 mm. For comparison, the average annual rainfall at this location is 1300 mm. Despite maintaining relatively high leaf water potentials, stomatal conductances declined exponentially as the saturation deficit of the air increased from zero to 4.5 kPa. This indicated effective stomatal control of transpiration by the trees, especially during the dry season when the saturation deficits were large. These observations provide no support to suggest that *E. grandis* is 'a large consumer of water'.

The same authors later reported the results of a similar study on water use, but this time by *Eucalyptus tereticornis* at two sites, planted at different densities (1800 and 1090 stems ha^{-1}) in the Kerala State Forest in south-west India (annual rainfall 2500–3000 mm). During the pre-monsoon period, transpiration estimates ranged from 0.6 to 1.2 mm h^{-1} at site 1 and from 0.2 to 0.6 mm h^{-1} at site 2. Totalled over a year on rain-free days, these equate to annual transpiration totals of 1560 (high density) and 850 mm (low density) respectively. Although with this species the stomata did not close completely even at high saturation deficits (5 kPa), these are realistic transpiration totals for a perennial tree crop (Kallarackal and Somen, 1997b).

In Brazil, Cabral *et al.* (2010) used the eddy-covariance method to measure the water use of *Eucalyptus* (*grandis* × *urophylla*) hybrids over a two-year period. The average daily evapotranspiration (*ET*) rates were 5.4 (± 2.0) mm in the summer and 1.2 (± 0.3) mm in the winter. Annual evapotranspiration totalled 1120 mm in year 1 (82% of the annual rainfall) and 1235 mm in year 2 (96%). A detailed study in China on the Leizhou peninsula produced similar estimates of annual evapotranspiration, which averaged 1070 mm over two years at two sites on contrasting soil types (range 970–1150 mm). These values represented from 50% to 70% of the annual rainfall. Transpiration did not exceed 600 mm in either year. Soil evaporation was 15–26% of the total evapotranspiration. During the dry season *ET* matched or exceeded rainfall.

Water Table

In Brazil, a water-use model was developed and tested in a nine-year-old eucalypt plantation (*Eucalyptus grandis*) (Soares and Almeida, 2001). The upward flux of water from the water table was about 82 mm (over a 12 month period), and piezometric measurements showed a 2.5 m decline in the height of the water table over the same period. At the end of a long dry season, the upward flux of water into the root zone was about 1 mm d^{-1}. Total transpiration for the year was 1100 mm, with 150 mm intercepted by the foliage, and another 80 mm evaporated from the soil surface. When taken together, the total annual evapotranspiration was 1345 mm, compared with a rainfall total of about 1400 mm. The transpiration deficit (i.e. the difference between potential evapotranspiration, calculated with the Penman–Monteith equation, and actual evapo-transpiration) was 125 mm.

In the State of Victoria, Australia, Heuperman *et al.* (1984) monitored the effects of a 2.4 ha mixed-species (five, including *E. grandis*) plantation of eucalypts, established with irrigation, on the water table beneath the plantation and in adjacent areas of irrigated pasture. The water table in the land surrounding the plantation was at a depth of 2–3 m below the surface, and more than 4.5 m deep below the trees. The influence of the trees only extended 8 m into the pasture. Piezometric levels were higher than the water tables both inside and outside the plantation. In shallow aquifers, closely spaced trees will lower the water table, even when high piezometric pressures maintain the water table.

Water-Use Efficiencies

Stape *et al.* (2004) compared the growth of 14 stands of (fertilised) *Eucalyptus* hybrid (*grandis* × *europhylla*) situated in a geographic gradient across north-east Brazil. There was a fourfold difference in productivity along this gradient. Water was identified as the most important limiting factor. For each 100 mm increase in rainfall there was an increase in above-ground biomass production (net) of 2.3 t ha^{-1}. The most efficient stands produced 3.2 kg m^{-3} transpired water or 1.14 kg GJ^{-1} of solar radiation. At a regional level, a high productivity stand could produce the same amount of wood in a six-year rotation with half as much water as a low productivity stand on twice the land area.

In south-western Australia, Mendham *et al.* (2011) were concerned that eucalypts planted on agricultural land may deplete soil water stored at depth (down to 8 m) during the first rotation and, should this be the case, the trees would then suffer from a shortage of water during the second and subsequent rotations, unless the water was replenished between rotations. To test this theory, Mendham *et al.* (2011) used a combination of field experiments (*E. globulus* at three contrasting sites) and modelling. At all three sites, water was depleted at all depths, but it was only at the wettest site that the soil profile was fully recharged – but only after the first ratoon had been harvested. The maximum soil water deficits recorded at each site were similar (about 800 mm), regardless of the crop management practice imple-mented. On the sites that were responsive to nitrogen, the trees used more of the stored soil water each year, but also produced more wood. Modelling indicated that at most of the sites with a soil depth greater than 4 m the stored soil water was unlikely to be replenished during the second rotation. Wood yields in the second rotation are therefore likely to be less than those obtained from the first rotation. Later rotations would be affected even more. Plantation managers will need to do some innovative thinking if wood yields in the second and subsequent rotations are to match those obtained in the first. For example, fallowing between rotations would allow surplus rainfall to recharge the profile, or run-off water could be directed from elsewhere on the farm to the trees for storage in the soil profile (a practice known as run-off farming).

Shelter from Wind

In the late 1960s, the widespread removal of shade trees in most tea estates in East Africa meant that the tea bush was more exposed to wind than previously. It then became fashionable to plant *Hakea salicifolia* (previously known as *H. saligna*) hedges as shelter belts across the direction of the prevailing wind. It has since become common practice to remove them. What evidence is there for the net benefit of *Hakea* hedges to tea productivity, or, on balance, are the costs incurred in maintaining the hedges greater than any benefits?

To try to answer these questions, yield measurements were made in a commercial area of tea, comprising six blocks of young clones surrounded on all sides by 4–6 m tall *Hakea* hedges. The fresh weight of young shoots was recorded at successive distances from the hedges, which were planted across the direction of the prevailing wind. Ancillary measurements were made during the year of the plant water status, soil water content, stomatal opening and open water evaporation also at successive distances from the hedges. The sequence of wet/dry, warm/cool seasons was the same as at the Ngwazi Tea Research Station, which was about 20 km distant (see Box 3.7). The shelter experiment was however situated close to the 600 m Uzungwa escarpment (hot air rising from the plains below, but cooling rapidly), and the rainfall was higher than at Ngwazi. The experiment was initially rain-fed. The results obtained were not easy to interpret: responses to shelter by individual clones were influenced in part by differences in their sensitivity to (a) drought, (b) a stem canker (*Phomopsis theae*) and (c) the planting date/age of the bushes. In summary, the observations during three years of measurements were as follows:

1. During the rainy season and the cool dry months that followed, shelter had a beneficial effect on yield, namely that yields declined with increasing distance from the hedge. In addition, rates of refoliation after pruning were faster during the dry season in sheltered areas.
2. Towards the end of the extended dry season, when it was warm, shelter had an adverse effect on green leaf production. Tea plants in the lee of the shelter belt came under greater water stress than tea that was exposed to the wind. Apart from the rows immediately adjacent to the hedges, there was no evidence that competition between a hedge and the tea for water was the reason for these unexpected observations (Box 16.2).
3. When the adverse effects of drought on wind-sheltered tea bushes close to the hedges were severe, these plants took longer to come back into production after the start of the rains. As a result, yields during the rains remained low in the lee of the hedge.

Because of the different responses to shelter depending on the sequence of seasons, and the carry-over effects from one season to another, together with the stage of development of the tea plants, it is not easy to make a single recommendation (Carr, 1985). The answer, as ever, is 'it all depends … '

Box 16.2 Shelter from Wind

Possible mechanisms responsible for the seasonal differences in responses to shelter include:

- *Cool dry season* – a 1 °C increase in mean air temperature in the lee of the shelter represents a 20% increase in the effective temperature (base temperature = 12.5 °C, mean ambient temperature = 17–18 °C). This is enough to explain the differences in yield between sheltered and wind-exposed tea during the winter months.
- *Warm dry season* – the adverse effects of shelter during the long dry season (in the absence of irrigation) can be explained by the relative changes in the ratio of the canopy (associated with the degree of stomatal opening) and aerodynamic resistances (a function of wind speed) on actual rates of evapotranspiration. At the same time, in the wind-exposed areas, the wind will cool the leaves and by so doing reduce the saturation deficit (the driving force for transpiration) between the saturated vapour pressure at the leaf surface and the actual vapour pressure in the surrounding air (Carr, 1985).

Contour Hedgerows

Most tea in Sri Lanka is grown on sloping terrain in areas of high rainfall (> 2500 mm yr^{-1}). The loss of soil nutrients in the eroded soil, as well as the removal of nutrients 'exported' in the processed crop results in the depletion of nutrients from the soil profile. The continued cultivation of tea is unsustainable if these losses continue without adequate replenishment. One way to control erosion is to plant hedges on the contour within the tea, but there is a risk that the hedges will compete with the tea for resources both above (e.g. light) and below ground (e.g. water). De Costa *et al.* (2005) reported the results of a detailed study of the impact of several potential hedgerow species on tea in a mid-altitude (945 m) humid zone of Sri Lanka. The mean annual rainfall over the 50 months of the study was 2900 mm. There were no prolonged periods without rain. The soil was classified as a haplic alisol with a clay loam texture on the surface and loam to clay in the subsoil. The experimental field had a 30–35% slope (Figure 16.6).

The impact on the tea (clone TRISL-2023) of each of six hedgerow shrub species, all of which were planted on the contour, was assessed. The six species were *Calliandra calothyrsus, Senna spectabilis, Eupatorium inulifolium, Flemingia congesta, Gliricidia sepium* and *Tithonia diversifolia*. The tea was pruned at three-year intervals, and the hedgerows every four months. As with experiments on shade (from the sun) and shelter (from wind), interpretation of the results was not straightforward. Total tea yields from the hedgerow intercrops were, with one exception, always below those from the sole tea crop. Addition of the hedgerow prunings as a mulch to the tea increased yields from both young (six months after planting at the start of the experiments) and mature tea

Figure 16.6 Watering young tea plants to aid establishment in the field. Note the woody plants planted across the slope to control run-off and soil erosion (Sri Lanka, MKVC).

(six years). Some hedgerow species reduced yields from the young tea more than from the mature tea. In other cases it was the reverse. Removing tree root competition by cutting a trench increased tea yields by 11–19%. The one exception was with *Eupatorium* with mature tea plus mulch. This was the only treatment to show a significant yield improvement (+18%) over the control. Providing the right hedgerow species can be identified, there may still be opportunities for the use of contour hedgerows in Sri Lanka.

According to Gnanapragasam and Sivapalan (2004), soil degradation as a result of excessive erosion and the consequent loss of soil fertility, and profligate use of agrochemicals, were all 'causes for concern' in Sri Lanka. To ensure the sustainability of the tea industry, tea production should be concentrated on the best soils, and the steep, eroded areas used for fuelwood planting. Terracing, surface drainage and establishing fast-growing, mulch producing, leguminous contour hedgerows which can protect existing land and also help to rehabilitate degraded fields on sloping terrain at mid- and high altitudes, should all be encouraged (although the economics are not considered).

Agroforestry Systems

There are opportunities to promote agroforestry systems that include tea intercropped with another crop, as noted in the following examples.

In China, interplanting tea with another tree crop is a long-established cultural practice. An estimated 100,000 ha of tea, 10% of the total planted area, was intercropped

Figure 16.7 The stems of these *Grevillia robusta* shade trees are providing support for black pepper, a profitable partner (South India, MKVC).

in 1991. Associated tree species include a number of fruit crops such as pear, plum, peach and grape, and commercial crops like rubber and *Paulownia* (for timber). Research has concentrated on studying the effects of shade on the microclimate and the effect of these changes on the growth and physiology of the tea (Huang *et al.*, 1991; Yu *et al.*, 1991). Of particular interest is a rubber (*Hevea brasiliensis*) and tea agroforestry system that has been developed for application on degraded soils in South China. It has apparently contributed significantly to slowing soil erosion, improving soil quality, and providing economic benefits to farmers (Pareham, 2000).

A tea/tallow intercrop is also promoted in China. The tallow tree (*Sapium sebiferum*) increases soil fertility, which in turn benefits the tea. Certain enzymes are increased in the soil, including invertase, urease and phosphatase, as is the organic-matter content of the soil (Huang and Ding, 1989).

In South India and Sri Lanka black pepper is cultivated as an intercrop in rubber, coconut and areca nut gardens and climbing on shade trees in coffee and tea plantations (Figure 16.7). For smallholders, intercropping can help to spread the risks associated with monoculture, increase farm incomes and provide additional employment opportunities.

17 A Nice Cup of Tea

Is It Made in the Field or in the Factory?

Good tea is made in the field, value is added in the factory – an old tea planter's saying

One cannot write about the agronomy of tea without considering the impact of field practices on the quality of the end product. Since the objective is to produce a 'nice cup of tea' it is important to understand how what is done in the field can influence the processing procedures in the factory and the value of the final product. It is also reasonable to expect the field manager to appreciate the challenges faced by the factory manager.

Although the emphasis in this chapter is on black tea, the basic processes for producing the whole range of teas available on the market are summarised in a chart (Figure 17.1) and in Boxes 17.1 (black tea), 17.2 (green tea, oolong tea and pouchong tea), 17.3 (instant tea and ready-to-drink teas) and 17.4 (white, puerh, compressed, flavoured and decaffeinated teas). The details are abstracted from a range of sources.

What Is Quality?

We start the questions with an easy one! What do we understand by the word 'quality'? Tea producers rely on the discrimination of professional tea tasters to determine quality and therefore price. The tea tasters judge quality based on appearance (mainly – one exception is that first flush teas in Assam are judged on flavour, although appearance is also important) and taste. Judging these criteria can be very subjective, with two tea tasters often disagreeing on a valuation, but not always – sometimes good tasters agree (Corley and Chomboi, 2005). Since quality means different things to different people, failure to agree can lead to a certain level of cynicism amongst those with a less discerning palate. To add more confusion, when tea is scarce the price differential for quality is less than when the supply of tea on the world market is good. There can also be a lag in time between when the tea is processed and when it is sold, during which the quality criteria may have changed as well as the quality deteriorating.

Quality in black tea is determined partly on the basis of the appearance of the processed tea leaf, including its colour (usually, the blacker the better); particle size, since this affects the 'cuppage' (that is, the smaller the particles the greater the number of cups of tea that can be obtained from 1 kg of made tea); size uniformity; and general

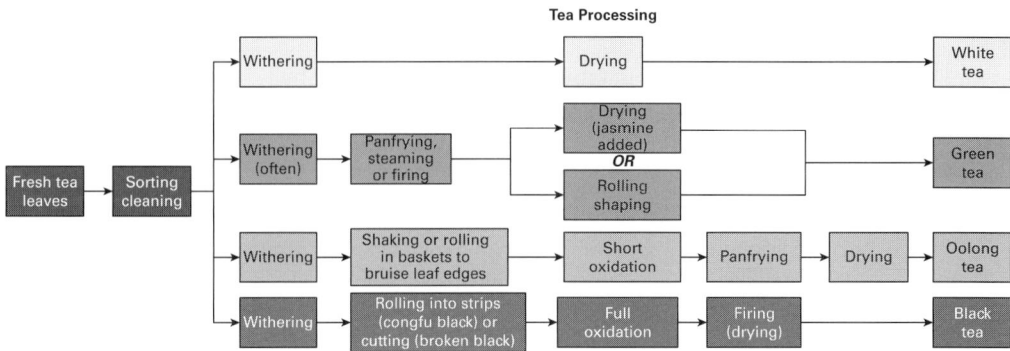

Figure 17.1 Tea processing chart: fresh green leaves enter the factory on the left-hand side, and after a range of processes leave the factory on the right-hand side as a specialty tea.

Box 17.1 Black Tea

The basic process for producing black tea involves five stages: withering, rolling or cutting, oxidation, drying and sorting. *Withering* allows the harvested leaf to lose moisture over a period of 18 hours or so, while the leaves become soft and malleable. This also assists in breaking down the cell walls to facilitate the rupturing process. The leaves are then ready for the next stage, which can be either *rolling* (orthodox manufacture) or *cutting* (CTC manufacture). Both these processes rupture the leaf cells, allowing the juices to mix and, in the presence of air, to initiate *oxidation* of the *catechins* (a process facilitated by the release of the enzyme *polyphenol oxidase* from lignified tissue). This results in the synthesis of complex flavonoids known as *theaflavins* and *thearubigins*, a process often referred to (wrongly) as *fermentation*. Oxidation is allowed to continue for 1–3 hours, depending on conditions. During this time the *aroma* and *flavour* develop, and the *colour* of the macerated leaf becomes darker. In order to arrest the oxidation process, the leaf is then *dried*. This reduces the leaf water content to 2–3%, and fixes the theaflavins and thearubigins, the quantity and proportions of which determine the colour and taste of the tea liquor, to the black tea particles. The dry product is then *sorted* into different particle size fractions (grades) by sieving, and any fibre is removed (Ellis and Nyirenda, 1995). Bulking and packing are also important, although not strictly part of the manufacturing process.

Box 17.2 Green Tea, Oolong Tea and Pouchong Tea

Green tea
- There is no oxidation process involved and hence no direct chemical reactions.
- Processing methods vary from country to country and from region to region, but the basic manufacture includes the following components:

Box 17.2 Continued
- a short period of withering (not always), followed by
- steaming (or pan firing), which deactivates most of the enzymes responsible for changes in the aroma complex, and at the same time makes the leaves soft and supple, then
- a series of rollings (traditionally done by hand) to macerate the leaf, and firings to shape and dry the leaf. These give each tea its own characteristic appearance.
- In Japan, the whole process is mechanised.

Oolong tea is partially oxidised tea traditionally made in China and Taiwan but now also made elsewhere.

- There are two processing methods. One produces dark, open-leafed oolongs; the other results in greener, rolled or 'balled' oolongs.
- Both methods involve withering, which starts the oxidation process, and then turning or tumbling the leaf.
- In the case of the open-leafed oolongs the leaf is turned for 5–10 minutes when oxidation is 70% complete before drying.
- For rolled oolongs, the leaves are heated when oxidation is 30% complete, then dried.
- The following day the leaf is wrapped into large cloths and rolled, unwrapped, and then rewrapped many times until the leaves are formed into rough green pellets. They are then dried.

Pouchong tea is another type of oolong tea, but with an even shorter period of oxidation, making it closer to green tea.

Box 17.3 Instant and Ready-to-Drink Teas

Instant tea – scientists have been working for many years, certainly since the 1960s, to produce an instant tea that can compete with instant coffee. But, for the connoisseurs, they have yet to succeed in their aim.

- The process begins by creating an infusion of tea, separating the solid material, and then alternately heating and freezing the solution before filtering through a special membrane.
- After separation, the solids are freeze-dried and the powder is packed into glass jars. Much of the instant tea produced is used in ready-to-drink teas.

Ready-to-drink tea – a vast range of plain and fruit-flavoured ready-to-drink 'teas', bottled or canned, is now being marketed. They are based on either black or green tea. It is an expanding market, having been pioneered in Japan and the USA during the 1990s.

Box 17.4 White, Puerh, Compressed, Flavoured and Decaffeinated Teas

White tea was originally only made in China from the terminal buds of young shoots (before they start to unfurl). White tea is now made elsewhere, using the new bud together with one or two young open leaves. After harvesting, the buds/leaves are dried. When infused they produce a very pale liquor (see Box 17.5).

Puerh tea – a traditional tea from Yunnan Province in south-west China. There are two types: 'raw puerh', which takes many years to mature, during which time true fermentation occurs; and 'cooked puerh', a modern version designed to replicate the mature earthiness of raw puerh by a faster method.

Compressed tea – a traditional way of storing and transporting tea in China in the form of bricks or balls.

Flavoured tea – any type of tea can be scented or flavoured with flowers, fruits, spices or herbs. The additional flavourings are normally blended in at the end of the manufacturing process. Perhaps the best-known flavoured tea is Earl Grey, in which the essential oil of a Chinese citrus fruit (bergamot; *Citrus aurantium* subsp. *bergamia*) is added to the processed leaf.

Decaffeinated tea – when tea is drunk, the caffeine (a stimulant) is released into the bloodstream slowly (because of the controlling effects of other ingredients in tea). It can take 15–20 minutes to be absorbed. Caffeine from coffee enters into the bloodstream much more quickly. There are three different methods by which caffeine can be removed from tea: (1) carbon dioxide is used as an organic solvent – no chemical residues are left in the tea; (2) methylene chloride – this is the most widely used method; and (3) ethyl acetate – this extracts other components as well as caffeine, but is difficult to remove.

Box 17.5 Hairy Leaves

Wight (1932) described the presence and importance of hairs on the back of leaves of tea, particularly on the terminal bud where they play a role in tea manufacture, particularly those teas that include a high content of 'tips' (e.g. white tea). The average length of a hair is 0.015–0.030 mm, and on an average second leaf (of two and a bud) there are about 170 hairs mm^{-2} on a good jat and about 50 mm^{-2} on a poor one. Yes, someone counts them!

---

appearance. There are two markets: (1) *loose leaf* must be 'black, clean, granular and neat', and (2) *tea bags* – modern tea-bag machines need dense tea. The colour of the liquor is also important, while what is perceived as flavour comes from smell. Countries rank and weight these quality criteria in different ways. For the trader selling cups of tea in the street in Mumbai, cuppage is important. At the other end of the market, it is the aroma that is important in what are known as 'flavoury teas'. 'Plain' teas are not considered to be flavoury, while 'medium' teas, which the British tend to prefer, have some aroma. When people in the UK are in a hurry, it is the speed at which the colour develops after the tea bag has been inserted into boiling water that is considered to be important.

The Raw Material

The market is constantly changing its preferences. For this reason managing the processing of green leaf into black tea is a continual challenge. A further complication is that the raw material (green leaf) coming to the factory from the field is not consistent. Its specification varies, depending on:

1. How it is harvested (e.g. fine or coarse plucking; selective plucking or not).
2. The year within the pruning cycle (e.g. 'tippings' – leaf produced in the first few harvests after pruning – give a black tea appearance but the liquor is 'thin').
3. The clone (e.g. shoot size, fibre content, rate of oxidation and flavour).
4. The fertiliser regime (e.g. an annual application of 300 kg N ha^{-1} makes a very different tea from 100 kg N ha^{-1}).
5. The season (e.g. wet or dry, cold or hot).
6. The time of day (e.g. with or without dew on the leaves, temperature).
7. How it is stored in the field (e.g. under cover or in the open) and how it is handled.
8. The time taken to transport it to the factory, including the state of the roads, which affects the condition in which the leaf arrives (Figure 17.2).

In addition to the challenge of handling a variable product at the point of entry to the factory, there is an infinite number of variables that can be adjusted in the factory at the discretion of the manager. For example, the duration of the withering process, the depth of leaf in the troughs, the degree of ventilation and the temperature of the air passing through the troughs (Figure 17.3). In addition, the amount of maceration can also be varied, for example by the number of CTC (crush, tear, curl) cuts (Figure 17.4), with or without a rotavane in the line, followed by the time allowed for oxidation, the temperature in the drier and the duration of the drying process. Finally, the grading of the leaf can be adjusted to suit the prevailing market. Some of these processes are now automated, but monitoring the external environment and adjusting the machines according to the condition of the green leaf on its receipt at the factory is a highly skilled job. Managers need to be mechanics and chemists as well as to have discerning taste buds. Watching a tea taster identify a tea that has been in the drier for five minutes longer than necessary or recognise the need to make some other relatively minor adjustment to the CTC machine is impressive (Figure 17.5).

Notwithstanding that many of these variables are outside the control of the factory manager, the market still expects a consistent product from the factory. The skill of the

Figure 17.2 The state of the roads affects the condition in which the leaf arrives at the factory (MKVC).

factory manager is to manage the processes involved in ways that add value to the product coming into the factory from the field. Compare this with the quality-control procedures that are imposed on the growers of peas for the frozen pea market in Europe.

Chemical Analysis

For obvious reasons there is a need to find a more objective and replicable way of scoring tea quality than relying on tea tasters alone. Research has focused on the links between quality and the chemical composition of the processed leaf. Roberts (1950, 1962) was the first person to discover that black tea contained two groups of phenolic substances, which he called *theaflavins* and *thearubigins*, and their significance for tea quality (Figure 17.6). They are produced when compounds known as catechins (found in the green leaf) are oxidised by enzymes during the oxidation stage of tea manufacture

Figure 17.3 The duration of the withering process, the depth of leaf in the troughs, the degree of ventilation, and the temperature of the air passing through the troughs are variables that the factory manager has to consider (MKVC).

Figure 17.4 The amount of maceration of the leaf can be varied, for example by the number of CTC cuts (MKVC).

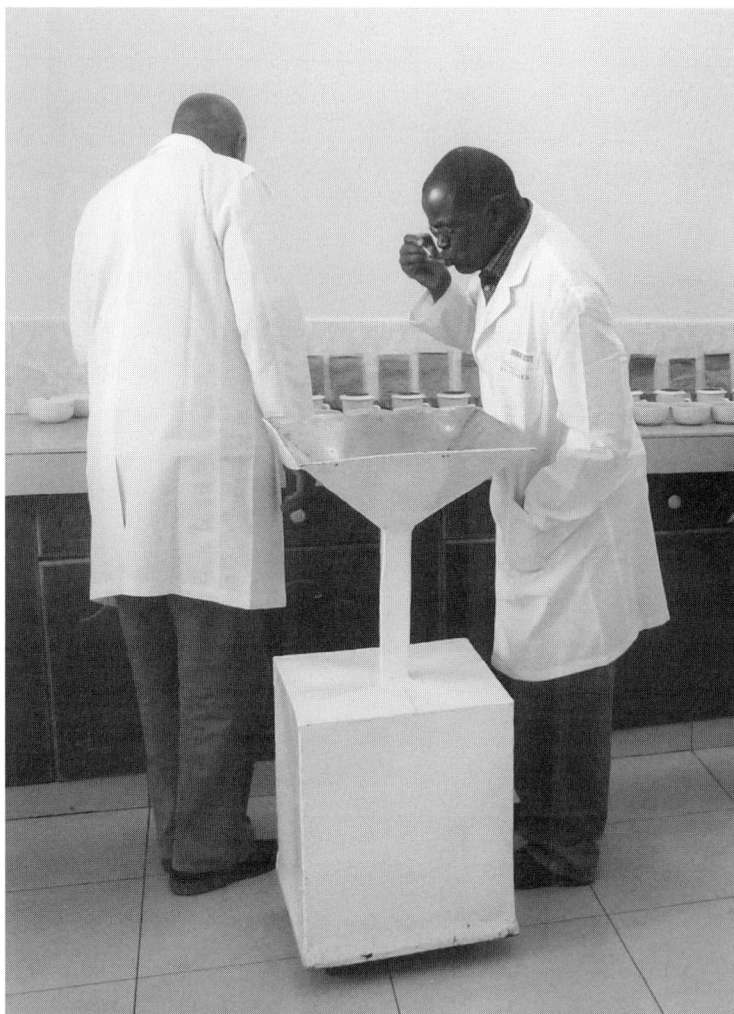

Figure 17.5 Tea producers rely on the discrimination of professional tea tasters to determine quality and therefore price. Tea tasters judge quality based in part on appearance and taste (MKVC).

(McDowell and Owuor, 1992). The intense bright red colour of the (four) theaflavins gives tea a quality called 'brightness'. The same molecules contribute to what the tasters call 'briskness' and 'freshness'. These words are used to describe how tea feels in the mouth. These qualities are essential components of 'breakfast teas', particularly those from Kenya and Assam. The four theaflavins do not contribute equally to quality, and the ideal mix is not yet known, neither is the ideal combination of their precursors, the catechins, in the green leaf.

Like the theaflavins, the thearubigins are coloured phenolic compounds produced during the oxidation process. It is these molecules that are responsible for what tasters call 'body', 'depth of colour', 'richness' and 'fullness'. Catechins are oxidised to

Figure 17.6 Theaflavins and thearubigins are coloured phenolic compounds produced during the oxidation process. The former give tea 'brightness', 'briskness' and 'freshness', while the latter are responsible for the 'body', 'depth of colour', 'richness' and 'fullness' of the tea liquor (MKVC).

theaflavins, which are then further oxidised to thearubigins, so the longer the duration of the oxidation process during black tea manufacture the more thearubigins are produced and the fewer theaflavins remain (Figure 17.7). Unlike the theaflavins, the chemical structure of the thearubigins is not known. Tea tasters are looking for teas that combine the highest levels of thearubigins with briskness. But these two groups of compounds are

Figure 17.7 (a, b) Oxidation is allowed to continue for 1–3 hours, depending on conditions. During this time, the aroma and flavour develop and the colour of the macerated leaf becomes darker. The leaf is then dried. The quantity and proportions of theaflavins and thearubigins determine the colour and taste of the tea liquor (MKVC). *A black and white version of this figure will appear in some formats. For the colour version, please refer to the plate section.*

not the only molecules that give tea its taste. Substantial quantities of catechins that survive the oxidation process contribute to astringency of black tea; in green teas they provide the characteristics of 'astringency' and 'bitterness' (Robertson, 1992).

In Malawi and North-East India, good relationships have been established between theaflavin contents and prices or tasters' evaluations. But no such relationships have been established in Sri Lanka or Kenya (Owuor *et al.*, 1986).

Aroma

All these molecules (theaflavins, thearubigins and catechins) are non-volatile. Other, more volatile, molecules are responsible for the aroma to be found in the more expensive teas. Some of the aroma molecules are present in the green leaf, while others are produced during processing in the factory. There are more than 500 such molecules (Robinson and Owuor, 1992). At specific times of the year, depending on the climatic conditions, flavoury teas occur in Darjeeling (India) and in the high-altitude (above 1200 m) areas of Sri Lanka. Specific weather conditions lasting for at least two weeks uninterrupted by rain prior to the onset of the monsoon are needed to stimulate the production of the aroma complex. Climatic conditions similar to these occur in January

Table 17.1 Grades of black tea – the key grades for CTC and orthodox manufacture are listed in descending order of size

CTC grades (crush, tear and curl)	Orthodox grades (this is not a complete list)
Brokens	**Whole leaf**
BP1 – broken pekoe one	TGFOP – tippy golden flowery orange pekoe
BP – broken pekoe	FOP1 – flowery orange pekoe one
Fannings	OP – orange pekoe
PF1 – pekoe fannings one	**Brokens**
PF – pekoe fannings	TGBOP – tippy golden broken orange pekoe
Dusts	BOP1 – broken orange pekoe one
PD – pekoe dust	BP – broken pekoe
D1 – dust one	**Fannings**
D – dust	BOPF – broken orange pekoe fannings
	PF – pekoe fannings
	Dusts
	PD – pekoe dust
	D – dust

and February at high altitude in the Dimbula District (Sri Lanka), and in August and September in the Uva District. Flavoury teas also occur during the first and second flushes in Darjeeling in India when similar climatic conditions prevail. The aroma associated with semi-oxidised teas such as oolong from China and Japan and pouchong from Taiwan is associated with *methyl jasmonite*, an important constituent of jasmine flowers. Some odours are unpleasant, but low concentrations of these may be essential components of the overall aroma. High-grown Darjeeling teas contain five times as many floral odour molecules as low-grown Assam teas. The characteristic flavour of Darjeeling tea is believed to come from geraniol.

Grades

Confusingly, flavour has nothing to do with the way teas are graded. The traditional names, from 'orange pekoe' through to 'dust', reflect the size and appearance of the leaves, nothing more (Table 17.1). Particle size is important for cuppage – hence the relatively high value of dust in South India, for the tea shops. Dust is also important for rapid infusion in tea bags.

Caffeine

Tea is a mild stimulant. This is due principally to the presence of caffeine. The highest concentration is in the terminal bud. As the leaf develops and matures the caffeine content declines. Fine plucking tea will therefore produce a tea with a higher caffeine content than will coarse plucking. For leaves of the same age each clone has a different caffeine content. There are also large seasonal variations. For all these reasons, it is not possible to consider including caffeine content as a parameter in any international standard designed to ensure that the product is authentic tea (Owuor and Chavanji, 1986).

Interactions

Assessing 'quality' with any degree of reliability and robustness is not easy, while many factors contribute to yield. It is therefore not always easy to demonstrate cause and effect. Owuor *et al.* (2011) summarised the state of understanding in Kenya of the impact of environmental factors on tea quality. These included:

- **Soil type** – There was no evidence in Kericho, Kenya, to show that the productivity and quality of different clones varied with (a very limited range of) soil type. Where soils have degraded over time, nutrient depletion may influence productivity, but any variations in black tea quality have not yet been documented.
- **Altitude** – This is essentially a surrogate for temperature, Owuor *et al.* (2011) provide some (limited) evidence to show that the 'quality' of aromatic black teas as measured by the caffeine content, the 'flavour index' and 'sensory evaluation' were all positively correlated with increases in altitude (attributes associated with slow growth rates), but not all the genotypes that were tested responded in the same way (Owuor *et al.*, 1990).
- **Location and seasons** – Cooler weather and slower shoot growth rates are linked with improved quality. Since the switch from dry to wet conditions, and from cool to warm weather, can lead to a large flush, it is critically important to have sufficient harvesting and processing capacity available at the right time, otherwise crop is wasted and quality will definitely be reduced.
- **Genotypes** – There are genotypes that yield well in one location but relatively less well in another (Owuor *et al.*, 2011). The same genotype × environment interaction can occur with tea quality parameters. For example, the theaflavin content of black tea from clone TRIEA-31/8 (a standard in East Africa) varied between 21.8 µmol g^{-1} at one site in Kenya, where it was ranked 14th out of 20, and 25.7 µmol g^{-1} at another site, where it was ranked 7th. By contrast, another clone (TRFK-56/59) performed equally badly at each of three sites, being ranked 20th, 19th and 20th (Owuor *et al.*, 2010a).
- **Harvesting** – The *timing* of each harvest is largely a function of temperature, which varies between sites and seasons. The plucking *standard* is also influenced by the plucking *intensity* (defined by the number of leaves left on the plucked stump after harvest).[1] This reflects a trade-off between yield and quality. For example, when the leaf standard was one leaf and a bud (clone TRIEA-6/8), the theaflavin content was 25.0 µmol g^{-1}, for two leaves and a bud it was 29.3 µmol g^{-1}, but when the leaf standard was five leaves and a bud it was only 19.0 µmol g^{-1}. For clone AHP-S15/10, theaflavin concentrations were 17.7, 19.2 and 13.4 µmol g^{-1}, respectively (Obanda *et al.*, 2002).

Box 17.6 Commentary: Quality Criteria

Cost saving is a panic reaction to an adverse situation. The long-term solution is to receive higher prices for your product. That is not possible as long as tea is treated as a commodity. It must be accepted as a consumer product, which is designed for those markets that can support higher prices. Product quality and specific attributes must be designed as per customer requirement. There is no better way of doing that than understanding the chemistry of tea. It can then be manipulated to make a designer product that will fetch the price you want.

These were Nigel Mellican's opening remarks as a session chairman at the Third International Conference on Global Advances in Tea Science, Kolkata, November, 2003. They remain valid today.

- **Nutrition** – The recommended economic rates of nitrogen fertiliser to apply to tea for yield vary from country to country, but are commonly between 100 and 300 kg N ha^{-1} a year (see Chapter 12). This is subject of course to affordability and availability. In Kenya, increasing the amount of nitrogen applied from 100 kg N ha^{-1} to 600 kg N ha^{-1} reduced the 'flavour index' from 0.94 to 0.62 and the 'sensory evaluation' from 40 to 34. There was also a small reduction in the theaflavin content (from 11.6 to 10.3 μmol g^{-1}) (Owuor et al., 1997a). Similar results were obtained from an experiment on rates of nitrogen repeated at five sites across Kenya. Raising the level of nitrogen applied from 0 to 300 kg N ha^{-1} reduced the 'sensory evaluation' index from 96 to 26, the theaflavin content from 25.4 to 22.1 μmol g^{-1}, and the thearubigin content from 17.5 to 15.1 μmol g^{-1}. There were also significant differences between some of the sites in the sensory evaluation and in the theaflavin content (Owuor et al., 2010b). Leaf analysis shows that genotypes differ in their capacity to absorb nutrients. There is no published evidence to suggest that phosphorus or potassium influence quality (Owuor et al., 2011).

Ideally, different genotypes require individual processing treatments, while processing practices need to be location-specific if a uniform tea is to be produced. For example, different genotypes subjected to the same conditions during withering have different optimal oxidation ('fermentation') times. But such adjustments are only worthwhile if the amount of green leaf for an individual clone (and its market value) justifies the changes needed.

There is always a trade-off between yield and quality, but since both the price obtained for the processed tea and the premium for quality are inherently volatile, it is not always easy to make the correct choice (Box 17.6).

Comparisons Between Sensory Evaluations and Chemical Analysis of Black Tea

In Kenya, the quality of the black tea manufactured (CTC) from leaf produced by three popular, high-quality commercial clones, as judged by sensory evaluation and

Figure 17.8 (a) Ready to go, and (b) on its way to port (MKVC).

by chemical analysis, was assessed. The leaf was harvested according to four standards (one leaf and a bud, two leaves and a bud, three leaves and a bud, four leaves and a bud). Differences between the clones in quality were greater than the influence of the leaf standard. According to Kamunya *et al.* (2012), this could mean that clones that produce high-quality tea could be harvested at an inferior standard (even four leaves and a bud) and still produce teas of a comparable standard to clones of inherently moderate to low quality when plucked at a comparable stage. It is understood that one commercial company is harvesting one clone (AHP-S15/10) at the four-leaves-and-a-bud stage. A detailed evaluation of a range of biochemical compounds found in black tea was undertaken in Sri Lanka by Kottawa-Arachchi *et al.* (2014). A total of 35 germplasm accessions, representative of the three main taxa (Assam, Cambod and China), were selected from the field gene-bank collection at the Tea Research Institute of Sri Lanka. Leaf samples (two leaves and a bud) from each clone were processed under standard conditions using the Teacraft mini-manufacturing system.[2] The tea leaves were then infused in boiling water and after cooling analysed for the important biochemical compounds. The results were compared with the organoleptic evaluations made by four professional tea tasters. There were significant differences among the 35 accessions in all the biochemical parameters tested. Only a small proportion of this genetic diversity has so far been used as parents in tea breeding programmes.

When the results of the biochemical analysis were combined with the tasters' evaluations, the 35 clones could be separated into four groups, distinguished on the basis of the origin and ancestry of the genotypes. The most important compounds for characterising this germplasm were theaflavins, thearubigins, total polyphenols, total catechins and caffeine. These assessments supplemented the organoleptic evaluations made by the tea taster based on leaf colour, liquor colour, strength and liquor quality. There were also seasonal differences (between wet and dry seasons) in the contents of all but one (theaflavins) of these five compounds. Cultivars with low caffeine contents were identified.

Residues

Another aspect of quality of current concern is that of 'chemical residues' in the processed tea. This is particularly important in India, where it is necessary to use pesticides to control the many pests and diseases. Pesticides are toxic chemicals, and there is international pressure to reduce, minimise or preferably avoid their use. Increasingly, legislation now restricts the use of chemicals that were previously allowed. For each chemical the government of India has specified the maximum recognised upper limit (MRL) for several of the pesticides used in tea. Some of the limits have been set very low, but the levels are constrained by the precision/sensitivity of the analysis available. In reality, this means zero tolerance (Barooah, 2010).

The consumers of tea in the West are also imposing their own controls on the use of pesticides in tea. For teas aimed at export to the European Union there are, in effect, only three pesticides (possibly five) that can be used. Japan has also set strict limits. Research at Tocklai has however shown that only a small fraction of the pesticide residue in the processed tea is transferred into the tea liquor. The amount depends on the relative solubility of the residue in water. The acceptable level for a residue also varies from one country to another. For example, endosulfan, an insecticide, has an MRL of 24 mg kg^{-1} in the USA, 30 mg kg^{-1} in Japan and the EU and 5 mg kg^{-1} in India (Barooah, 2010).

Despite the uncertainty about harmful levels, the pressure to reduce the use of pesticides in tea will not go away. As concerns about risks to health and adverse environmental impact grow, this pressure can only increase. Research on integrated pest management will need to continue. Tea producers in Africa are very fortunate in that they have little need to use pesticides on a regular basis. Ironically, it was herbicides that, by controlling the weeds, enabled the canopy to shade out most of the weeds, especially couch grass, so that herbicides are no longer much needed in well-managed tea areas.

Box 17.7 Milk First or Second?

And to answer the question which you have all been waiting to ask . . .

If you take milk with your tea, should the milk go into the cup before the tea, or does the tea go in first?

Answer 1 – Citing George Orwell, Mair and Hoh (2009) say add tea to milk (that way you know how much milk you are using).

Answer 2 – Hereward Corley says: Add milk to tea and you can see by the colour how much to add. If you put milk in before the tea bag it adversely affects the infusion – lower temperature.

Now you know! *It all depends.*

Notes

1. Intensity depends on a combination of leaf standard and the harvest interval. If the interval is too long, then for fine plucking it will be necessary to leave leaves on the stump, so plucking will be light (unless breaking back is done).
2. http://teacraft.com.

18 Fair Trade?

Smallholders Are Beautiful

Growing tea has not always had a good press in the Western media. Film makers keen to make a name for themselves quickly, and at low cost, concentrate on the sensational, descending on a tea estate with their film crew and recording the less attractive side of the industry. They rarely pay attention to the role that tea plays in the rural economies by creating employment opportunities and generating wealth, often in locations with few viable alternatives. Tea creates a cash economy which leads, in turn, to a demand for other products, be it a bicycle, a radio or, now, the ubiquitous mobile phone. Skills are needed at every level, for which training is required, whether those skills are associated with the running and maintaining of a complex tea factory, maintaining the vehicles and the associated road infrastructure or propagating a tea plant from a leaf-bud cutting.

That is not to say there are no 'rogues' amongst the tea fraternity. Sadly there are those who only take, in terms of crop, what they can get. They make no investment, apply no fertiliser, and their labourers are not paid a living wage and live in sub-sub-standard housing. At the other extreme are those international companies that, for example, attempt to impose health and safety regulations designed for a developed economy on a poor country. This is not always possible, although their intention is an honourable one.

The majority of the employees on an estate are pluckers. This is a very responsible job, if a monotonous one. Protective clothing is essential because of the high rainfall. The protective capes are not always comfortable when it is hot. In some areas lightning is a risk to life, and harvesting should then cease until the storm is over. Factories also vary in their level of safety awareness. Often, the moving parts of machinery are not covered. There is a lot that can and should be done to implement safety procedures appropriate to the rural sector and to improve the general welfare facilities for staff and their families.

The usual argument for not doing these things is 'cost', and a view that the price of tea is never high enough to justify expenditure on what some judge to be 'non-essential' items. That is where organisations like the Fairtrade Foundation and the Rainforest Alliance have a role to play (see below).

In this chapter, examples are presented to illustrate a number of issues which affect the socioeconomic wellbeing of millions of people involved in the tea industry, both employees and small-scale tea farmers. These examples cover employment laws in India, corporate social responsibilities, including gender issues, ethical trading and poverty alleviation. The objective is to protect and to raise the standard of living of all those involved, from the farmer to the consumer. A good starting point is to consider the topic from a historical perspective.

A Historical Perspective

Tea is a labour-intensive industry, and tea is often grown in remote areas of a country away from the attractions of urban living. Communication is often difficult, and with poor roads transport is not easy. To attract labourers and managers of the right calibre, the terms of employment need to be good (relative to the economy of the country and competing industries). This means providing all the services found in a town. These include clean drinking water, sanitation, health services, hospitals and education facilities. In some places, including parts of India, food is also still provided where there is no land available for employees to grow their own. Managing a tea estate is akin to running and maintaining a small town (Figures 18.1 and 18.2). In some places the social situation is further complicated by the presence of different ethnic groups, some of whom came to the tea areas to work over 100 years ago and settled for good. In some cases, these migrants are treated as second-class citizens – as for example the Tamils in Sri Lanka and Hindus in Bangladesh; often, such treatment leads to sociopolitical unrest.

Attempts to address poverty in the tea industry have been attempted before. For example, good intentions prevailed in the 1950s in Tanzania, with mixed outcomes, although one should not pass judgement on the success or otherwise of the actions of yesterday by the standards of today (Box 18.1).

Figure 18.1 A successful business needs to attract and retain good staff. This means providing services that would be expected living in an urban area. Managing a tea estate is like running a small town, with all of the responsibility that goes with it (Sri Lanka, MKVC). *A black and white version of this figure will appear in some formats. For the colour version, please refer to the plate section.*

Figure 18.2 The provision of housing on tea estates is improving, in some cases (Tanzania, MKVC). *A black and white version of this figure will appear in some formats. For the colour version, please refer to the plate section.*

Box 18.1 Commentary: Good Intentions, but Who Was Right?

I think it was Doris Lessing who said that it can be a mistake to judge the actions of yesterday by the standards of today.

My personal experience of the tea industry only goes back to 1955, but one hears stories.

1. The tea plantations of Mufindi in what was then Tanganyika were for the most part planted by German settlers between the wars. At the outbreak of the Second World War, they were rounded up and incarcerated in South Africa. The Custodian of Enemy Property advertised for someone to look after the plantations, and the firm I subsequently went out to work for got the job.

   Workers were needed, and were recruited from remote highland areas where there were plenty of people but little if any work. When they arrived, many of them were suffering from serious malnutrition. Recognising this, the company provided them with a daily meal.

   People we might now regard as agitators tried to persuade the workers to reject the food, on the grounds that it was only being provided so that the company could get more work out of them.

   Who was right?

Box 18.1 Continued

2. The immigrant workers often arrived with their families. Mum and Dad went out to work, together with the older children – there were no schools in the area in those days, and the children were learning useful skills. The younger children stayed on the field boundaries and looked after the toddlers, while Mum carried the baby on her back. Shades of *The Darling Buds of May*?

    Now, of course, all that is illegal, and employers have to be vigilant to ensure that no one under the legal age is employed. But with the AIDS epidemic sweeping the area, this often means that the parents are gone, and the oldest child is left to look after and support the others. Denied the ability to work, the whole family suffers.

    So measures intended to protect children from exploitation end up causing real hardship.

    But no employer dare make an exception on humanitarian grounds for fear of exposure to the media. And it must be acknowledged that much of an employer's corporate social responsibility agenda is dictated by the need to project a favourable image.

3. A friend of mine once remarked that in his opinion, in Tanzania, the only organisation more corrupt than the police was the Civil Aviation Authority. One has to assume he had dealings with both.

    But operating in an environment where corruption is rife does pose problems for corporate entities. Ethically, of course, no inducements, facilitating payments, or non-refundable deposits are ever made, no matter what.

    I remember a time when government salaries were fixed, and inflation was running at around 30%, year after year. The purchasing power of salaried officials was catastrophically reduced. People will fight tooth and nail to maintain some standard of living, and government servants are no exception. It was in these circumstances that an acquaintance of mine arranged to import some camping equipment. When the goods arrived in port, import duty of 100% became payable, on top of which sales tax was due. The imposition of these dues would have made the goods prohibitively expensive, and if there had been any intention of paying them, the goods would not have been ordered in the first place.

    A facilitating payment to the appropriate customs official got the goods released.

    My acquaintance got his goods. The manufacturer made a sale. The customs official got enough to live on. And if Adam Smith is right, the country was better off by the value of the imports.

    Who lost?

T. C. E. Congdon

Conditions of Employment in India (an Example)

In India, the Plantation Labour Act governs the conditions of employment of plantation labour, jointly with the rules promulgated by the state governments. The Act requires the employer to provide the workers with medical facilities, housing, sickness and maternity benefits, and other forms of social security. It also provides for the setting up of canteens, crèches, recreational facilities, suitable accommodation and educational facilities for workers and their families in and around the work places in the estate. The standard of these, however, varies in different states depending upon the state laws. Additionally, the Act states that no worker shall be employed for more than 48 hours a week, and child labour is banned. Every worker is entitled to a day of rest at the end of a six-day period.

The wage rates are generally fixed by tripartite agreement between representatives of the employers, the employees and the state government. In Kerala they are fixed at the level of the minimum wage prescribed by the state. In addition to the wages, plantations in north-eastern India provide subsidised food grains, and some plantations provide fuel as well. This is meant to offset the lower wage rates in comparison to those prevailing in South India. The social costs of employment represent about 5–8% of the average sale price for the tea. These social security obligations add another 30–40% to the daily wage rate. This means that the total employment costs are now considerably in excess of the minimum wage. This in turn leads to pressure to mechanise the field operations, particularly harvesting, in order to reduce the labour requirement. The introduction of mechanical harvesting has also been encouraged as a result of a shortage of workers willing to pluck tea in certain areas (see Chapter 11).

Corporate Social Responsibilities

In addition to these legal requirements in India there are also 'corporate social responsibilities'. These embrace economic and environmental issues as well as social ones. The pressures to address these concerns may come from within a company wishing to be seen as ethically superior to its competitors, from consumers concerned about human rights or from mainstream buyers, including non-government organisations (NGOs), driven by ethical principles which may subsequently be adopted by schemes such as the UK Fair Trade Codes and Certification scheme. These schemes result in additional costs being imposed on those in the tea industry who wish to meet these criteria of good practice. To check on compliance, a new regulatory industry has emerged.

Corporate social responsibilities have largely remained confined to the large tea companies, who can ill afford not to comply. By contrast, small farmers cannot afford to take part, as to do so would involve additional costs. Through their involvement in the processing, blending and packaging of tea, the larger companies can influence tea trade policies in tea-producing countries and beyond. In addition, they can influence, for example, (1) fertiliser prices through bulk purchases and associated economies of scale and (2) the costs of road and rail transport and shipping. Small farmers usually get paid only for the green leaf they harvest from their fields (less deductions), unless they are members of a cooperative or shareholders in the factory, when they may also get a share of the sale value of the processed tea.

Corporate social responsibility should extend to women, who represent the majority of workers in both large and small tea gardens in India. Whether from a business, social justice or stakeholder perspective, the question of gender equality needs to be addressed. For example, most smallholders hire both male and female workers, but some farmers employ only women. Since females are also the dominant source of family labour, it is probably true to say that most smallholder tea is produced by women. This disparity can lead to conflict between the women doing the work and the head of household, usually male, over the control of the proceeds from the sale of the green leaf to the factory. Owuor *et al.* (2007) emphasised the need to take gender balance into account when planning the Kenya Tea Development Agency (KTDA) extension services.

Ethical Trading: a Marketing Opportunity

Concern about the ethical dimension of the way tea is produced amongst the consuming public, particularly in the richer countries of the world, has led to the creation of a number of NGOs.

The Ethical Trading Initiative (ETI) is an alliance of companies, NGOs (including Oxfam, the Fairtrade Foundation and Save the Children), high-street companies and trade unions working to promote and improve the implementation of corporate codes of practice. ETI is a UK-based partnership. Certification schemes are offered which aim to ensure that internationally recognised labour standards apply throughout the supply chain (Box 18.2).

Box 18.2 Rainforest Alliance and Fairtrade Certified Labels

Information from www.rainforest-alliance.org *and* www.fairtrade.org.uk.

The Rainforest Alliance and Fairtrade are both international organisations committed to improving the lives of farmers and farm workers in the developing world. The Rainforest Alliance Certified seal can be found on an array of farm goods – from coffee and bananas to flowers and ferns – as well as timber, paper and other forest-derived products. The Fairtrade Certified label appears on an assortment of agricultural goods, including tea, sugar, coffee and vanilla. While the Rainforest Alliance and Fairtrade share similar missions and goals, they differ in focus and strategy.

Agricultural expansion is responsible for 70% of global deforestation, and is the single greatest threat to tropical forests. In these biodiversity-rich regions, farmers are often responsible for soil erosion, water pollution and wildlife habitat destruction. Rainforest Alliance certification encourages farmers to grow crops and manage ranchlands sustainably. Because the certification system is built on the three pillars of sustainability – environmental protection, social equity and economic viability –

Box 18.2 Continued

and no single pillar can support long-term success on its own, the Alliance helps farmers improve in all three areas.

The Fairtrade Foundation was formed in 1992 to help small-scale producers and workers in developing countries to improve their quality of life by paying them a premium for their produce, in this instance, tea. The workers and management agree jointly how this premium should be spent to improve the living standards of the workers against specified criteria. Small-farmer associations operate in a similar way.

Fairtrade certification ensures that small-scale producers and workers are treated fairly and receive wages greater than or in line with the legal minimum wage. Tea workers from Fairtrade-certified estates and smallholder organisations also benefit from the Fairtrade premium paid directly to the worker committees or the producers for community improvements.

Examples

In 2010 Unilever saw a marketing opportunity, if it could be perceived by the tea-buying public as a responsible company concerned about:

- the health and wellbeing of its employees
- reducing the (adverse) environmental impact of its operations worldwide
- enhancing livelihoods

The company set itself the challenge of sourcing all its supplies from 100% sustainable sources. This included Lipton, Unilever's 3.5 billion Euro tea brand. The company's goal was to have all the tea in Lipton's tea bags sourced from Rainforest Alliance certified tea farms by 2015, and every kilogram of Unilever tea sustainably sourced by 2020.

One challenge for Unilever was how it could transform a supply chain that was not only geographically diverse but also highly fragmented. In many of the markets, smallholders produced the majority of the tea (Figure 18.3) (Henderson and Nelleman, 2011). Unilever, in association with Rainforest Alliance, successfully certified their own tea estates (see below) and other large estates, but found it difficult to convince smallholders of the value of the scheme's certification. There was some success in convincing an influential group of farmers in Kenya of the value of this approach, using a Farmer Field School (FFS) approach. This was run as a partnership between Lipton tea (Unilever) and the Kenya Tea Development Agency (KTDA). KTDA manages 66 tea factories, serving 565,000 smallholder tea growers. The 720 farmers who registered for the initial FFS became good examples of the value of Rainforest Alliance certification and indeed acted as extension agents to the remaining 38,000 farmers supplying the four tea processing factories chosen for involvement in this pilot project (Mitei, 2015).

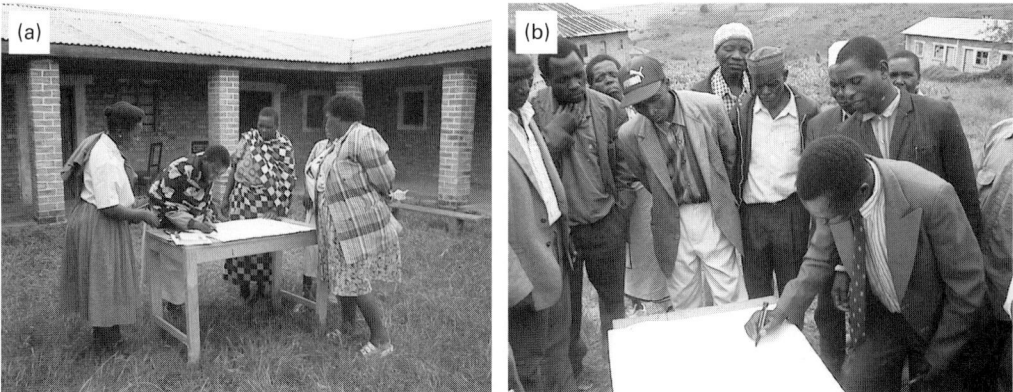

Figure 18.3 (a, b) In many countries, tea production by smallholders exceeds that from estates, providing a livelihood for millions of people. There are mutual obligations from the smallholder and the factory processing their leaf, including regular payment (as shown here in Tanzania, MKVC).

A second challenge was to see if Unilever could obtain a market advantage from its move to 'sustainable tea', especially in 'emerging markets'. There were issues on the sourcing side. For example, in India there is no central agency like KTDA to take a lead in promoting the certification and to make a case against the continued use of child labour (and certain pesticides). Could consumers in countries such as Turkey, Russia and India be persuaded to pay more for certified tea (Henderson and Nellemann, 2011)?

One doesn't have to be a born-again cynic to see such initiatives as opportunism in action, with a bandwagon leading the way. All the politically correct phrases are being employed, including those whose meaning is difficult to define. For example, what is really meant by 'sustainable tea'? This issue is discussed further in Chapter 20 under the title *How sustainable is sustainability?* But assuming all the motives are honourable, how well has this ambitious initiative progressed (Figure 18.4)?

Sustainable Tea

Unilever has produced a mission statement in which four principles for sustainable agriculture were specified together with 10 (later 11) indicators. This was in the context of increasing consumer concerns about the food chain, and the threat that these concerns posed to existing markets and brands (e.g. Lipton tea). In 2010 Unilever sold nearly 350,000 t of tea, with about 90% coming from external suppliers. The balance came from Unilever-owned tea estates in Kenya (Kericho), Tanzania (Mufindi) and India (north and south). In an initial self-assessment for Rainforest Alliance certification, a wide range of activities was identified as indicators of 'sustainable tea', as summarised in Box 18.3 (abstracted from Pretty *et al.*, 2008), all of which may be well-intentioned in theory, but how they work in practice is uncertain. Many are merely what a good farmer should be doing anyway. Hanging a label on something helps to focus attention on the issue, at least in the

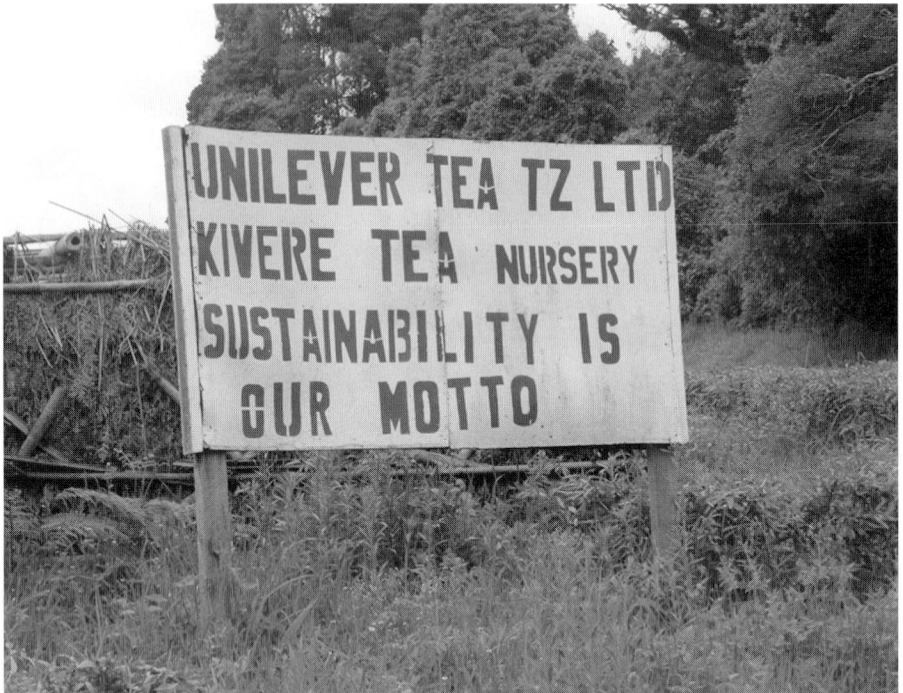

Figure 18.4 Sustainability is everywhere, but how sustainable is sustainability? (Tanzania, MKVC). *A black and white version of this figure will appear in some formats. For the colour version, please refer to the plate section.*

Box 18.3 Rainforest Alliance Certification and Sustainable Sourcing

The following are examples of the range of activities put forward to justify certification by the Rainforest Alliance for sustainable sourcing of tea in Kenya, Tanzania and India.

Kenya – Sustainable practices include using prunings as a mulch; no pesticides are used on mature plants; the indigenous forest and conservation area are maintained and enhanced, including a dedicated monkey sanctuary; over 400,000 indigenous trees have been planted in cooperation with neighbouring communities; fertiliser trials are used (1) to fine-tune applications to each field, and (2) to evaluate the use of factory ash and lime to restrict further acidification of the soil; 97% of energy consumption now comes from renewable resources (hydroelectric and fuelwood), while drying wood before burning and replacing older factory boilers with energy-efficient models has reduced fuel consumption further; a dedicated hospital facilitates education and health programmes, including an HIV/AIDS peer-group counselling, prevention and treatment course.

Tanzania – Where the growing area is flat the tea is mechanically harvested; improved machinery alleviates soil compaction created during harvest; there is

Box 18.3 Continued

limited use of fertilisers and herbicides; no pesticides are applied to mature plants; rainwater harvesting and storage in on-site reservoirs provides water for irrigation during the six-month-long dry season; supporting biodiversity is critical; forest fires and other systems of land clearing are a constant threat to the indigenous 'Eastern Arc' forest, which is especially rich in bird and plant species; some 14,000 ha of indigenous forest are maintained within the property; a biodiversity action plan aims to promote biodiversity through a range of projects, including the protection of identified key habitats and associated education projects; an environmental impact assessment is undertaken before any new developments are started; a comprehensive programme to combat HIV/AIDS is also under way.

India – This project operates in two climatically distinct areas (see Chapter 3), Tea Estates India Division (TEI) in the mountains of South India and Assam in the north; farmers in TEI focus on soil loss and fertility, pest management and biodiversity; leguminous cover crops planted between young tea plants help to prevent soil erosion and also fix nitrogen; rainwater harvesting helps young plants to survive in drought-prone areas in the south; renewable energy (wood from trees grown on the plantation) is used to dry tea, and two windmills supply almost 10% of the electricity; a rainforest species nursery was established with 75,000 plants of 60 different species; local communities are being trained in environmental conservation, particularly in the surrounding Anamala rainforest; prunings remain in the field to improve soil fertility; nitrogen and potash applications have been reduced by a third; fungal agents and pheromone traps are replacing inorganic pesticides; part of TEI has been certified for organic tea cultivation; water table management and pest management are priorities in Assam.

short term, but it can lead to mistaken priorities during implementation. Was reducing nitrogen and potash application in South India (also in Tanzania) really a good thing? Evidence please!

Assuming that 'sustainability is indeed sustainable', the challenge for a tea company is how best to encourage others to improve the sustainability of the supply chain, whether it is supplying a factory with green leaf or a consumer with a branded tea.

Poverty Alleviation

Owuor *et al.* (2007) highlighted the role that KTDA smallholder tea has played in the redistribution of wealth in Kenya. It has provided meaningful year-round employment to nearly half a million farmers on their own farms. In addition, there is also regular wage employment for farmers' relatives in modern tea-processing factories, located in rural areas. These require highly skilled artisans as well as unskilled labourers. The employment of drivers and mechanics for maintaining transport for collecting green leaf has, in turn, led to improvements in the rural road networks. In short, tea has been a catalyst for development.

Further expansion in Kenya is however constrained by a shortage of labour for harvesting. With the yields greater than originally expected – and they could be a lot more – the farming family often finds it difficult to cope, especially as child labour is illegal in Kenya. Shears or other mechanical aids will continue to play an increasing role in harvesting, particularly during peak seasons.

The least that can be said is that concern about the image of the tea industry is being addressed. Hopefully this will mean that any criticism from the media will be constructive and supportive of the initiatives being taken by the leaders in the industry.

19 The Agronomist's Report

A Synthesis

The focus of this penultimate chapter is on a proposed new tea project somewhere in eastern/southern Africa. A detailed evaluation is to be undertaken of a possible site where a large (up to 1000 ha) modern tea estate with factory could be developed. It is expected that a smallholder tea scheme will be created alongside the estate, with the green leaf going to the same factory. To initiate the project, an agronomist has been employed by the Southern Africa Tea Development Agency to evaluate the site, and to prepare a proposal for an economist to determine the viability of the project. The consultant has been instructed to adopt an innovative approach to the task by challenging, at every opportunity, conventional wisdom on how a tea project should be managed, and not to feel constrained by what has gone before. The challenge to the consultant is to know what questions to ask and where to find the answers, and to judge whether those answers are realistic.

This chapter is written as if it is the report written by the agronomist for consideration and evaluation by the Directors of the Southern Africa Tea Development Agency. It therefore provides a synthesis of many of the topics covered in the preceding chapters. As such, its style differs from that used elsewhere in the book. As stated at the beginning of Chapter 1, this is not a recipe book that one is expected to follow word-for-word, rather it is a book that challenges the reader to seek to justify the cultural operations practised on the tea farm/estate.

The agronomist's evaluation is reported in the following sequence:

1. Location
2. Soil suitability
3. Climate constraints
4. Water management
5. Nutrition
6. Clones
7. Crop establishment
8. Potential yields
9. Pests and diseases
10. Harvesting systems
11. Fuelwood.

For this proposed new tea enterprise a possible location has been identified, but its suitability for intensive (black) tea production has yet to be determined. The agronomist will seek to make recommendations based on (1) an analysis of published and unpublished reports, (2) consultation with a local research station, (3) comparisons with other tea-growing areas in the region and elsewhere, and (4) visits to the proposed location.

A detailed list of the issues to be addressed in such an analysis, and the decisions to be taken, are given in Box 19.1. This report does not claim to be complete. It merely gives some examples of how data from miscellaneous sources can be interpreted. Answers to many of the outstanding questions can be found elsewhere in this book.

Box 19.1   Some of the issues to be addressed and decisions to be taken before a new project can be fully implemented (the chapter numbers in brackets indicate where a particular issue is discussed in more detail)

Part 1: New Projects

Identifying a Market
- Which type of tea? (Chapter 17)
- CTC or orthodox-manufacture black tea (plus possibly some green tea) (Chapter 17)
- Organic? (Chapter 14)
- Fair trade? (Chapter 18)
- Local market or traded internationally? (Chapter 17)
- What is the demand?
- Will it pay? Sensitivity to price changes
- Is it sustainable? (Chapter 20)

Selecting a Site
- Availability of land
- Access to water
- Access to ports
- Labour availability
- Opportunities for outgrowers

Detailed Evaluation
- Suitability of soils (Chapter 3)
- Suitability of climate (Chapter 3)
- Limiting factors: drought, high/low temperatures, dry air, frost, hail, wind (Chapter 3)
- Climate change risks (Chapter 3)
- Potential yields (Chapter 10)

Box 19.1 Continued

Getting Started: Crop Establishment
What to plant: the criteria
- Yield and/or quality? (Chapters 6 and 17)
- Evaluate evidence for the best most recently released clones/composites (or those in late stages of evaluation) (Chapter 5)
- Drought resistant and/or responds well to irrigation (Chapters 5 and 15)
- Resistant to major diseases
- Ease of plucking (Chapter 10)
- Suitable for mechanical harvesting (Chapter 11)
- High or low base temperature for shoot extension/development (Chapter 7)
- Spreading or upright posture (Chapter 7)

Planting
- How many different clones to plant (Chapter 6)
- Area of each clone (Chapter 6)
- Plant density/spacing (must suit mechanical harvesting) (Chapter 7)
- Total number of plants needed (net and gross) (Chapters 5, 6 and 7)
- Area to be planted each year (Chapters 7 and 11)
- Time period available for planting (Chapter 3)
- Size of nursery, location (Chapter 6, Box 6.1)
- Number of cuttings needed (single leaf/bud or 'pluckable shoots')
- Number of mother bushes to set aside (Chapter 6)
- Planting method (Chapter 6; e.g. post-hole, Box 19.4)
- Method of bringing into production (Chapters 6 and 9)
- Shade trees? (Chapter 16)

Part 2: Existing Tea

Plucking (Chapter 10)
- Plucking intensity
- Quality assessment (of harvested leaf)
- Scheduling harvests
- Coping strategies for peak harvests
- Gang plucking
- Scheme plucking
- Monitoring creepage

Mechanical Harvesting (Chapter 11)
- Harvesting aids (e.g. shears)
- Machine assisted

Box 19.1 Continued

- Choice of tea harvester
- Criteria for selection
- Management of machines
- Maintenance of machines
- Training of staff

Nutrition (Chapters 13 and 14)

- Observation of deficiency symptoms in the field
- History (previous cropping)
- Soil/leaf analysis
- Research institute data
- Care in interpreting the data
- Remedial applications
- Routine applications
- Organic (special considerations)
- Young tea (special case)

Water Management (Chapter 15)

- Water conservation
- Drought mitigation
- Irrigation
- Deficit irrigation
- Drainage

Pruning (Chapter 7)

- Duration of pruning cycle
- Time of prune
- Height of prune
- Avoid skiffing
- Retention of prunings in field

Replanting (Chapter 6)

- Taking a difficult decision
- When to replant?
- What with?
- How fast?
- Plan ahead

Figure 19.1 (a, b, c, d) A prerequisite for establishing a successful tea estate includes an assessment of the indigenous vegetation, a soil survey, assessment of the water resources, a topographic survey, and reliable weather data (MKVC). *A black and white version of this figure will appear in some formats. For the colour version, please refer to the plate section.*

Location

The proposed tea farm is located in eastern/southern Africa at latitude 12°S. The average altitude is about 1000 m. The climax vegetation is miombo/*Brachystegia* woodland (Figure 19.1).

There are sufficient people living in the locality to provide labour for the project, some with a technical knowledge of tea.

The total area of the farm, known as Rashidi, is 2300 ha, but, owing mainly to topographic restrictions, the planned area of tea is about 800–1000 ha together with up to 400 ha of eucalypts for fuel. About 200 ha has already been cleared for arable crops, and this area could be planted with tea first. The farm is bordered on its eastern boundary by a river, which flows throughout the year, and to the north by a mountain (1600 m). Rashidi farm is bisected by several streams, which also flow throughout the year. A dam has been constructed on one of the streams, and there is a site for a second dam if needed.

Soil Suitability

Based on existing topographic and geological maps and aerial photographs (to delineate soil boundaries), together with recent soil auger borings and soil profile pits (all to depths of 1.5–2.0 m), the physical and chemical characteristics of the predominant soils have been described by the consultant according to the Food and Agriculture Organization (FAO) guidelines. Three mapping units were identified based on landform and slope. The largest part of the farm (950 ha) consists of flat to nearly flat ridges with very deep, well-drained clays. On the gently sloping sides of the mountain (300 ha) the topsoils are coarser (sandy clay loam). The soils of the valley bottoms (450 ha) are poorly drained peaty topsoils over loam or clay subsoils. Field observations confirm the prevailing view that the soils have been developed *in situ* as the weathering products of the underlying bedrock (granite). The soils of the ridges and the valley bottom slopes are classified (FAO) as acrisols (acid, low base status, strongly leached, but less so than a ferrasol).

Analysis of the soils in the National Laboratory in Nairobi confirmed that the soils on Rashidi farm are all acid, with pH values in the range 5–6, and therefore suitable for tea. Organic matter contents are low (range 1.1–3.6%), especially in previously cultivated fields (mean 2.2%), compared with natural vegetation (mean 2.9%). The C:N ratios are around 15, indicating decomposed, good-quality organic matter. The cation exchange capacities (CEC) are low in the topsoils, with values between 5.5 and 12.4 me 100 g^{-1} soil, confirming the low organic matter content. The CEC of the clay fraction is below 10 me 100 g^{-1} clay, suggesting low-activity kaolin-type clays (as at the local research station). The soil is well supplied with exchangeable potassium, with levels ranging from 0.2 to 0.5 me 100 g^{-1} soil. Available phosphorus levels are highly variable, while exchangeable calcium and magnesium levels are both low (range 1.5–3.7 and 0.7–1.8 me 100 g^{-1} soil respectively). Of the micronutrients, the sulphur content is very low (< 0.01%), suggesting the need for sulphur. Copper and zinc levels are both low to moderate, being in the range 20–70 and 30–40 mg kg^{-1} respectively.

The soil profile is porous and friable throughout. The available water capacity is however low (because of the relative pore size distribution), averaging about 10% by volume, or 100 mm m^{-1} depth. Although this is a relatively low value for a clay soil, it is similar to that recorded on similar soils on a nearby coffee farm. These properties have been confirmed by visual inspection of profile pits by the consultant during a visit to the farm.

Summary: Soil Suitability

- The soil physical and chemical properties are similar to those found in other success-ful tea-growing areas in the region.
- Soils are deep and well-structured clays (but behave more like loams).
- Soil pH is within the optimum range for tea (5–6).
- Organic matter content is low, as is CEC.
- Low nutrient status (except K).
- Low available soil water content.

Climate Constraints

Reliable weather data are required so that the suitability of the site for tea can be evaluated. In particular temperature and rainfall data are needed over a long period (at least a 10-year time span). Unfortunately, the data are not all available or, as with rainfall, are unreliable due to poor siting of a non-standard rain gauge (it is too close to the roof of the office). A maximum/minimum thermometer, sited in a rudimentary screen near the office, has been recorded intermittently over the previous five years. As there are successions of days when the maximum temperature remained the same (suggesting the thermometer had not been reset between readings) these data too are suspect. There are similar problems with the wet and dry thermometer, with the wet bulb reading the same as the dry bulb on numerous occasions. On inspection, it was noted that the wet bulb was dry and the wick was very dirty. The humidity values could not therefore be trusted. As is often the case, the day-to-day recording of the data has been delegated to the most junior person in the office, without any training. The importance of the job of recording is only appreciated when there is a requirement for good data to support a project proposal.

At the nearby research station, the records are again incomplete. The hours of bright sunshine have not been recorded for several years because there are no recording cards. The evaporation tank leaked, and bees had occupied one of the cups on the anemometer.[1]

The nearest reliable data are those recorded at the airport 60 km away and at an altitude of 600 m, that is 400 m below the altitude of the farm, but still within the influence of the mountain.

From this miscellaneous collection of data, an attempt has been made by the agronomist to obtain something useful, which could be used as a basis for planning.

Rainfall

The mean annual rainfall total at Rashidi is about 1500 mm, with very distinct wet (December to April) and dry seasons (June to October), with November and May being transition months. Irrigation will be needed from May/June to October/November.

The start and end of the rains and hence the duration of the single rainy season are particularly difficult to define. During the rains, rainfall normally exceeds evapotranspiration on a weekly basis. Using data from the nearby airport and adjusting for differences in distance from the mountain, the duration of the dry season on a 1-in-20-year expectation was estimated to be 20–24 weeks. There are no records of hail, and wind speeds are low.

Temperature

After allowing for the differences in altitude between the airport and the farm, and assuming a lapse rate of –0.6 °C for each 100 m increase in altitude, the mean daily air temperature during the rains (December to May) at 23–24 °C is judged to be close to the

Table 19.1 Rashidi farm: mean monthly air temperatures 2004–2008

	Max °C	Min °C	Mean °C	Days > 30 °C
Jan	29.9	19.0	24.5	15
Feb	30.0	18.8	24.4	13
Mar	29.7	18.8	24.2	10
Apr	29.0	18.5	23.7	11
May	28.2	16.4	22.3	9
Jun	27.2	14.0	20.6	4
Jul	26.3	12.5	19.4	0.3
Aug	27.7	12.6	20.1	4
Sep	28.7	14.4	21.6	11
Oct	31.1	16.6	23.8	21
Nov	31.7	18.6	25.1	26
Dec	30.8	19.2	25.0	16
Mean/total	29.2	16.7	22.9	140

optimum for shoot extension (*c.* 25 °C), but the mean daytime maxima during September to November, the hot dry season, were close to, or in excess of, 30 °C. This means that on individual days temperatures would often exceed 30 °C. High temperatures like these can restrict shoot growth and limit rates of photosynthesis. How important this is will depend on the corresponding saturation deficits. In the warm dry season these are likely to exceed 2.0 kPa, a critical value which, if exceeded, will limit the response to irrigation at that time. This will result in large peaks of production in late December/January after the start of the rains when the atmospheric humidity increases (Table 19.1).

Over the period 2004–2008, inclusive, the mean number of days at Rashidi farm on which the daily maximum temperature exceeded 30 °C in a year totalled 106 (data set not complete). This broke down as follows: June to August = 1; September to November = 70; December to February = 30; March to May = 5.

Saturation Deficit/Humidity

Using actual vapour pressures recorded at the airport (1944–1964) and saturated vapour pressures as 'measured' on the farm (based on the maximum air temperature, 2004–2008) the calculated mean monthly saturation deficits (mid-afternoon) are close to 2.0 kPa or above throughout the year, reaching 3.0 kPa in October. On individual days they will be substantially in excess of these mean values (e.g. 4.1 kPa when the air temperature is 35 °C in October). These values are similar to those recorded in the south of Malawi when dry-season yields from irrigated mature seedling tea are in some years restricted by dry air.

Evaporation

Potential evapotranspiration (*ET*c) has been estimated from evaporation pan data (after allowance for the poor siting of the pan) and by analogy with similar climates elsewhere.

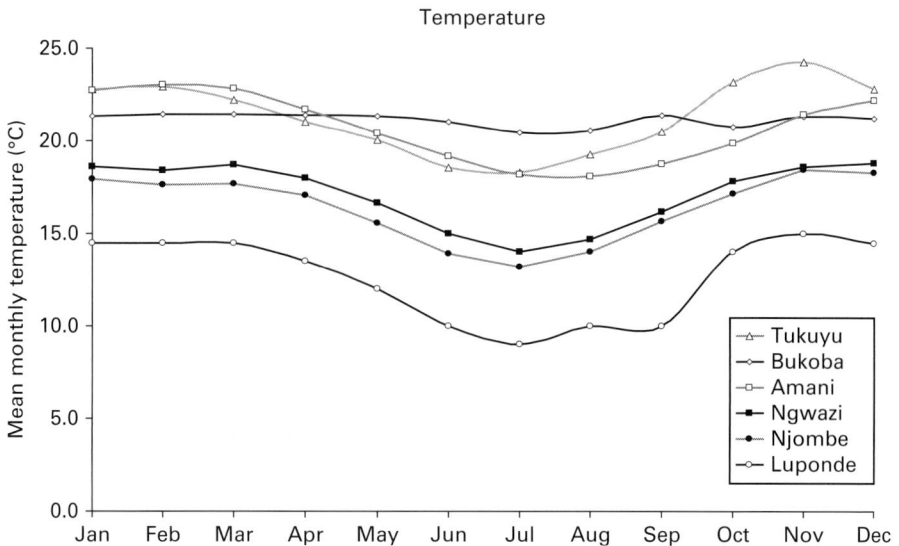

Figure 19.2 Mean monthly air temperatures in the principal tea-growing regions of one country, Tanzania: the effective temperature, assuming a base temperature of 12.5 °C, ranges between 0 and 2 day °C in Luponde and from *c.* 7 to 12 day °C in Tukuyu. This translates into two complete shoot replacement cycles at Luponde, and between five and six at Tukuyu (if other factors are not limiting).

ETc was estimated to be about 3–4 mm d^{-1} during the winter months and during the rains, and up to 5–6 mm d^{-1} during the hot dry months. This gives an estimated potential soil water deficit over the 20–24-week dry season of 800–900 mm, when no/little rain is expected.

Regional Comparisons

Within the eastern/southern Africa region, mean monthly air temperatures compare most closely with those experienced at Mulanje in southern Malawi. Similar high temperatures occur in September and October in Zambia. Two high-altitude sites (1800 m) in Tanzania are substantially cooler throughout the year than Rashidi, Njombe and Luponde (Figure 19.2). It is substantially wetter at Rashidi than it is at several other locations where tea is grown in the region. The dry season in Rashidi is long.

Summary: Climate Constraints

- Rainfall during the months December to April is greater than evaporation. This is followed by a cool dry season, and then a hot dry period.
- From May to November (inclusive) potential evapotranspiration exceeds rainfall by 800 mm.
- Mean monthly air temperatures are close to the optimum during the main growing season for tea.

- The nearest equivalent location within the region is Mulanje, Malawi.
- Maximum air temperatures can exceed 30 °C on 140 days a year.
- Mean monthly daytime saturation deficits of the air are close to 2.0 kPa throughout the year and can exceed 3.0 kPa in October.
- High air temperatures and dry air may limit yields (even with irrigation).
- Climate change could make this worse.

Water Management

With an annual dry season lasting from four to six months, a policy decision has been taken to irrigate the whole area of tea, except possibly in the year of prune. The expectation is that drip irrigation will be the preferred method of applying water, but the option remains to use centre-pivot irrigators. Much will depend on the capital and running costs, as the price of energy varies, mostly upwards, by the day. Following successful field trials in Tanzania, drip irrigation is also a realistic option for small-holders. The layout of the fields and spacing of the rows of tea must take into account the method of irrigation to be used. (Easy access into the tea to check for leaks in the drip system will be needed, for example, especially if the lateral pipes are buried.) Some flexibility will be needed to allow for future changes. Water distribution from sprinklers is distorted by the prevailing wind (from the south-east for 10 months of the year), which is an argument for drip irrigation, unless low-pressure downward-facing sprinkler nozzles are used on the centre-pivot, but there are then run-off implications owing to the high water application rate. This means planning the infrastructure early in the process, with pumping stations and buried main pipes supplying a drip irrigation system.

Stream flow records obtained from the Ministry of Water show that it will be necessary to build a dam to store surplus water during the rainy season for irrigation in the dry season. Young tea will be fully irrigated throughout the dry season, but mature tea will be partly irrigated, so that by the end of the dry season there will be a potential soil water deficit of 200 mm (known as deficit irrigation).

The results of research in the region suggest that the annual yield response of mature tea to irrigation will be of the order 3–5 kg made tea for each (ha mm) of effective irrigation. For an annual application of 600 mm (net) this equates to a response of 1800–3000 kg ha^{-1}. For younger tea the yield response to irrigation will be greater than this, up to 15 kg ha^{-1} mm^{-1} or, put another way, yield will be lost due to drought once the soil water deficit exceeds a certain value. This critical soil water depletion level will need to be specified, as it will form the basis for scheduling irrigation. This is probably best done using a water balance approach in which rainfall and evaporation are both measured on site. In the case of young tea, crop cover will also need to be assessed in order to determine the value of the crop factor.

It is important to make sure that the rain gauge and evaporation pan are well sited and well maintained, and that the recorder is suitably trained. A Stephenson screen with maximum and minimum thermometers and wet and dry bulb thermometers also needs to be purchased. Mean daily air temperature can then be used to determine the duration of a

---

Box 19.2 Shoot Replacement Cycle: an Exercise

1. Use the mean monthly maximum and minimum air temperature data below to calculate:
   - the expected length of each shoot replacement cycle, beginning at the start of the cropping year, 1 July
   - the total number of shoot replacement cycles in one year
   - the recommended length for the plucking intervals in the 'winter' months (July and August), and during the spring (October and November), summer (January and February) and autumn (April and May)
2. Show what effect a 2 °C temperature rise in the mean monthly temperature, either due to climate change or at an estate 350 m lower in altitude, would have on each of these variables.

Mean monthly maximum and minimum air temperatures (°C)

	Jan	Feb	Mar	Apr	May	Jun	Jul	Aug	Sep	Oct	Nov	Dec
Max.	24.5	26.0	27.0	24.5	23.0	21.0	20.7	21.0	23.0	26.5	27.0	25.0
Min.	14.0	15.0	16.0	14.0	12.0	11.0	11.3	11.5	12.5	15.5	16.0	15.0

Assumptions:

- 475 day °C are required above a base temperature of 12.5 °C for a bud released from apical dominance to reach a pluckable size.
- There are no other limiting climatic factors, such as dry air or dry soil, or excessively high temperatures.
- The target shoot population is a 50:50 mix of two leaves plus bud and three leaves plus bud.
- Day degrees = (max. + min.)/2 minus 12.5 times number of days.

---

phyllochron (see the exercise and worked example in Boxes 19.2 and 19.3). A fully equipped agrometeorological station should be developed in due course as an aid to management.

Recommendations: Water Management

- The irrigation system must be designed by a qualified and experienced irrigation engineer, otherwise there may be false economies.
- For mature tea, a soil water deficit of 200–250 mm can be allowed to accumulate prior to the start of the rains without a loss in the *annual* yield.
- But this has an impact on crop distribution, with proportionally more crop produced in the rains and less in the dry season.

Box 19.3 Solution to Shoot Replacement Cycle Exercise

Month	Mean air temperature (°C)	Day degrees	Monthly total day degrees	Date and days when cumulative total reached 475 day °C (from 1 July)
Jul	16.0	3.5	109	
Aug	16.3	3.8	118	
Sep	17.8	5.3	159	03/9 (112 d)
Oct	21.0	8.5	264	10/10 (102 d)
Nov	21.5	9.0	270	
Dec	20.0	7.5	232	04/12 (55 d)
Jan	19.3	6.8	211	
Feb	20.5	8.0	224	
Mar	21.5	9.0	279	08/03 (60 d)
Apr	19.3	6.8	204	
May	17.5	5.0	155	13/05 (66 d)
Jun	16.0	3.5	105	

1. Duration of each shoot replacement cycle = 102, 55, 60, and 66 d
2. Number of shoot replacement cycles in 12 months: 4 completed (4.5)
3. Assuming four generations of shoots on the bush at any one time: winter (Jul/Aug) = 25 d; spring (Oct/Nov) = 14 d; summer (Jan/Feb) = 15 d; autumn (Apr/May) = 16/17 d.
4. Increasing mean air temperature by 2 °C:
   - Duration = 79 d (17/09); 49 d (05/11); 46 d (21/12); 52 d (11/02); 39 d (27/03); 69 d (25/05); 84 d(18/08)
   - 6 completed (6.5)
   - 20 d; 12 d; 13 d; 15 d

- For young tea (< 3 years) water use will be proportionally less, depending on ground cover, but sensitivity to drought will be greater because of the shallow and less dense root system.
- Young tea is very sensitive to water stress, and yields decline rapidly when the soil water deficit exceeds 200–250 mm.
- If sprinklers are used, irrigation will need to be applied more frequently (say every two weeks) compared with mature tea (every three weeks).
- For drip irrigation, daily application of water is a practical possibility, as is fertigation. Allocate people to monitor and repair leaks.
- By the time all the tea is planted, it may be necessary to build a second earth dam on one of the streams to cope with the peak water demands.

Nutrition

Because of the difficulty in obtaining sufficient quantities of certified organic matter within a reasonable distance of the estate, and the vagaries of the market, no attempt will be made to produce organic tea.

Inorganic fertilisers only will be used.

Soil analysis suggests that there is adequate K in the soil, but P and S may be deficient. Phosphate will therefore be mixed in the planting hole. Nitrogen will be applied as sulphate of ammonia, which should also provide the S needed for young tea. The quantity of nitrogen required will have to be based on prevailing recommendations, perhaps initially 200 kg N ha^{-1} as an annual application (to be confirmed after consultation).

A decision needs to be taken about whether or not the nitrogen application should be split into two or more events. Or indeed should it be applied in the irrigation water (if drip or centre-pivot irrigation is chosen)? The method of fertiliser application needs careful consideration in order to ensure that it is applied as uniformly as possible, and with minimum opportunities for any to 'go missing'.

Smallholders always make rational decisions based on their individual priorities. They will need to be convinced that the fertiliser is best applied to the tea and not to their maize.

Routine leaf analysis will begin at a later date.

Recommendations: Nutrition

- The expectation is that mature tea will respond (profitably) to annual applications of nitrogen of up to 250 kg ha^{-1}.
- This can be applied as two equal applications in July/August and December.
- The soils are deficient in phosphorus: remedial applications of phosphate are needed, initially in the planting hole.
- The soils are well supplied with potassium. Annual applications of 150–200 kg K ha^{-1} are needed to replace that lost annually in a high-yielding crop (6000 kg ha^{-1}).
- The soils are deficient in sulphur; this should be remedied by applying sulphate of ammonia as the source of nitrogen.
- There may be a need for zinc (and copper).

Clones

With a long-term crop such as tea, it is unwise to plant anything but the best cultivars available at the time. Most of the clones recommended in eastern and southern Africa have been around for a long time. Commercial companies have largely led the way in looking ahead and developing breeding programmes, which have resulted in new genotypes that show great promise. Unfortunately, these are not normally available for general release. Growers have to select from the products of government-funded and/or cess-funded tea research organisations.

Recommendations: Clones

- Proven high-yielding clones under irrigation include AHP-S15/10, TRFCA-PC81, AHP-SC12/28 and TRIEA-31/8.
- Others include TRIEA-6/8.
- Clones worthy of evaluating include TRFCA-PC117, PC123 together with PC168 and PC213 from TRFCA, Malawi.
- Possible clones from TRF Kenya include the 303 series and 301/5 and 301/6.
- Consideration should be given to planting composites.
- Most of these are long-established clones selected in eastern and southern Africa which are likely to be superseded in the near future by the products of current breeding progammes.
- If there are sufficient plants, consider high-density planting (20,000+ ha^{-1}).
- It is recommended that further enquiries are made concerning the most promising new clones best suited to the region, and that 10 are selected initially to provide enough cuttings for evaluating and multiplication on site.
- A planting schedule needs to be developed as soon as possible.

Crop Establishment

Recommendations

- Plan row spacing and field layout for machine harvesting.
- Follow standard procedures in the nursery and for planting in the field.
- Be innovative. For example, test the idea of using a post-hole borer for digging planting holes to a depth of 0.9 m to aid establishment (see Box 19.4).
- Try high-density planting (20,000+ ha^{-1}).
- Bring plants into production by 'bending the dominant shoot' and then 'tipping in'.
- Aim to achieve full crop cover by year 3 at the latest: first prune year 4 (possibly year 6).

Potential Yields

At the regional research station, the clones that have performed best in the early years when well irrigated have achieved yields in the range 7000–8000 kg ha^{-1} within six years after planting. In the second rank are a group of clones yielding an average over two pruning cycles of 6000–7000 kg ha^{-1}. The remaining clones have averaged a yield of about 5000–6000 kg ha^{-1}. By year six after planting a realistic commercial yield target is c. 6000 kg ha^{-1}.

Since the proposed site is away from the equator (latitude 12°S, alt. 1000 m) there will be a seasonal influence on yields. These will decline during the winter months due to

Box 19.4 Replanting Experiment ABC Tea Ltd

A replanting experiment is to be designed because there is a realisation at senior management level that a sustained programme of replanting is needed if ABC Tea Ltd is to remain profitable in the long term:

- Yields have remained static (or declined) in old seedling tea fields – needs confirmation.
- Visual evidence of excessive sun-scorch on plant frame.
- Pruning at a constant height (needs down-prune, means loss in yield).
- Spacing is not suited to mechanical harvesting.

Aim:
- To justify (or otherwise) the benefits of replanting 'moribund' seedling tea in financial terms.
- To specify the rate of replanting that is possible.

Need to be able to:
- Maximise the rate of replanting at a realistic cost.
- Minimise loss of yield between clearing and new crop coming into production.
- Minimise loss of plants (from drought) during the first four years after planting.[2]

Other requirements:
- Which clone(s) to plant? Choice is currently very limited. Needs immediate action.
- To be able to provide answers to questions posed as soon as possible.
- To predict the benefits ahead of the crop coming into full bearing.
- Use diameter of main stem as surrogate measure of dry matter production and yield.[3]
- A calibration curve will be needed relating stem diameter to total dry mass and root dry mass.
- Both relationships are expected to be linear over most of the range.

  Roots – a preliminary observation trial has shown the potential of using post-hole borers to excavate a planting hole to a depth of at least 0.75 m. This facilitates more rapid root penetration compared with a standard planting hole of only about 300 mm.

  Canopy – the same trial has shown the apparent benefits of 'bringing into bearing' by bending the main stem and then allowing the shoots that emerge to grow freely for two years before pruning.

These observations need to be tested in a systematic way in order to provide the quantitative evidence (including a cost–benefit analysis) needed to support the case for a replanting programme.

Box 19.4 Continued

Treatments:
1. Deep planting hole (D) × constant tipping (T)
2. Shallow planting hole (S) × T
3. D × bending (B)
4. S × B

Replication × 4 (randomised blocks):
50 bushes per plot = 800 bushes net (@ 1.2 m × 0.6 m) plus guard area of, say, 0.1 ha
   total. Plant April 2020.
Clone: to be decided.
Recording: to be agreed.

---

declining temperatures extending the duration of the shoot replacement cycle, which results in a build-up of shoots followed by a large peak in production when the temperature rises. There is no experimental evidence that this sequence of events is a photoperiodic effect. The corresponding decline in yields during the winter months in Assam is due partly to the effects of drought at that time, but it is probably made worse by the tradition of *skiffing* the tea (a light prune).

Skiffing is intended to stimulate the first flush, which has a reputation in Assam for being of a high quality. It is perhaps another case of a traditional practice being accepted without questioning the evidence that was the basis for the original introduction of the practice. In any event, there will be differences in the labour requirements during the year as a result of fluctuations in crop distribution (even with irrigation). This also impacts on the green leaf transport and factory capacity requirements. An attempt will have to be made to predict the likely month-by-month crop distribution pattern, based on the experience of other tea producers operating at similar latitudes.

One way of mitigating the influence of the winter season on crop distribution is to plant clones with a low base temperature for shoot extension and shoot development.

Recommendations: Potential Yields

- Potential yield of clone TRIEA-6/8 = 6000 kg ha^{-1}, and of clone AHP-S15/10 = 7000 kg ha^{-1} by year 6 or 7.
- If high temperature/dry air is limiting, potential yields are reduced by up to 20% (= 4800 and 5600 kg ha^{-1} respectively).
- If there is insufficient water, yields are reduced by 4 kg ha^{-1} mm^{-1} (TRIEA-6/8) or 5 kg ha^{-1} mm^{-1} (AHP-S15/10).
- Total production of 5 million kg made tea is possible by year 10 after the start of planting (from 800 ha).
- Yield distribution during the year will be uneven, with peaks in October and January, and low yields during the winter months (June to August).

Pests and Diseases

- With the possible exception of termites, there appear to be no major risks to the tea from pests and diseases, especially as the site is isolated from other tea crops.

Harvesting Systems

It is a fallacy to imagine that machine harvesting is easier to manage than hand plucking. It is also a fallacy to believe that machines can be successfully deployed in a previously hand-plucked field without any field preparation. Nevertheless, machine harvesting of tea is the future. Therefore, plant on the assumption that all the tea will be harvested by machine.

Recommendations: Harvesting Systems

- Chose a machine that is the most appropriate match for the objectives and the terrain.
- Use the concept of programmed plucking, based on the phyllochron, to determine when to harvest (see Boxes 19.2 and 19.3).
- Evaluate the effect on yield of 2.5 phyllochron intervals between (machine) harvests (compared with 2 phyllochrons for hand plucking).
- Judge the quality of the harvested leaf on the basis of number of shoots per kilogram of green leaf.
- Be prepared for large yield peaks in September/October and again in January.
- Monitor (minimise) the rise in canopy height regularly, keeping it below 80 mm per year.
- Bear in mind that machines require more management than hand plucking.
- Machines are unreliable unless they are well maintained and there are sensible replacement policies.

Fuelwood

The maximum annual production of made tea from a 1000 ha estate and 500 ha of smallholder tea is estimated to be 650 t. When fully operational, the tea factory will require 2200 m^3 of dry *Eucalyptus grandis* wood a year. If the trees are well managed (like a commercial crop) and coppiced in year five after planting and thereafter at five-year intervals, the total area of trees required will be at least 220 ha (net of fire breaks and roads) (Box 19.5). It is suggested that a proportion of the *Eucalyptus* seedlings are purchased on contract from neighbouring farmers, who would also eventually provide a proportion of the firewood. The sale proceeds would be valuable additions to the annual incomes from their farms.

Box 19.5 Fuelwood Calculation Question

Imagine you are responsible for planting the trees needed to supply fuel for a factory to process tea from a new estate. At maturity there will be 1000 ha of estate-grown clonal tea that will need to be processed on site. In addition there will be 500 ha of rain-fed smallholder tea feeding into the same factory.

Specify the area of trees (*Eucalyptus grandis*) that will be needed to provide the energy needed in the factory to produce 'black tea'. State any assumptions that you make when answering this question.

An answer (there will be others)
- Area of tea and target yields of black tea (maximum expected, averaged over a four-year pruning cycle):
  - Estate tea 1000 ha @ 5000 kg ha^{-1}.
  - Smallholder tea 500 ha @ 3000 kg ha^{-1}.
  - Based on current practices, the maximum annual production of made tea is expected to reach 0.65 million kg (650 t).
- 1 m^3 of dry wood when burnt will provide enough energy to process 300 kg of black tea.
- Hence, 2200 m^3 of wood will be needed annually (650 t tea divided by 300 kg m^{-3}).
- If coppiced in year 5, and afterwards at five-year intervals, when the mean annual increment (MAI) is 50 m^3 ha^{-1}, the area of trees to harvest each year is 44 ha.
- The net area of *E. grandis* required is therefore *c*. 220 ha.
- Remember to allow extra space for fire breaks and roads.

Recommendations: Fuelwood

- Farmers to provide a high proportion of the *E. grandis* seedlings on contract.
- Farmers to supply a proportion of the fuelwood to the factory, again on contract, the proportion increasing as mutual trust grows.
- Fuelwood growing to be treated like any other commercial crop, with attention to detail.

Risks

Potential risks to the viability of the project are listed in Box 19.6.

Box 19.6 Risks to the Project

- **Climate change**. Time scale = decades. An increase in the mean air temperature could be beneficial at high altitudes (> 1900 m), but damaging at low altitudes (< 1200 m); compensatory growth after a drought ends can compensate for loss in yield during the dry season; rainfall is already erratic in many tea-producing areas.
- **Soils**. Time scale = years/decades. Increased acidification occurs over time, but no problems in the immediate future are anticipated. Need to maintain carbon content of the soil.
- **Water**. Time scale = five years. Abstraction licensing is enforced; energy cost to pump water for irrigation becomes prohibitive. Self-sufficiency in energy (including factory processing) is desirable. There are opportunities from hydro, solar and fuelwood.
- **Labour**. Time scale = monthly/annual. A shortage of pluckers is a strong possibility; need to plan ahead to mechanise harvesting.
- **Scale of production**. Time scale = up to 5–10 years. Trend towards smallholders (with lower green-leaf production costs) to continue, with corresponding decline in the traditional role of estates as 'outdoor factories' producing green leaf.
- **The market**. Time scale = daily, monthly, next year? The big unknown! How much will someone pay for a kilogram of tea? Need to add value in producing country, and consume more.

At first analysis, tea production practices in most countries appear to be very robust and resilient to change, except for sudden shocks such as internal strife or currency changes, and therefore sustainable over a 10–50-year time scale. The tea plant itself is robust.

Notes

1. These are not exaggerated statements. All have been observed. One wonders how relatively small temperature changes over time can be distinguished (in the context of climate change) when there is so much variation in the data for the reasons stated.
2. The period when there is an imbalance between the water demand by the atmosphere through the crop canopy (transpiration) and the rate of supply of water from the soil through the roots to the leaf.
3. Assuming a constant value for the harvest index.

20 Support Services

How Sustainable Is Sustainability?

The word *sustainability* is to be found everywhere, but what it means is an open question. No doubt, it means different things to different people. In the context of agriculture, and development work in particular, the S-word is associated, for example, with 'conservation agriculture', in which tillage is perceived as being incompatible with conservation, as it leads to a loss of soil from one field to the next, or transfer within a field (Andersson and Giller, 2012). The S-word is also linked to organic agriculture, but how sustainable is that? The example given earlier concerning the import of organic chicken manure by road from South Africa to Tanzania to produce organic tea exposes some of the irrational nonsense that is perpetrated under this 'politically correct' banner. Sustainability now also includes environmental and social aspects, where it sometimes seems to mean 'not changing anything'.

In this chapter, sustainability is discussed in terms of the contributions from support services to the tea industry. Specifically, the roles of research, training and extension are described using three case studies: namely agronomy research in India and Tanzania, training courses for estate managers in Africa and elsewhere, and an innovative extension service for smallholders in Tanzania. All three of these ancillary services have as an (unwritten) objective the desire to ensure the sustainability of the tea industry and the livelihoods of those people dependent on it being successful. Finally, the need for reliable data with which to monitor sustainability is considered.

Research and Sustainability

Agricultural research has a critical role in developing and sustaining agricultural production systems, whether they are described as small-scale or commercial. A core element of agricultural research is agronomy, defined by Sumberg and Thompson in their book *Contested Agronomy: Agricultural Research in a Changing World* (2012) as a science that affects:

the biological, ecological, physical, socio-cultural and economic bases of crop production and land management.

Or, more simply:

the application of plant and soil science to crop production.

Given that the series in which this book is published is entitled *Pathways to Sustainability*, it should be possible to find a clear definition of what is meant by the word *sustainability*. However, despite the S-word pointing to 10 page references in the index, there is only one attempt to provide a clear definition, namely that sustainability is:

the maintenance of unspecified features of systems over time.

The word *resilience* is perhaps clearer to understand than this definition of sustainability. Thus, resilience is:

the ability to maintain form and function.

Both words have entered the vocabulary of mainstream agriculture, leading to their 'casual rhetorical usage that has been allowed to mask any real commitment to change' (Thompson and Sumberg, 2012). This means (I think!) that the word has become shorthand for something that is difficult to define, but must be a good thing. For example, donors like to see it used in their project proposals. *Robustness* is used in a similar way to resilience.

More success has been achieved in defining sustainability in the context of livelihoods:

A livelihood comprises the capabilities, assets and activities required for a means of living ... A livelihood is sustainable when it can cope with and recover from stresses and shocks and maintain or enhance its capabilities and assets both now and in the future, while not undermining the natural resource base (DFID, 1999).

The obvious missing parameter in these definitions is time. No doubt dinosaurs thought they were sustainable after surviving 180 million years on this planet, but then came the ultimate shock! Tea has been cultivated for 4000 years, which implies sustainability as well. What therefore is an appropriate time scale for sustainability to be considered sustainable in the tea industry. Is it one year, a pruning cycle, a decade, or a lifetime?

Research Institutes: Large or Small

Research institutes serving an industry such as tea have similar problems when attempting to answer questions like these on sustainability. For example, the producers who are funding the research, often through a statutory levy, as well as donors, usually want answers to short-term problems that are often time- and location-specific. At the same time, researchers are meant to serve an industry located in diverse ecological areas, with both large-scale commercial (for profit, Figure 20.1) and small-scale producers (for livelihoods), often with conflicting and constantly changing priorities and limited resources. Fundamental information is needed to enable results of experiments to be extrapolated from one location to another where conditions may be very different. It is also necessary for researchers to anticipate future problems and not to concentrate only on the immediate challenges facing the industry. Somehow a balance has to be struck within the prevailing financial and skill constraints.

Figure 20.1 A potential new site for mechanised irrigated clonal tea (MKVC). *A black and white version of this figure will appear in some formats. For the colour version, please refer to the plate section.*

Let us consider what agronomic research has contributed to the sustainability of the tea industry, to tea productivity, and to the enhancement of livelihoods. Take as one example a large well-established research station (the Tocklai Tea Research Institute, formerly known as the Tocklai Tea Experimental Station) serving a major tea industry in North-East India. Then, compare this with a small recently established research institute serving a small, but diverse, tea industry in a geographically large East African country (Tanzania). The Tanzanian government and the tea industry together, and with help from donors, are struggling to support an independent research programme (or should that be written the other way round – namely, the research institute is struggling to support the tea industry (Figure 20.2)?

In their early years, the well-established institutes made important contributions to the tea industry worldwide (e.g. clonal selection and vegetative propagation are important everywhere), and continue to do so. A list of activities at the Tocklai Tea Research Institute in Assam since 1910, and achievements, can be seen in Box 20.1. It is an impressive list! Agronomic research has clearly played a major role in ensuring the sustainability of the tea industry in India.

There are a large number of other national and regional research institutes serving the tea industry. Some are funded by the industry through a cess or levy based on the production of processed tea by individual growers/producers. Those serving the tea industries in the Indian subcontinent are listed in Box 20.2, while tea companies and universities also undertake their own research.

Figure 20.2 Research stations have an obligation to communicate the results of their research in the most effective way to all stakeholders. The line-source irrigation system provided an excellent visual picture of the effect of drought and irrigation on the yield of different clones. (Ngwazi Tea Research Station, Tanzania, MKVC). *A black and white version of this figure will appear in some formats. For the colour version, please refer to the plate section.*

Box 20.1 Research Activities and Outputs from the Tocklai Tea Research Institute, Assam, North-East India, 1910–2010

(abstracted from Hazarika and Muraleedharan, 2011)

1910–1930
- Soil pH analysis standardised.
- Effect of lime and sulphur on soil pH ascertained.
- Insects recorded in different districts.
- Systematic recording of weather data began.
- Ammonium sulphate most efficient source of nitrogen.

1931–1940
- Ammonium sulphate role confirmed.
- Higher yield, lower quality: inverse relationship.
- Need for fertiliser when replanting noticed.
- Green crops for two years when replanting.
- Tea breeding initiated (resulted in release of clones).
- Science behind shoot growth characteristics investigated.

1941–1950
- Systematic experiments on pruning, plucking, manuring and shade laid out.
- Responses to 90 kg N ha^{-1} as ammonium sulphate, and up to 180 kg ha^{-1}.

Box 20.1 Continued

- Young tea with NPK response to 40–80 kg N ha^{-1}.
- Pit mixture when planting (4.5 kg cattle manure + 28 g superphosphate).
- Sleeve size for cuttings standardised (200 mm × 175 mm lay flat)
- Three-year pruning cycle tried for first time.
- Tipping height best at 100 mm.
- Role of maintenance foliage understood.
- Optimum time of (light) pruning was November/December.
- Chemical analysis of whole leaf initiated.
- Release of polyclonal seed (stock 303, Gaurishanka; since discarded).
- First clones released in 1949 (TV1, TV2 and TV3). TV1 continues to dominate the tea industry in North-East India.

1951–1960

- Nomenclature proposed for tea species.
- Winter dormancy explained (?).
- Patents obtained for Rotorvane, Continuous Tray Drier, and Rotorvane Attachment.
- Photosynthesis in tea in relation to light intensity and temperature explained.
- Suitable shade trees identified: *Albizia chinensis* and *A. odoratissima.*

1961–1970

- Longer pruning cycles evaluated.
- Paraquat and 2.4-D herbicides introduced.
- Rehabilitation with *Mimosa invisa* and Pusa giant hybrid grass after uprooting tea recommended.
- Long-term effects of ammonium sulphate on soil investigated.
- Eight clones (TV10–TV17) and biclonal seed (Nanda Devi) released.
- Life cycles of insect pests and their predators described.
- More met. stations established in the regions.

1971–1980

- Three more biclonal seed stocks (TS449, TS450 and TS397) released, plus seven clones (TV18–TV24).
- Garden selection scheme (to widen genetic base) introduced.
- Machines patented included Boruah Continuous Roller, Continuous Withering Machine and Rotary Type Continuous Tea Roller.
- Chemical composition of Assam and Darjeeling teas differentiated.
- Fate of nitrogen in soil quantified; leaching loss 40%; split applications recommended.
- Pigment profile used to detect flush characteristics.
- Factory floor fermentation test developed.

Box 20.1 Continued

1981–1990

- Importance of potassium recognised: NPK compound recommended.
- Clones respond differently to nitrogen.
- New machines included Tea Breaker cum Stalk Separator, Green Leaf Storage System, Green Leaf Spreader and CTC Screen.
- Four biclonal seed stocks and four clones released.
- In addition, 92 garden clones released.
- Radioactive carbon isotopes used to trace carbohydrate flow and storage in plant.
- Necessity for drainage and irrigation recognised.
- Life cycle of tea mosquito bug and *Agistemus* spp. a predatory mite studied in tea ecosystem.

1991–2000

- Withering, rolling and fermentation practices for the production of colour, quality and flavour in made tea standardised.
- Pheromones studied in tea/bacteria pest complex.
- Reducing pesticide loads initiated.

2001–2010

- Emphasis on sustainability.
- Integrated pest management.
- Integrated nutrient management.
- Drainage development.
- Catchment development.
- Molecular breeding.
- Drought-tolerant and blister-blight-tolerant genes identified.
- Model Tea Factory – a unique automated processing facility.
- Climate change: minimum temperature risen by 2 °C in last 80 years at Jorhat; rainfall reduced by 200 mm.

Box 20.2 Tea Research Institutes Serving the Indian Subcontinent

- The *Tocklai Tea Experimental Station*, now known as the *Tocklai Tea Research Institute* ($26°43'$N $94°$E, alt. 50 m: annual rainfall *c*. 2000 mm; established in 1911 at Jorhat), serves 760 member estates. In total there are about 60,000 holdings in Assam (43,000), West Bengal (9000) and elsewhere in northern India (9000). Average yields in North-East India are about 1800 kg ha^{-1}.
- The *Darjeeling Tea Research Centre* (founded in 1976) is based at Kurseong at an altitude of 1240 m ($26°55'$N $88°12'$E), where the annual rainfall often exceeds 3000 mm.

Box 20.2 Continued

- South India also has its own tea research institute, known as the *United Planters Association of Southern India* (or UPASI). It is situated at Valparai (11°21'N 76° 08'E) at an altitude of 1050 m, where the average annual rainfall is close to 4000 mm. It reaches 7500 mm in some places. UPASI serves 143 member estates and 12,000 smallholders (classified as farmers who own 8 ha of tea or less).
- The headquarters of the *Tea Research Institute of Sri Lanka* is located in the central highlands at Talawakelle. It was founded in 1925. There are about 400,000 holdings (average size 0.5 ha). Smallholders contribute about 74% of the national production. The national average yield is about 2000 kg ha^{-1}.
- The *Bangladesh Tea Research Institute* (BTRI) was founded in 1972 soon after the War of Liberation had led to the creation of a separate state called Bangladesh. It provides research and extension support to 159 registered tea gardens. Because of the proximity of Assam, technical support until 1947 came from the Tocklai Tea Experimental Station in Jorhat, but in 1957 the Pakistan Tea Research Station was formed at Srimangal in Sylhet (24°18'N 91°44'E, alt. 23 m).

By way of contrast, the recent history of tea research in Tanzania is summarised in Box 20.3. Since the mid-1950s, when research on tea began in East Africa, there have been continual disruptions to funding reflecting, in part, the ebb and flow of political priorities. The lack of continuity in Tanzania has been due to the break-up of regional alliances, changes in policies at government level, vagaries in the profitability of the tea industry, a failure of research to convince the stakeholders of its value to their industry, failure of government organs to truly understand how an appropriate research institute can function and to truly understand the practical economics of agriculture. Each of these has played a role. For a perennial crop like tea, long-term sustainable funding is essential

Box 20.3 Tea Research in Tanzania: a Short History

Tanzania is a small player on the world scene, producing 30,000–35,000 kg of processed teas annually, virtually all of which are black teas. As has been pointed out in Chapter 3, tea growing is concentrated in the Southern Highlands, where there is an extended dry season, lasting up to six months. Tea is also grown in the Usambara Mountains in the north-east of the country, and to the west of Lake Victoria in Kagera, where there are two dry (or less wet) seasons. So tea researchers in Tanzania face the challenge of a small industry with its tea widely scattered geographically as well as ecologically. There is also a balance to be struck between short-term and longer-term goals.

From the late 1950s until 1978 tea growers depended for their technical advice on the staff of the Tea Research Institute of East Africa. With its headquarters in

Box 20.3 Continued

Kericho, in Kenya, this institute, which was largely funded directly by the industry, served the three East African countries, Kenya, Tanzania and Uganda. The break-up of the East African Community in 1980 meant that cooperation between the three countries ceased. For the next five or six years, there was no effective research undertaken in Tanzania. That is, not until a major tea company decided to take unilateral action and contracted Cranfield University in the UK to undertake some very specific field studies on tea irrigation and nutrition.

From a very humble start, the research programme was slowly consolidated and its funding sources diversified. The UK Department for International Development (DFID) provided some support, despite its officials blowing hot and cold over whether a government aid agency should be seen to be contributing funds to the private sector and, more specifically, to the tea crop. This, predictably, led to conflict with the government of Tanzania, which had its own plans and priorities for agricultural research, as indeed did the World Bank. The Bank at that time was funding, through the Tanzanian government, the provision of infrastructure (particularly laboratories) for crop research. Tea had previously been identified as a priority crop. This decision eventually resulted in laboratories and other facilities being built at a site entirely unsuited for forward-thinking, innovative tea agronomy research to be undertaken.

Eventually, it was accepted that the headquarters of the newly formed (1998) Tea Research Institute of Tanzania (TRIT) should be at Ngwazi, with a substation known as Marikitanda serving the north of the country from its base in the East Usambara Mountains (Carr, 1999).

With its predictable climate and extended dry season (see Box 3.7), Ngwazi is akin to an outdoor controlled environment facility. Southern Tanzania is where the majority of new tea projects are being developed, with irrigation on flat land suited to mechanical harvesting. This is where the future of tea in Tanzania will be decided, not on steep slopes in rainforests in remote locations, and it is here where the research needs to be centred.

TRIT has since been very successful in attracting funds from donors, in particular the European Union, for a diverse range of activities largely targeted at the small farmer. These include establishing tea nurseries, and mobile farmer training centres in villages across the country. In the view of some stakeholders (particularly the major cess payers) TRIT has now become too large and has lost its original focus, and with it their support. No doubt some redirection was needed, but it was never intended that TRIT should become large.

Donors can take some responsibility. They saw TRIT as a successful, well-led organisation where relatively large sums of money could be dispersed with minimum risk. But what happens to TRIT when EU priorities change? Is it sustainable? You decide!

if progress is to be made, while at the same time the industry needs to be confident that the research outputs will be of value (Box 20.4).

Box 20.4 lists the issues that research managers face when seeking to prioritise their activities in an attempt to keep their numerous stakeholders happy. The Ngwazi Tea Research Unit (NTRU), a precursor to TRIT, started with a simple formula: prioritise short-term issues of concern, understand the mechanisms responsible for the observed responses, and communicate the results to the industry and the wider scientific community in the most appropriate way.

Box 20.4 Sustainable Tea Research

Aim: to ensure the long-term *sustainability* of the tea industry by undertaking and reporting high-quality agronomy research.

The issues that need to be recognised by researchers seeking to achieve this aim include: the diversity of the stakeholders (small- and large-scale tea producers with different requirements), the diversity of ecological areas where tea is grown, even within one country, and the necessity to seek secure funding, and to attract and retain good staff.

Small-scale tea producers trying to maintain and/or enhance their livelihoods need to develop successful partnerships between tea producer and tea processor in order to seek to receive regular, reliable and *fair* payments for green leaf.

Large-scale tea producers seeking to remain profitable or to become (more) profitable need to reduce unit costs of production.

Agronomy research can help large-scale producers providing that researchers understand commercial realities, and can ensure that 'value' is not lost or left behind in the field.

Agronomy research can help small-scale producers providing that researchers recognise the technical, financial and social constraints to increases in productivity, by delivering training and extension services and, in so doing, enhancing the self-confidence of farmers.

Who is best qualified to judge the value of agronomy research? Small- and large-scale producers, consumers represented by government, donors, other researchers, self-assessment, whoever is paying the cess!

The value of agronomy research is judged on the basis of its developmental impact (e.g. rapid uptake, increased profitability, improved livelihoods), scientific impact (e.g. new knowledge, international recognition), value for money.

When agronomy research is judged depends on whether it is addressing short-term problem solving and/or long-term strategic issues, when the next annual report is published, or when someone says we have to!

The contributions of agronomy research with obvious long-term commercial value include: vegetative propagation, clones, NPK fertilisers, herbicides, shade.

Box 20.4 Continued

Small independent tea research organisations wishing to be sustainable need:
- to earn and retain the confidence of the tea industry
- to focus on problem solving
- to be supported by good-quality science
- to assess the financial worth of any recommendations
- to communicate the results to the industry in the most appropriate ways in order to facilitate uptake

Risks
- If the tea research organisation becomes too large, and forgets its primary function,
  it becomes unsustainable.
- Similarly, if donors encourage rapid expansion in infrastructure, when the money stops,
  it becomes unsustainable.

The contributions that NTRU/TRIT has made to the tea industry over a short period with limited funds is shown in Box 20.5. It serves as an example of what a small research unit, operating out of a simple wooden building for many years, can achieve. This formula has also worked to the extent that the NTRU was the basis on which TRIT was formed. But the question remains – For how long can this level of research output and uptake be maintained? Is it sustainable?

Small can be beautiful, but it does not always match the aspirations of those who seek to compete for prestige with large sister organisations (Figure 20.3)! The pressure is to expand, to get bigger, sometimes too big (often with the encouragement of the donor who is under pressure to spend money quickly), and hence to become unsustainable when funds become scarce, as they always do. All single-crop research stations face similar problems. In the view of some stakeholders, TRIT has already become too big and is not sufficiently focused on issues of concern to the commercial sector.

The perception is that large, long-established tea research institutes may not be as productive as small ones.

Management Training and Sustainability

Both at research stations and on tea estates, opportunities for staff development are key to sustainability. The following statement described the state of the research at one research institute. Many readers will identify with what is written, as it applies to many other single-crop, underfunded research organisations, not just those devoted to tea.

The . . . . . . is constrained in its work by its links with the Tea Board and the Ministry responsible: in particular this relates to questions of remuneration and promotion of research staff. The existing

---

Box 20.5 Impact of Agronomic Research at the Tea Research Institute of Tanzania, 1985–2005

Outputs from the research include:

Development

1. Two modern, irrigated, machine-harvested, clonal estates established (and expanding).
2. Intensification of existing tea; average estate yield increased from 2300 to 3500 kg ha^{-1} over the same period. Target commercial yields now 6000 kg ha^{-1}.
3. Extension service provided to 10,000 farming families (in association with district extension staff), commercial contract.
4. Village-based tea nurseries facilitated nationally (EU support).
5. National production increased from 15,500 t in 1985 to 32,800 t in 2012 (FAOSTAT, 2013).
6. Extra employment opportunities.
7. Improved livelihoods.

Science

1. Understanding of the effects of climate and water on development and productivity of tea has improved.
2. International recognition through refereed scientific papers.
3. Staff development, research theses.

Figure 20.3 Some tea research institutes have their priorities right, with their own cricket ground (Sri Lanka, MKVC). *A black and white version of this figure will appear in some formats. For the colour version, please refer to the plate section.*

departmental structure is also too rigid, with not enough emphasis on problem solving. The staff is academically isolated from other research organisations, with very limited opportunities to travel and without access to relevant up-to-date journals.

It has not been common in the tea industry to train managers and others in a systematic way. Rather it has been a question of 'learning on the job' (or learning how *not* to do it on the job!). In the late 1980s, a leading international tea company identified the need for training of its staff working in the tea sector if their businesses were to remain sustainable. A study was commissioned to determine the training needs at different levels in the management structure in tea businesses in Africa.

Training in this context was defined as:

Any systematic process aimed at equipping people with the knowledge and skills required to undertake a present or future job.

The types of knowledge or skills required by different levels of staff fell into two main categories:

- **Management** – associated with the general management or supervision of resources (e.g. personnel management, financial management, resource management and control).
- **Technical** – associated with the technology required to undertake a specific current or future job (e.g. field agronomy, factory processing or computer skills).

In addition, the study listed four categories of training according to purpose:

- **Preparatory or pre-entry training** – to determine the suitability for employment of potential recruits to the industry
- **Induction training** – to equip newly recruited management trainees with basic knowledge and skills
- **Performance training** – to improve performance of current staff
- **Developmental training** – to prepare high-potential staff for broader and more senior management responsibilities.

This analysis resulted in the development of a series of training programmes for staff at all levels. As a prerequisite, training began with managers at the highest level in the businesses (namely the directors). It then continued with senior managers, followed by junior managers. These courses were then repeated over a period of several years at an international training centre established for this purpose in Kenya. On completion of the course, participants returned to their places of work 'uplifted'. They took with them previously prepared action plans, the implementation of which was later evaluated by a tutor. The key to the success of this programme was that it had the support of the chief executive and it was implemented by people who had earned the confidence of the tea industry.

Following this experience, a similar training needs assessment was undertaken for the tea industry in southern Africa, specifically in Malawi and Zimbabwe (Harding *et al.*, 1993). Over the same 12-year period management courses were also run in the UK for senior managers from all over the world, including many working for Indian tea companies. This coming together of managers, used to working in relative isolation in their own country, exposed them to new ideas from course participants from elsewhere in the world.

Unfortunately systematic training programmes for plantation managers of the sort just described are no longer offered internationally, although they served a valuable function.

At around the same time (late 1980s/1990s) a major training programme, funded by what is now known as the UK Department for International Development (DFID) was being implemented for the tea industry in Bangladesh. This project had its own challenges, in particular the question of who were its principal beneficiaries – the consultants? the trainees? the profit-sharing shareholders in the tea companies? wage-earners? Or those people living on the non-viable, underperforming tea gardens – the so-called 'poorest of the poor', the declared target beneficiaries of British aid programmes at the time? These issues, which are common to many aid projects, are well described as a lesson to all project managers by Sullivan in his book *The Making of a One-Handed Economist* (2011).

Extension and Sustainability

Mutual trust between participants is central to the sustainability of any enterprise. This applies in particular to the relationship between the smallholder producing green leaf and the factory owner who processes that leaf and sells the tea. If the process is perceived not to be fair, and the means by which the proceeds are shared are not transparent, the relationship will soon break down, and plucking tea will not be a priority activity for the farmer. This was the situation in 1999 when TRIT was awarded a three-year contract to coordinate all the tea extension services for smallholders in the Rungwe District of southern Tanzania. This contract has since been renewed several times and the service is continuing to this day (2017), which must demonstrate to some extent how successful it has been. The initiative followed years of neglect by the authority responsible for smallholder tea at that time, the privatisation of the two tea factories supposedly serving the smallholders, and the creation of a company in which the smallholders became substantial shareholders. TRIT runs the contract independently of its core business, which initially was funded by a compulsory cess from the industry. Unfortunately this is no longer the case. Government is once again involved. It is very important that the two income streams are seen to be separate and that there is no hidden support of what is a private company with cess money. The contract is with the managers of the company. It is self-funding.

Following a baseline geographic survey (with GIS) of the area in 2001, a total of 3626 ha of smallholder-owned tea was identified, of which 2233 ha were being plucked and 658 ha were in the process of being rehabilitated, while the balance had been abandoned. Recruitment of nine staff by TRIT to advise the smallholders took place in 2000. The responsibilities of TRIT, the company and any third parties (e.g. haulage) were clearly defined so as to establish the links. Their initial task was to establish a smooth green-leaf buying and collection system. Collection centres were renovated at central and accessible points across the district and linked with village tea committees.

Special attention was paid to aligning any government-employed district extension staff into the process. This was despite their limited knowledge at the outset of tea growing. One issue that had to be handled with particular care was that, although TRIT had no line-management control over government employees, it was very dependent on their goodwill. Assistance was provided to the extension staff such as bicycles and fuel allowances. Motorbikes, means of communications and incentives were provided for all the extension staff as the area to be serviced was/is vast.

A close working relationship was also necessary with the green-leaf haulage company that was employed to deliver leaf to the factory. An early challenge was to gain the trust of the tea growers, who for many years had been exploited (or robbed). This included accurate (and transparent) weighing of the green leaf at each collection point (including quality control), timely delivery of fertilisers, regular and reliable leaf collections, transparent systems for payments (and charges), and fair sharing of profits through dividends. Trust once gained had to be maintained. In addition, TRIT has established demonstration plots and implemented farmer training programmes.

A contract between a fully commercial company and a research organisation, such as the one described, to run extension services in a developing country is unique. With TRIT controlling and having full access to an extension network, it makes it much easier to link research output directly to a total of about 12,000 farmers in the Rungwe District. This means of technology transfer also ensures that feedback from farmers reaches researchers directly (Simbua and Nyanga, 2001). This model for extension is now being repeated elsewhere in Tanzania.

Statistics and Sustainability

Over-production of tea on a world scale is something to be avoided if the tea industry is to be sustainable. The International Tea Committee (ITC) was established in 1933 to provide the tea industry with statistical information on a regular basis in what turned out to be an unsuccessful attempt to control production. (In a way, its role was similar to that of OPEC for oil-producing countries.) In the early 1920s, world exports of tea were around 310,000 t each year. Of this total, about 75% came from British-owned plantations in India and Ceylon (Sri Lanka) and 9% from Dutch-owned plantations in the Dutch East Indies (Indonesia). At this time the UK absorbed about 60% of the world's tea exports. By the end of the 1920s, tea production in the Dutch East Indies had doubled from 35,000 t in 1921 to 72,000 t. Fearful of a world surplus and a consequential decline in prices, the British and Dutch producers agreed a Regulation Scheme in 1929 to restrict, on a voluntary basis, tea production and exports. But, because of disagreements between the two parties, this agreement was abandoned two years later. Production continued to grow and, because the world market was unable to absorb the extra tea, prices fell.

Subsequent negotiations led to the formation of the International Tea Committee by representatives of tea growers in India, Ceylon and the Dutch East Indies, with the twin tasks of (1) administering the Regulation Scheme under the terms of the International Tea Agreement, and (2) collecting and compiling data on tea production, exports, consumption and stocks. This scheme stayed in place until 1955, after which ITC became responsible only for the statistical component of its remit, a task it continues to this day (ITC, 2013). It is a non-partisan, non-profit-making organisation funded by many of the major tea-producing and tea-consuming countries in the world. Good data are an essential prerequisite for any decision making that concerns planning for the future.

Conclusion: Transparency and Fairness

Is sustainability sustainable? That was the question posed at the start of this chapter. Is the tea industry sustainable? We are back though to definitions, and the question of an appropriate time scale for tea. It is probably decades or at least several pruning cycles. (On that scale, I am between 14 and 17 pruning cycles old!) Well-focused and directed research is always a necessity if tea industries are to survive and to operate at the highest level. There is also a continuous need for training, providing managers are allowed to manage. If a tea company is to make progress and to be successful, it needs well-qualified and well-motivated employees. Extension services, when effective, can contribute towards the long-term sustainability of the livelihoods of the stakeholders, particularly smallholders. The transparency and fairness with which the rewards are shared need to be monitored. Tea is a long-term investment complete with its own insurance policy, since the plant is virtually indestructible. It can be abandoned for years but then recovers quickly once normal service is resumed (as happened in Tanzania and Uganda: Figure 20.4). Tea is a very forgiving plant. Tea is sustainable

Figure 20.4 The tea industries in several countries have been constrained for many years as a result of military adventures. For example, the coup led by Idi Amin in Uganda in 1970 led to the abandonment of tea estates for several years before they were rehabilitated (MKVC).

Box 20.6 What Has Changed?

Following a visit from South India (UPASI) to Assam, Sharma and Swaminathan (1976) compared and contrasted tea cultural practices in Assam with those practised in South India at that time. It makes interesting reading 40 years later.

Differences highlighted included: the flat terrain in Assam (compared with South India); the occurrence of paddyfields between tea fields; a high water table, at depths from the surface ranging between 1.20–1.50 m (January) and 150–300 mm (July); alluvial soils; marked seasonal production; replanting with clones at a rate of 2–2.5% per annum; use of biclonal and polyclonal seed; clones selected on commercial estates; double-hedge planting $120 \times 75 \times 75$ cm (13,675 plants ha^{-1}) and also single-hedge 120×60 cm (13,888 plants ha^{-1}); use of mulch in newly planted areas; bringing into production by bending and then tipping, cut-across prune in year 3, followed by pruning on a three-year cycle – this policy was questioned when used at high densities (bending is expensive), but 500 kg tea ha^{-1} was obtained from single-hedge in year 2 after planting and 2200 kg ha^{-1} from double-hedge; dense (excess) shade from leguminous trees (*Albizia* spp.): noticed that efficiency of pluckers declines in fields without shade; policy of applying fertiliser to mature tea in single dose (April/May, 90–135 kg N ha^{-1}, 40 kg K$_2$O ha^{-1} annually, plus 20 kg P$_2$O$_5$ ha^{-1} every three years) was questioned; potash (and zinc) deficiencies suspected; poor frames due to repeated high prune (skiffing); excellent plucking standards, seven-day rounds for 9–9½ months; black rot and red rust were major problems; weed control variable.

Perhaps things haven't changed as much as we thought.

(whatever that means!), providing there continues to be a market at the right price for the teas produced and over-production is avoided.

We began in Chapter 1 by referring to things that had changed in the tea industry over the last few decades. But examination of a report on a visit to North-East India by Sharma and Swaminathan (1976) from South India suggests that perhaps not so much has changed after all (Box 20.6).

A useful three-word descriptor or motto for an effective tea research institute might be:

Solve (the problems), **understand** (the processes) and **communicate** (the results)

The tea industry, and its supporting research institutes, might then both be sustainable. Sustainability can indeed be sustainable.

Figure 20.5 Tea revives – it is the sovereign drink of pleasure and health. It is time for tea. *A black and white version of this figure will appear in some formats. For the colour version, please refer to the plate section.*

Postscript

If you think education is expensive, try ignorance (seen written on the back of a bus in Dar es Salaam).

Glossary

This glossary is based on a paper prepared for the Tea Research Institute of Tanzania (Carr, 2000). Reference has been made to a number of sources, including the *Tea Growers' Handbook* (4th edition), published by the Tea Research Foundation of Kenya (TRFK, 1986), and the *Tea Planter's Handbook*, published by the Tea Research Foundation of Central Africa (TRFCA, 1990).

Acceptable leaf: a term used to describe whether the relative proportions of shoots at different stages of development and size within a freshly harvested sample meet the factory specifications.

Apical dominance: a term that describes the controlling influence of the apical or *terminal bud*, whether active or dormant, over the development of the *axillary buds* lower down the stem.

Assam-type: this term is used to describe cultivars of tea that show vegetative characteristics tending towards those of *Camellia sinensis* var. *assamica* (Masters), namely with large, horizontal, broad, mostly non-serrated and light green leaves (cf. *China-type*).

Axillary bud: the vegetative bud in the angle between a leaf and the stem.

Banjhi: the (Indian) name given to a tea shoot with a dormant terminal bud, recognised by its relatively small size (2–3 mm long). There is *'hard' banjhi*, where the upper leaves and stem have become coarse through age, which is unsuitable for processing, and *'soft' banjhi*, where the subtending leaves and stem are young, that can be successfully used in tea production.

Basal population density: the number of (small) shoots per unit area of bush, or ground, remaining after harvest (cf. *harvested population density* and *total population density*).

Breaking back: the process in which old leaves unsuitable for processing are removed and discarded in order to maintain a level *plucking table*.

Cataphyll: a reduced, small leaf (see *Scale-leaves*).

China-type: this term is used to describe cultivars of tea that show vegetative characteristics tending towards those of *Camellia sinensis* var. *sinensis* (L.), namely with small, erect, narrow, serrated and dark green leaves (cf. *Assam-type*).

Crop distribution management: the term used in Malawi for techniques recommended to mitigate the problems of managing large crop peaks, particularly those that follow the start of the rains. These include (1) removing 30–40 mm from the top of bushes (*skiffing*) 17–27 days after the start of the rains or (2) *tipping* pruned tea 50 mm lower than the recommended height, then *plucking* normally until the peak occurs when one round of plucking is missed. Both practices are claimed to reduce the size of the crop peak without any loss of crop.

Field capacity: the maximum amount of water a soil can hold against the forces of gravity (that is when free drainage ceases) (mm).

Flush: the word used to describe the state of the crop when there is a large number of visible, actively growing tea shoots per unit area of the bush surface.

Fordham effect: the name given (after the scientist who first described the process) to the large peak in production that occurs after a limiting factor, such as low temperatures or drought, has ended and allowed the accumulated buds of many ages to develop together. Once these shoots have been harvested there is then a decline in production until the next generation of shoots has developed, followed by a second but smaller peak. The first large peak, such as that which follows the start of the rains in Malawi, can cause large logistical problems in the field, for transport, and in the factory.

Gang plucking: the number of *pluckers* needed to harvest a given area of tea is decided on a daily basis by management depending on the availability of crop and labour; the pluckers then move across the field in a group or gang.

Green leaf: a collective name commonly used to describe the harvested leaf before processing begins in the factory.

Harvested population density: the number of harvested shoots per unit area of bush, or ground (cf. *basal population density* and *total population density*).

Immature leaf: a term used to describe shoots that are less developed than those specified for harvest, typically those with one unfolded leaf, or with a partially open second leaf, and an active *terminal bud*.

Leaf appearance rate: a measure of the rate of unfolding of individual leaves on a tea shoot (leaf/day); the reciprocal of a *phyllochron* $(1/P)$.

Leaf handling: this term covers the movement of *green leaf* within the field, weighing and storage, and transport from the field to the factory. The positioning of weighing points, and ease of access from the field, affects the productivity of the *pluckers*. Similarly, bad handling of the green leaf in the plucking basket and poor storage conditions after weighing and during transport can have adverse effects on the condition and tea-making quality of the leaf entering the factory. Since leaf handling extends from the time that an individual shoot is plucked in the field to the onset of the formal withering process in the factory, it is often the neglected part of the total tea production process, being the full responsibility of neither the field nor the factory manager.

Leaf pose: a measure of the inclination of a leaf as indicated by the angle between the leaf petiole, or (better) the leaf tip, and the vertical stem or axis. Tea plants with large, horizontal ($> 70°$) leaves are considered to be of the *Assam-type*, while those with small, vertically inclined ($< 50°$) leaves tend towards the *China-type* of bush.

Maintenance foliage: the name given to the canopy of leaves below the level at which the tea shoots are harvested. These are the leaves that intercept sunlight and, through the process of photosynthesis, manufacture carbohydrates that contribute to the growth of the tea bush. For large yields it is essential to ensure that the leaf canopy completely covers the ground surface for as large a proportion of the *pruning cycle* as possible.

Mature leaf: the name given to leaves that have stopped expanding in area and become hard with ageing. They are unsuitable for processing.

Mother leaf: the name given to the leaf, the bud in the axil of which develops into a shoot or, in the case of a leaf-bud cutting, a plant. In India, the term 'mother leaf' is used to describe the first true foliage leaf to unfold on the expanding shoot.

Node: the point of attachment of a leaf to the stem; by comparison the *internode* refers to the length of stem between adjacent nodes.

Origin: see *plucked origin*.

Phyllochron: the time interval (days) for two successive 'true' leaves to unfold from the terminal bud. Harvest intervals can be estimated using the relationships that exist between mean daily air temperature and a phyllochron. On the assumption that, after harvest, the shoots remaining on the bush that will form the basis for the next harvest have, on average, one unfolded leaf, and that the target shoot is *'three leaves and a bud'*, the interval to the next harvest will be two phyllochrons. The reciprocal of a phyllochron is known as the *leaf appearance rate* (leaf/day).

Pieces per kilogram: this index can be used as a measure of both the 'quality' of *green leaf* after plucking, and also of *plucker* productivity (see Box 10.3). The target number of *pieces per kilogram* varies with stage in the pruning cycle, and the type of tea plant or clone. The higher the number, the more small shoots in the sample. In Kenya the target is about 1000 pieces kg^{-1} for seedling tea in the second and third years after pruning. This is considered to represent a 50:50 split between shoots with *two leaves and a bud* and *three leaves and a bud*.

Plucked origin: a tea shoot that has developed from the *axillary bud* of the leaf immediately below the point of abscission of a harvested shoot, following the removal of *apical dominance*. By comparison, a shoot that has developed from a bud without the removal of apical dominance is said to have come from an *unplucked origin*.

Pluckers: the name commonly used to describe people who harvest (pluck) shoots by hand from tea bushes growing in the field.

Plucking: the skilful process of removing targeted tea shoots by hand. Various terms are used to differentiate types or intensities of plucking. For example, in *selective plucking* the pluckers are encouraged to harvest only shoots of a specified maximum length and *stage of development*. Thus *fine plucking* refers to the selective removal of *two leaves and a bud* and *soft banjhi* only. By contrast, in *non-selective plucking* the pluckers are expected to remove all the shoots above the level of the table regardless of size or stage of development. This is also sometimes referred to locally as *plucking black*, a reference to the dark colour of the *maintenance foliage* when all the young shoots have been removed, or *hard plucking*. Shoots unsuitable for processing should always be discarded.

Plucking round: the interval (days) between one harvest and the next.

Plucking table: the level surface of the leaf canopy which *pluckers* are charged to maintain after *tipping*, or allowed to increase in height at a certain rate. A bamboo *plucking wand* facilitates this process. The faster the table rises the shorter the interval between one pruning and the next. One important role of management is to ensure that the rate of increase in the height of the table, above ground level or the last prune, is kept below an acceptable maximum. This should be measured on a regular basis. Shoots left on the bush represent lost crop.

Plucking wand: a length (2.5–3.5 m) of bamboo, with a diameter of 10–20 mm, used by *pluckers* as an aid in maintaining a level surface on the top of tea bushes.

Primordia: embryonic leaves (or flowers) that are initiated in buds that subsequently may develop into shoots.

Programmed plucking: the term used to describe a system of management in which the harvest interval is determined on the basis of the time taken for a *shoot generation* to develop to a stage suitable for plucking. In Malawi this is taken to be equivalent to one-quarter of the duration of the *shoot replacement cycle*, which, in the main growing season, is about 42 days. The recommended harvest interval is therefore alternately 10 and 11 days.

Pruning cycle: the time interval between one pruning of a tea bush (once the formative prunes have been completed) and the next. In eastern Africa commercial practice is usually to prune every three or four years, but this can be extended if the height and level of the *plucking table* allow the *pluckers* to continue to harvest efficiently. The terms *first*, *second*, *third* and *fourth year after pruning* refer to the successive 12-month periods following the prune. In Malawi the three years in the pruning cycle are referred to as *pruned* (P), *unpruned one* (UP1) and *unpruned two* (UP2).

Saturation deficit: a measure of the dryness of the air. Namely, the difference between the saturation vapour pressure of the air (e_s, kPa) and the actual vapour pressure (e_a, kPa). For comparison, the relative humidity is $e_a/e_s \times 100$.

Scale-leaves: two or sometimes three very small smooth-edged leaves (or cataphylls) that can be found at the base of a tea shoot. Previously enclosing the bud, they usually fall off as the shoot matures. An upper, much larger, scale leaf is known as the *janum*, while the term *fish-leaf* is sometimes used to describe a similar leaf, serrated on one side, situated above the *janum*. These two appendages are not always easy to distinguish, and are of little known practical importance, although buds, which can develop into shoots, form in their axes.

Scheme plucking: a management system in which each *plucker* (or a family) is given an area of tea for which he or she is responsible throughout the season. The area of each plot, representing the number of bushes that can be harvested comfortably in one day is, commonly, about 0.05–0.1 ha, depending on the yield and the harvesting interval. Sometimes two (half-size) plots are allocated each day so that it is easier to accommodate a five-and-a-half-day working week, and to introduce flexibility in the system. Plots are demarcated by stakes at each corner or by allowing a shoot of tea to grow to a length of about 0.5 m. The responsibility for applying fertiliser, weeding and pruning on each plot can also be delegated to individual pluckers. Systems that suit local practices need to be developed to suit individual estates. *Programmed scheme plucking* is the term used to describe the combination of *programmed plucking* with scheme plucking.

Shoot generation: the term used to describe a population of shoots that develops from (mainly) *axillary buds* following the removal of *apical dominance* by *plucking*.

Shoot replacement cycle: the time taken (days) for an *axillary bud*, when released from *apical dominance* after plucking the main shoot, to develop into a shoot suitable for harvesting. In Malawi this is taken to be when the third true leaf has unfolded.

Skiffing: a very light prune intended to level the *plucking table* if it has become uneven, or sometimes to extend the length of the *pruning cycle* in a mature crop; usually a sign that management has failed to control the harvesting process.

Soil water deficit: a measure of how much water is needed to bring the soil profile back to field capacity (mm). It is analogous to a bank overdraft.

Stage of shoot development: a term that signifies the developmental stage that a shoot has reached, for example, a shoot having one, two or three unfolded leaves with an expanding terminal bud or with a dormant (*banjhi*) bud. By contrast, *growth* represents an increase in the dry weight of a shoot, and *size* would commonly be a measure of its length.

Terminal bud: the name of the vegetative bud found at the apex of a shoot from which new leaves unfold when it is actively developing, but which, if left to grow, becomes dormant (or *banjhi*) after a certain number of leaves have unfolded. Later the cycle is repeated.

Tipping: the name given to the first and second harvests after pruning when a new leaf canopy is being established and a level *plucking table* is being formed at a specified height above the ground or prune.

Total (shoot) population density: the total number of shoots (*basal* and *harvest*) per unit area of bush, or ground. This is a major determinant of yield in tea. Values can range from 100–200 to 1000–1200 m^{-2}.

Two/three leaves and a bud: a shoot with two or three unfolded 'true' leaves (ignoring any *scale-leaves*) and an actively growing *terminal bud*, > 10 mm in length. These are shoots at *stages of development* considered to be suitable for processing.

References

Allen, R.G., Pereira, L.S., Rees, D. and Smith, M. (1998). *Crop Evapotranspiration: Guidelines for Computing Crop Water Requirements*. FAO Irrigation and Drainage Paper 56. Rome: Food and Agriculture Organization.

Anandacoomaraswamy, A., De Costa, W.A.J.M., Shyamalie, H.W. and Campbell, G.S. (2000). Factors controlling transpiration of mature field-grown tea and its relationship with yield. *Agricultural and Forest Meteorology* 103: 375–386.

Andersson, J.A. and Giller, K.E. (2012). On heretics and God's blanket salesmen. In *Contested Agronomy: Agricultural Research in a Changing World* (ed. J. Sumberg and J. Thompson). Oxford: Earthscan, Routledge, 22–46.

Anon. (1997). Tea in Georgia. Business Information Service for the Newly Independent States (BISNIS). http://www.itaiep.doc.gov/bisnis/country/ggtearpt.htm (accessed September 2013).

Anon. (2003). Tea statistics: performance of tea in China. *International Journal of Tea Science* 2 (3): 24–30.

Anon. (2012). Manuring. Tea Research Association, Tocklai. http://www.tocklai.net/activities/tea-cultivation/manuring/ (accessed June 2014).

Anon. (2013). Vietnam tea overview. www.vietnam-tea.com/overview.html (accessed September 2013).

Apostolides, Z., Nyirenda, H.E. and Mphangwe, N.I.K. (2006). Review of tea (*Camellia sinensis*) breeding and selection in southern Africa. *International Journal of Tea Science* 5 (1): 13–19.

Arifin, M.S., Sanusi, M. and Subarna, N. (1991). The Indonesian tea industry. In *World Tea: Proceedings of the International Symposium on Tea Science*, Shizuoka, Japan, 1991, 58–64.

Ariyarathna, H.A.C.K., Gunasekare, M.T.K., Kottawa-Arachchige, J.D. *et al.* (2011). Morpho-physiological and phenological attributes of reproductive biology of tea (*Camellia sinensis* (L.) O. Kuntze) in Sri Lanka. *Euphytica* 181: 203–215.

Baillie, I.C. and Burton, R.G.O. (1993). Ngwazi Estate, Mufindi, Tanzania. *Report on Land and Water Resources with Special Reference to the Development of Irrigated Tea. Part 1: Soils and Land Suitability*. Soil Survey and Land Research Centre, Cranfield University.

Balasaravanan, T., Pius, P.K., Kumar, R.R., Muraleedharan, N. and Shasany, A.K. (2003). Genetic diversity among south Indian tea germplasm (*Camellia sinensis, C. assamica* and *C. assamica* spp. *lasiocalyx*) using AFLP markers. *Plant Science* 165: 365–372.

Balasuriya, J. (1998). Effects of altitude on the productivity of clonal tea in Sri Lanka. MPhil thesis, Cranfield University, UK.

Balasuriya, J. (2000). The partitioning of net total dry matter to roots of clonal tea (*Camellia sinensis* L.) at different altitudes in the wet zone of Sri Lanka. *Tropical Agriculture (Trinidad)* 77: 163–168.

Banerjee, B. (1992a). Botanical classification of tea. In *Tea: Cultivation to Consumption* (ed. K.C. Willson and M.N. Clifford). London: Chapman and Hall, 29–51.

Banerjee, B. (1992b). Selection and breeding of tea. In *Tea: Cultivation to Consumption* (ed. K.C. Willson and M.N. Clifford). London: Chapman and Hall, 53–86.

Barbora, B.C. (1991). Progress of R & D and its application by the tea industry in N. E. India. In *Proceedings of the International Symposium on Tea Science*, 494–498.

Barbora, B.C., Jain, N.K. and Baruah, U. (1982). Root studies in young tea. *Two and a Bud* 29: 52–55.

Barman, T.S. and Saikia, J.K. (2005). Retention and allocation of C14 assimilates by maintenance leaves and harvest index of tea (*Camellia sinensis* L.). *Photosynthetica* 43: 283–287.

Barman, T.S., Baruah, U. and Sarma, A.K. (1993). Effect of light and shade on diurnal variation of photosynthesis, stomatal conductance and transpiration rate in tea. In *Proceedings of the International Symposium on Tea Technology*. Calcutta: Tea Research Association, India, 208–218.

Barman, T.S., Baruah, U. and Saikia, J.K. (2008). Irradiance influences tea leaf (*Camellia sinensis* L.) photosynthesis and transpiration. *Photosynthetica* 46 (4): 618–621.

Barooah, A.X. (2010). Present status and use of agrochemicals in tea industry of eastern India and future directions. *Science and Culture* 77: 385–390.

Barua, D.N. (1963). Uses and abuses of clones. *Two and a Bud* 10: 3–6.

Barua, D.N. (1966). The interdependence of top and root growth in tea. *Two and a Bud* 13: 140–144.

Barua, D.N. (1969). Seasonal dormancy in tea (*Camellia sinensis* L.). *Nature* 224: 514.

Barua, D.N. (1970). Light as a factor in metabolism of the tea plant (*Camellia sinensis* L). In *Proceedings of the Symposium, Physiology of Tree Crops* (ed. L.C. Luckwill and C.V. Cutting). London: Academic Press, 307–322.

Barua, D.N. (1989). *Science and Practice in Tea Culture*. Calcutta-Jorhat, India: Tea Research Association.

Baruah, A., Phukan, I.K. and Bhagat, R.M. (2011). Comparison of soil and leaf nutrient status under organic and conventional teas. *Two and a Bud* 58: 155–159.

Baruah, R.D. and Bhagat, R.M. (2012). Climate trends of Northeastern India: a long-term pragmatic analysis of tea production. *Two and a Bud* 59 (2): 46–49.

Bezbaruah, H.P. (1987). Use of interspecific hybrids in tea breeding. *Two and a Bud* 34: 1–4.

Bezbaruah, H.P. and Dutta, A.C. (1977). Tea germplasm collection at Tocklai Experimental Station. *Two and a Bud* 24: 22–30.

Bishir, J. and Roberts, J.H. (1999). On numbers of clones needed for managing risks in clonal forestry. *Forest Genetics* 6 (3): 149–155.

Blackie, J.R. (1979). The water balance of the Kericho catchments. *East African Agriculture and Forestry Journal* 43: 55–84.

Bond, T.E.T. (1942). Studies in the vegetative growth and anatomy of the tea plant (*Camellia thea* Link.) with special reference to the phloem. I. The flush shoot. *Annals of Botany* N.S. 6 (24): 607–633.

Bond, T.E.T. (1945). Studies in the vegetative growth and anatomy of the tea plant (*Camellia thea* Link.) with special reference to the phloem. II. Further analysis of flushing behavior. *Annals of Botany* N.S. 9 (34): 181–216.

Bonheure, D. and Willson, K.C. (1992). Mineral nutrition and fertilizers. In *Tea: Cultivation to Consumption* (ed. K.C. Willson and M.N. Clifford). London: Chapman and Hall, 269–329.

Booth, T.H. and Pryor, L.D. (1991). Climatic requirements of some commercially important Eucalypt species. *Forest Ecology and Management.* 43: 47–60.

Bore, J.K.A. (1996). A review of problems of old tea fields. *Tea* 17 (1): 27–33.

Bore, J.K.A. (2008). Physiological responses of grafted tea (*Camellia sinensis* L.) to water stress. PhD thesis, Jomo Kenyatta University of Agriculture and Technology, Nairobi, Kenya.

Bore, J.K.A. and Ng'etich, W.K. (2000). Mechanical harvesting of tea. 1. Yield and leaf standards. *Tea* 21(1): 19–23.

Bore, J.K.A. and Njuguna, C.K. (1995). Chip budding in tea (*Camellia sinensis* (L.) O. Kuntze) and its effect on yields and quality. *Tea* 16: 9–13.

Bore, J.K.A., Njuguna, C.K. and Magambo, M.J.S. (1998a). Effect of high plant population densities on harvest index and yields: final report. *Tea* 19: 66–70.

Bore, J.K.A., Ng'etich, W.K. and Njuguna, C.K. (1998b). Performance of clonal tea in rehabilitated and replanted fields 1. Yields. *Tea* 19: 38–42.

Boyd, D.A. and Needham, P. (1976). Factors governing the effective use nitrogen. *Span* 19 (2): 68–69.

BTRI (1986). *Mature Tea Pruning.* Circular no. 79. Srimangal: Bangladesh Tea Research Institute.

Buckingham, J. (1885). Papers regarding the Sau tree and its remarkable influence on the tea bush. Indian Tea Association, Calcutta.

Burgess, P.J. (1992) Responses of tea clones to drought in southern Tanzania. PhD thesis, Silsoe College, Cranfield University, UK.

Burgess, P.J. (1993a). Irrigation scheduling for mature and young tea. *Ngwazi Tea Research Unit Quarterly Report* 11: 3–15.

Burgess, P.J. (1993b). Economic analysis of irrigation for mature tea. *Ngwazi Tea Research Unit Quarterly Report* 14: 8–12.

Burgess, P.J. (1994). Methods of determining the water requirements of mature tea. *Ngwazi Tea Research Unit Quarterly Report* 17: 11–21.

Burgess, P.J. (1996). Experiment N9. Responses of clonal tea to fertiliser and irrigation. In *Ngwazi Tea Research Unit Annual Report for 1993/94,* 6–7.

Burgess, P.J. and Carr, M.K.V. (1996a). Responses of young tea (*Camellia sinensis*) clones to drought and temperature. I. Yield and yield distribution. *Experimental Agriculture* 32: 357–372.

Burgess, P.J. and Carr, M.K.V. (1996b). Responses of young tea (*Camellia sinensis*) clones to drought and temperature. II. Dry matter production and partitioning. *Experimental Agriculture* 32: 377–394.

Burgess, P.J. and Carr, M.K.V. (1997). Responses of young tea clones (*Camellia sinensis*) to drought and temperature. III. Shoot extension and development. *Experimental Agriculture* 33: 367–383.

Burgess, P.J. and Carr, M.K.V. (1998). The use of leaf appearance rates estimated from measurements of air temperature to determine harvest intervals for tea. *Experimental Agriculture* 34: 207–218.

Burgess, P.J. and Myinga, G.R. (1992). Planning harvest intervals from phyllochrons and/or air temperature. *Ngwazi Tea Research Unit Quarterly Report, Dar es Salaam, Tanzania* 7/8: 17–25.

Burgess, P.J. and Sanga, B.N.K. (1993). Soil nutrient contents below a fertiliser experiment. *Ngwazi Tea Research Unit, Quarterly Report* 12: 3–7.

Burgess, P.J. and Sanga, B.N.K. (1994). Dry weight and root distribution of six-year old tea. *Ngwazi Tea Research Unit, Quarterly Report* 18: 12–16.

Burgess, P.J. and Stephens, W. (2010). Climate, climate change and tea production in Kenya, Tanzania and Malawi. Paper presented at Infini-Tea Conference: Ensuring the Future of Tea. New Delhi, India, May 2010.

Burgess, P.J., Carr, M.K.V., Mizambwa, F.C.S. *et al.* (2006). Evaluation of simple hand-held mechanical systems for harvesting tea (*Camellia sinensis* L.). *Experimental Agriculture* 42: 165–187.

Cabral, O.M.R., Rocha, H.R., Gash, J.H.C. *et al.* (2010). The energy and water balance of a *Eucalyptus* plantation in southeast Brazil. *Journal of Hydrology* 388: 208–216.

Calder, I.R., Hall, R.L. and Prassana, K.T. (1993). Hydrological impact of *Eucalyptus* plantation in India. *Journal of Hydrology* 150: 635–648.

Cannell, M.G.R., Njuguna, C.K., Ford, E.D. and Smith, R. (1977). Variation in yield among competing individuals within mixed genotype stands of tea: a selection problem. *Journal of Applied Ecology* 14: 969–986.

Carr, M.K.V. (1968). Report on research into the water requirements of tea in East Africa. Paper presented at a meeting of the Specialist Committee in Applied Meteorology, Nairobi, Kenya.

Carr, M.K.V. (1969a). The water requirements of the tea crop. PhD thesis, University of Nottingham, UK.

Carr, M.K.V. (1969b). Hydrological investigations. In *Annual Report for 1968/69, Tea Research Institute of East Africa, Kericho, Kenya*, 60–61.

Carr, M.K.V. (1970). The role of water in the growth of the tea crop. In *The Physiology of Tree Crops* (ed. D.C. Luckwill and C.V. Cutting). London: Academic Press, 287–305.

Carr, M.K.V. (1971a). The internal water status of the tea plant (*Camellia sinensis*): some results illustrating the use of the pressure chamber technique. *Agricultural Meteorology* 9: 447–460.

Carr, M.K.V. (1971b). An assessment of some results of tea-soil-water studies in Southern Tanzania. In *Water and the Tea Plant* (ed. M.K.V. Carr and S. Carr). Kericho, Kenya: Tea Research Institute of East Africa, 21–48.

Carr, M.K.V. (1974). Irrigating seedling tea in southern Tanzania: effects on total yields, distribution of yield and water-use. *Journal of Agricultural Science (Cambridge)* 83: 363–378.

Carr, M.K.V. (1976). Methods of bringing tea into bearing in relation to water status during dry weather. *Experimental Agriculture* 12: 341–351.

Carr, M.K.V. (1977a). Changes in the water status of tea during dry weather in Kenya. *Journal of Agricultural Science (Cambridge)* 89: 197–207.

Carr, M.K.V. (1977b). Responses of seedling tea bushes and their clones to water stress. *Experimental Agriculture* 13: 317–324.

Carr, M.K.V. (1985). Some effects of shelter on the yield and water use of tea. In *Effects of Shelter on the Physiology of Plants and Animals*. (ed. J. Grace). Lisse: Swets and Zeitlinger. *Progress in Biometeorology* 2: 127–144.

Carr, M.K.V. (1988a). *Integration of Research and Development, Advisory and Training Resources. Bangladesh Tea Rehabiliation Project Phase III*. Unpublished report for the UK Overseas Development Administration, London, by Cranfield University.

Carr, M.K.V. (1988b). Tea in Tanzania. *Outlook on Agriculture* 17 (1): 18–23.

Carr, M.K.V. (1995). Unpublished report to Rwenzori Highlands Tea Company Ltd. Cranfield University, Silsoe College, UK.

Carr, M.K.V. (1996). Mechanical harvesting of tea in Tanzania: a way forward. *Ngwazi Tea Research Unit Quarterly Report* 25: 3–7.

Carr, M.K.V. (1999). Evaluating the impact of research for development: tea in Tanzania. *Experimental Agriculture* 25: 247–264.

Carr, M.K.V. (2000). Definitions of terms used in tea harvesting 1: Shoot growth; and II: Harvesting and bush management. Paper prepared for the Tea Research Institute of Tanzania. Dar es Salaam, Tanzania.

Carr, M.K.V. (2010a). The role of water in the growth of the tea (*Camellia sinensis*) crop: a synthesis of research in Eastern Africa. 1. Water relations. *Experimental Agriculture* 46: 327–349.

Carr, M.K.V. (2010b). The role of water in the growth of the tea (*Camellia sinensis*) crop: a synthesis of research in Eastern Africa. 2. Water productivity. *Experimental Agriculture* 46: 351–379.

Carr, M.K.V. and Othieno, C.O. (1972). Hydrological investigations. In *Annual Report for 1971*. Kericho, Kenya: Tea Research Institute of East Africa, 34–35.

Carr, M.K.V. and Stephens, W. (1992). Climate, weather and the yield of tea. In *Tea: Cultivation to Consumption* (ed. K.C. Willson and M.N. Clifford). London: Chapman and Hall, 87–135.

Carr, M.K.V., Dale, M.O. and Stephens, W. (1987). Yield distribution at two sites in eastern Africa. *Experimental Agriculture* 23: 75–85.

Chandra Mouli, M.R., Onsando, J.M. and Corley, R.H.V. (2007). Intensity of harvesting in tea. *Experimental Agriculture* 43 (1): 41–50.

Chapotoka, O.H.F., Mphangwe, N.I.K. and Nyirenda, H.E. (2001). Above- and below-ground dry matter production of young clonal tea plants raised from plucking shoots and conventional cuttings. *Quarterly Newsletter, Tea Research Foundation of Central Africa* 144: 17–21.

Chen, L. and Yamaguchi, S. (2002). Genetic diversity and phylogeny of tea plant (*Camellia sinensis*) and its related species and varieties in the section *Thea* genus *Camellia* determined by randomly amplified polymorphic DNA analysis. *Journal of Horticultural Science and Biotechnology* 77 (6): 729–732.

Chen L., Yu, F.L. and Tong, Q.Q. (2000). Discussions on the phylogenetic classification and evolution of Sect. *Thea. Journal of Tea Science* 20 (2): 29–34.

Chen, L., Yang, Y.J. and Yu, F.L. (2005). *Descriptors and Data for Standard Tea (Camellia spp.)*. Beijing: China Agriculture Press.

Chen, L., Ming-Zhe, Y., Xin-Chao, W. and Ya-Jun, Y. (2012). Tea genetic resources in China. *International Journal of Tea Science* 8 (2): 55–63.

Chen, Z. (2012). Tea in China. *International Journal of Tea Science* 8 (2): 3–15.

Cilengir, E. (2010). Turkey's tea industry – the long and winding road. www.tching.com/2010 (accessed 18 July 2013).

Clowes, M.St.J. (1986). Pieces per kilogram: a useful guide to management and productivity. *Quarterly Newsletter, Tea Research Foundation of Central Africa* 83: 13–18.

Clowes, M.St.J. (1991). *Mechanical Plucking and Special Considerations*. Unpublished report, Tea Research Foundation of Central Africa.

Cooper, H.R. (1926). Shade trees. *Quarterly Journal of the Indian Tea Association* 1926: 83–102.

Cooper, H.R. (1946). *Nitrogen Supply to Tea*. Tocklai Experimental Station Memorandum No. 6, 3rd edn. Indian Tea Association.

Cooper, J.D. (1979). Water use of a tea estate from soil moisture measurements. *East African Agricultural and Forestry Journal* 43: 102–121.

Cordell, D., Drangert, J.-O. and White, S. (2009). Global food security and food for thought. *Global Environmental Change* 19: 292–305.

Corley, R.H.V. and Chomboi, K.C. (2005). Tea tasting: a statistical evaluation of tasters' skills. *Tea* 26: 10–18.

DAFF (2011). *A Profile of the South African Black Tea Market Value Chain*. Department of Agriculture, Forestry and Fisheries, Republic of South Africa.

Dagg, M. (1970). A study of the water use of tea in East Africa using a hydraulic lysimeter. *Agricultural Meteorology* 7: 3203–3220.

Dale, M.O. (1974). The Japanese plucking machine. *Quarterly Newsletter, Tea Research Foundation of Central Africa* 34: 3–6.

Damayanthi, M.M.N., Mohotti, A.J. and Nissanka, S.P. (2010). Comparisons of tolerant ability of mature field grown tea (*Camellia sinensis* L.) cultivars exposed to drought stress in Passara area. *Tropical Agricultural Research* 22 (1): 66–75.

Darmawijaya, M.I. (1988). The advance of tea production in Indonesia. In *Recent Developments in Tea Production: Proceedings of the International Symposium* (ed. Tsai-Fua Chiu and Chie-Huang Wang). Taiwan Tea Experiment Station, 111–117.

Dasgupta, R. (2007). Tea in Vietnam. *International Journal of Tea Science* 6 (2): 19–20.

De Costa, W.A.J.M., Surenthran, P. and Atttanayake K.B. (2005). Tree–crop interactions in hedgerow intercropping with different tree species and tea in Sri Lanka. 2. Soil and plant nutrients. *Agroforestry Systems* 63: 211–218.

De Costa, W.A.J.M., Mohotti, A.J. and Wijeratne, M.A. (2007). Ecophysiology of tea. *Brazilian Journal of Plant Physiology* 19: 299–332.

DFID (1999). *Sustainable Livelihoods Guidance Sheets*. London: UK Department for International Development.

Dharmaraj, N. (2012). Mechanisation of harvesting in tea: a classic case of science and technology interface with management. *International Journal of Tea Science* 8 (4): 57–58.

Doorenbos, J. and Kassam, A.H. (1979). *Yield Response to Water*. Food and Agricultural Organisation of the United Nations, Irrigation and Drainage Paper 33, Rome: FAO.

Eden, T. (1931). Studies in the yield of tea: I. The experimental errors of field experiments with tea. *Journal of Agricultural Science (Cambridge)* 21 (3): 547–573.

Eden, T. (1944). Studies in the yield of tea: Part V. Further experiments on manurial response with special reference to nitrogen. *Empire Journal of Experimental Agriculture*. XII (48): 177–190.

Eden, T. (1965). *Tea*, 2nd edn. London: Longman.

Eldridge, K., Davidson, J., Harwood, C. and van Wyk, G. (1994). *Eucalypt Domestication and Breeding*. Oxford: Clarendon Press.

Ellis, R. (1997). Tea, *Camellia sinensis* (Camelliaceae). In *The Evolution of Crop Plants*, 2nd edn (ed. J. Smartt and N.W. Simmonds). London: Longman, 22–27.

Ellis, R. and Grice, W. (1976). Plucking policy and techniques. *Quarterly Newsletter, Tea Research Foundation of Central Africa* 41: 3–10.

Ellis, R. and Nyirenda, H.E. (1995). A successful plant improvement programme in tea (*Camellia sinensis* L.). *Experimental Agriculture* 31: 307–323.

Etherington, D.M. (1990). Economic analysis of a planting density experiment for tea in China. *Tropical Agriculture (Trinidad)* 67 (3): 248–256.

Etherington, D.M. and Forster, K. (1989). The resurgence of the tea industry in China: 'beware the tail of the sleeping dragon'. *Outlook on Agriculture* 18 (1): 28–37.

Etherington, D.M. and Forster, K. (1991). Taiwan's tea industry: a dynamic transition. In *World Tea: Proceedings of the International Symposium on Tea Science*, Shizuoka, Japan, 1991, 519–523.

FAOSTAT (2013). Data. http://faostat.fao.org/site/339/default.aspx (accessed 12 June 2013; 17 July 2013).

Flowers, C. (2013). Machine tea harvesting in east and central Africa. [A synopsis of the last twenty years.] Unpublished paper.

Fordham, R. (1970). Factors affecting tea yields in Malawi. PhD thesis, Bristol University, UK.

Fordham, R. (1971). Stomatal physiology and water relations of the tea bush. In *Water and the Tea Plant* (ed. M.K.V. Carr and S. Carr). Kericho, Kenya: Tea Research Institute of East Africa, 89–100.

Fordham, R. (1972). Observations on the growth of roots and shoots of tea (*Camellia sinensis* L.) in southern Malawi. *Journal of Horticultural Science* 47: 221–229.

Fordham, R. (1977). Tea. In *Ecophysiology of Tropical Crops* (ed. P. de T. Alvim and T.T. Kozlowski). New York: Academic Press, 333–349.

Fordham, R. and Palmer-Jones, R.W. (1977). Simulation of intraseasonal yield fluctuations of tea in Malawi. *Experimental Agriculture* 13: 33–42.

Forster, K. and Etherington, D.H. (1991). China's dynamic tea industry. In *World Tea: Proceedings of the International Symposium on Tea Science*, Shizuoka, Japan, 1991, 524–528.

Georgia About (2013). How a Georgian prince smuggled tea out of China. http://georgiaabout .com/2012/09/14/about-food (accessed 18 July 2013).

Ghosh, J.J., George, U. and Barpujari, N. (2006). Available sulphur status of Dooars tea soils. *International Journal of Tea Science* 5: 1–9.

Gnanapragasam N.C. and Sivapalan, P. (2004). Eco-friendly management of tea plantations towards sustainability. *International Journal of Tea Science* 3: 139–146.

Gokhale, N.G. (1955). Estimating the decrease in yield on ceasing to manure unshaded tea. *Empire Journal of Experimental Agriculture* 24: 96–100.

Gokhale, N.G. (1960). Estimating the probable change in yield with time on altering the level of manuring of tea. *Empire Journal of Experimental Agriculture* 112: 316–326.

Green, M.J. (1961). Some problems of tea breeding. Unpublished paper presented at Specialist Committee on Agricultural Botany, Nairobi, Kenya, September 1961.

Green, M.J. (1964). *Vegetative Propagation of Tea*. Pamphlet No. 20. Kericho, Kenya: Tea Research Institute of East Africa.

Green, M.J. (1971). An evaluation of some criteria in selecting large yielding tea clones. *Journal of Agricultural Science (Cambridge)* 73: 143–156.

Greenwood, D.J. (1982). Nitrogen supply and crop yield: the global scene. *Plant and Soil* 67: 43–59.

Grice, W.J. (ed.) (1979). *Fertilisers for Tea in Central and Southern Africa*, revised edn. Mulanje, Malawi: Tea Research Foundation of Central Africa.

Grice, W.J. (1990a). Section 3. Replanting. In *Tea Planters' Handbook* (ed. W.J. Grice). Mulanje, Malawi: Tea Research Foundation of Central Africa.

Grice, W.J. (1990b). Section 5. Planting. In *Tea Planters' Handbook* (ed. W.J. Grice). Mulanje, Malawi: Tea Research Foundation of Central Africa.

Grice, W.J. (1990c). Section 6. Nutrition. In *Tea Planters' Handbook* (ed. W.J. Grice). Mulanje, Malawi: Tea Research Foundation of Central Africa.

Grice, W.J. and Mkwaila, B. (1990). Section 7. Pruning, pruning cycles and skiffing. In *Tea Planters' Handbook* (ed. W.J. Grice). Mulanje, Malawi: Tea Research Foundation of Central Africa.

Griffiths, J. (2007). *Tea: The Drink that Changed the World*. London: André Deutsch.

Gunasekare, M.T.K., Ranatunga, M.A.B., Piyasundara, J.H.N. and Kottawa-Archchi, J.D. (2012). Tea genetic resources in Sri Lanka: collection, conservation and appraisal. *International Journal of Tea Science* 8 (3): 51–60.

Guokun Yao and Tiejun Ge. (1986). Effect of close planting on output and quality of tea and the eco-environment of tea garden (in Chinese). *Journal of Tea Science* 6 (1).

Hackett, C.A., Wachira, F.N., Paul, S., Powell, W. and Waugh, R. (2000). Construction of a genetic linkage map for *Camellia sinensis* (tea). *Heredity* 85: 346–355.

Hadfield, W. (1968). Leaf temperature, leaf pose and productivity of the tea bush. *Nature, London.* 219: 282–284.

Hadfield, W. (1974a). Shade in North-East Indian tea plantations I. The shade pattern. *Journal of Applied Ecology* 11: 151–178.

Hadfield, W. (1974b). Shade in North-East Indian tea plantations II. Foliar illumination and canopy characteristics. *Journal of Applied Ecology* 11: 179–199.

Hadfield, W. (1975). The effect of high temperatures on some aspects of the physiology and cultivation of the tea bush, *Camellia sinensis* in North East India. In *Light as an Ecological Factor, II. The 16th Symposium of the British Ecological Society* (ed. G.C. Evans, R. Bainbridge and O. Rackham). Oxford: Blackwell, 477–496.

Hainsworth, E. (1969). A system of soil management for tea. *Tea Magazine* 10 (3): 14–15.

Hainsworth, E. (1976). Tea productivity. *World Crops* (Nov/Dec).

Hajra, N.G. (2001). *Tea Cultivation: Comprehensive Treatise.* Lucknow, U.P., India: International Book Distributing Company.

Hall, M.N., Robertson, A. and Scotter, C.N.G. (1988). Near-infrared reflectance prediction of quality, theaflavin content and moisture content of black tea. *Food Chemistry* 27: 61–75.

Harding, T.J., Carr, M.K.V. and Baker, D. (1993). *Training Needs Assessment of the Tea Industries in Malawi and Zimbabwe for the Tea Research Foundation (Central Africa), Malawi.* Unpublished report, Cranfield University.

Harler, C.R. (1966). *Tea Growing.* Oxford: Oxford University Press.

Harvey, F.J. (1988). Rootstock selection programme. *Quarterly Newsletter, Tea Research Foundation of Central Africa* 91: 10–11.

Harvey, F.J. (1989). Composite tea plants: a review. *Quarterly Newsletter, Tea Research Foundation of Central Africa* 94: 12–15.

Hazarika, M. and Muraleedharan, N. (2011). Tea in India: an overview. *Two and a Bud* 58: 3–9.

Henderson, R.M. and Nellemann, F. (2011). *Sustainable Tea at Unilever.* Boston, MA: Harvard Business School.

Herd, E.M. and Squire, G.R. (1976). Observations on winter dormancy of tea in Malawi. *Journal of Horticultural Science* 51: 267–279.

Hernandez, A.P., Cock, J.H. and El-Sharkawy, M.A. (1989). The responses of leaf gas exchange and stomatal conductance to air humidity in shade-grown coffee, tea and cacao plants as compared with sunflower. *Revista Brasileira de Fisiologia Vegetal* 1: 155–161.

Hess, T.M. and Kimambo, E. (1998). Rainfall data analysis for Marikitanda Tea Research Station, Tanzania. Paper presented at the Second Annual Open Day, Marikitanda Tea Research Station (Tea Research Institute of Tanzania), Amani, Tanzania, 6 October 1998.

Heuperman, A.F., Stewart, H.T.L. and Wildes, R.A. (1984). The effect of eucalypts on water tables in an irrigation area of northern Victoria. *Water Talk* 52: 2–8.

Hilton, P.J. (1974). Interpretation of foliar analysis results. *Quarterly Newsletter, Tea Research Foundation of Central Africa* 35: 2–4.

Hilton, P.J. and Ellis, R.T. (1972). Estimation of the market value of central African tea by theaflavin analysis. *Journal of the Science of Food and Agriculture* 23: 227–232.

Hilton, P.J. and Palmer-Jones, R. (1973). Relationship between the flavanol composition of fresh tea shoots and the theaflavin content of manufactured tea. *Journal of the Science of Food and Agriculture* 24: 813–818.

Htay, H.H., Kawai, M., MacNaughton, L.E., Katsuda, M. and Juneja, L.R. (2006). Tea in Myanmar, with special reference to pickled tea. *International Journal of Tea Science* 5 (3/4): 11–18.

Huang, S., Pan, G. and Gao, R. (1991). Physiological and biochemical characteristics of tea plants interplanted with trees. In *Agroforestry Systems in China* (ed. Z. Zhu, M. Cal, S. Wang and Y. Jiang). Chinese Academy of Sciences, China and International Development Research Centre, Canada, 162–173.

Huang, X. and Ding, R. (1989). Characteristics of soil fertility in tea (*Camellia sinensis*) and tallow (*Sapium sebiferum*) complex ecosystem in south of Anhui province. *Journal of Tea Science* 2: 116–120.

Hudson, N.W. (1992). *Land Husbandry*. London: Batsford.

IPGRI (1997). *Descriptors for Tea (Camellia sinensis)*. Rome: International Plant Genetic Resources Institute.

ITC (2013). International Tea Committee. http://www.inttea.com/history.asp (accessed July 2013).

Iwasa, K. (1991). Tea production and consumption in Japan. In *World Tea: Proceedings of the International Symposium on Tea Science*, Shizuoka, Japan, 1991, 65–73.

Jain, N.K. (1991). Indian tea in retrospect and prospect and the impact of R & D. In *World Tea: Proceedings of the International Symposium on Tea Science*, Shizuoka, Japan, 1991, 45–57.

Jain, N.K. (2001/02). Tea statistics: performance of tea in Kenya. *International Journal of Tea Science* 1 (2/3): 38–43.

Jain, N.K. (2007). Tea in India. *International Journal of Tea Science* 6 (4): 1–13.

Kabir, S.E. (2002). A study on ecophysiology of tea (*Camellia sinensis*) with special reference to the influence of climatic factors on physiology of a few selected tea clones of Darjeeling. *International Journal of Tea Science* 1 (4): 1–9.

Kallarackal, J. and Somen, C.K. (1997a). An ecophysiological evaluation of the suitability of *Eucalyptus grandis* for planting in the tropics. *Forest Ecology and Management* 95: 53–61.

Kallarackal, J. and Somen, C.K. (1997b). Water use by *Eucalyptus tereticornis* stands of differing density in southern India. *Tree Physiology* 17: 195–203.

Kamau, D.M. (2011). Optimising productivity of ageing tea plantations: lessons learnt from Kenya. *Two and a Bud* 58: 48–57.

Kamau, D.M., Spiertz, J.H.J., Oenema, O. and Owuor, P.O. (2008). Productivity and nitrogen use in tea plantations in relation to age and genotype. *Field Crops Research* 108: 60–70.

Kamau, D.M., Oenema, O., Spiertz, H. and Owuor, P.O. (2012). Changes in soil properties following conversion of forests into intensively managed *Camellia sinensis* L. plantations along a chronosequence. *International Journal of Tea Science* 8 (3): 3–12.

Kamunya, S.M. and Wachira, F.N. (2003). Alternative means for rapid multiplication of tea: use of shoots in tea propagation: preliminary indications. *Tea* 24: 66–68.

Kamunya, S.M. and Wachira, F.N. (2006). Two new clones (TRFK 371/3 and TRFK 430/90) released for commercial use. *Tea* 27: 3–14.

Kamunya, S.M., Muoki, R.C., Wachira, F.N. and Pathak, R.S. (2007). Inheritance of yield, drought tolerance and quality traits in tea (*Camellia sinensis* (L.) O. Kuntze). *Tea* 28: 17–25.

Kamunya, S.M., Wachira, F.N., Nyabundi, K.W., Kerio, L. and Chalo, R.M. (2009a). The Tea Research Foundation of Kenya pre-release purple tea variety for processing health tea product. *Tea* 30 (2): 3–10.

Kamunya, S.M., Wachira, F.N., Pathak, R.S. *et al.* (2009b). Quantitative genetic parameters in tea (*Camellia sinensis* (L.) O. Kuntze): I. combining abilities for yield/drought tolerance and quality traits. *African Journal of Plant Science* 3: 93–101.

Kamunya, S.M., Chalo, R., Korir, R., Kiplang'at, J. and Wachira, F.N. (2010a). Assessment of genetic similarities among popularly cultivated tea clones in Kenya using RAPD markers. *Tea* 31 (2): 11–16.

Kamunya, S.M., Wachira, F.N., Pathak, R.S. *et al.* (2010b). Genomic mapping and testing for quantitative trait loci in tea (*Camellia sinensis* (L.) O. Kuntze). *Tree Genetics and Genomes* 6: 915–929.

Kamunya, S.M., Lang'at, J.K., Wachira, F.N. *et al.* (2011). Mapping quantitative trait loci influencing pest damage and quality trait in tea (*Camellia sinensis* (L.) O. Kuntze). *Tea* 32: 75–87.

Kamunya, S.M., Muoki, R.C., Maritim, T. and Wachira, F.N. (2012). TRFK set to release a new clone 371/8 for commercial use. *Tea* 33 (1): 3–4.

Karunaratne, P.M.A.S., Wijeratne, M.A. and Sangakkara, U.R. (1999). Osmotic adjustment and associated water relations of clonal tea (*Camellia sinensis* L.). *Sabaragamuwa University Journal* 2: 77–85.

Kathiravetpillai, A. (1993). Soil erosion and conservation in new plantings. *Tea Bulletin, Tea Research Institute of Sri Lanka* 13 (2): 26–38.

Katikarn, K. and Swynnerton, R. (1991). Rejuvenation of abandoned tea smallholdings. In *World Tea: Proceedings of the International Symposium on Tea Science*, Shizuoka, Japan, 1991, 529–533.

Kayange, C.W. (1988). A modified method of chip budding tea (*Camellia sinensis*). *Acta horticulturae* 227: 110–112.

Kayange C.W. (1990). Cleft grafting on unrooted tea cuttings in the nursery. *Quarterly Newsletter, Tea Research Foundation of Central Africa* 98: 9–11.

Kayange, C.W., Scarborough, I.P. and Nyirenda, H.E. (1981). Rootstock influence on yield and quality of tea (*Camellis sinensis* L.). *Journal of Horticultural Science* 56 (2): 117–120.

Kekana, V.M., Kamau, D.M., Tabu, M., Nyabundi, K.W. and Wanyoko, J.K. (2012). Effect of varying ratios of enriched manures on nutrient uptake, soil chemical properties, yields and quality of clonal tea. *Tea* 33 (1): 18–29.

Kerfoot, C.W. (1961). Tea root systems. *Pamphlet, Tea Research Institute of East Africa, Kericho, Kenya* 19, 61–72.

Kerfoot, C.W. (1962). Tea root systems. *World Crops* 14: 140–143.

Kibblewhite, M.G. (2011). Harnessing the soil system as an engine for sustainable tea production. *Two and a Bud* 58: 29–32.

Kibblewhite, M.G., Prakash, S., Hazarika, M., Burgess, P.J. and Sakrabani, R. (2014). Managing declining yields from ageing tea populations. *Journal of the Science of Food and Agriculture* 94: 1477–1481.

Kigalu, J.M. (1997). Effects of planting density on the productivity and water use of young tea (*Camellia sinensis* L.) clones in southern Tanzania. PhD thesis, Cranfield University, UK.

Kigalu, J.M. (2002). Experiment N12: effects of plant density and drought on clonal tea productivity. In *Tea Research Institute of Tanzania, Annual Report 2000/2001*, 16–24.

Kigalu, J.M. (2007). Effects of planting density on the productivity and water use of tea (*Camellia sinensis* L.) clones. Measurement of water use in young tea using sap flow meters with a stem heat balance method. *Agricultural Water Management* 90: 224–232.

Kigalu, J.M. and Nixon, D.J. (1997). Responses of young clonal tea to planting density. 1. Annual yields. *Ngwazi Tea Research Unit Quarterly Report* 28: 3–12.

Kigalu, J.M., Kimambo, E.I., Msite, I. and Gembe, M. (2008). Drip irrigation of tea (*Camellia sinensis* L.) I. Yield and crop water productivity responses to irrigation. *Agriculture Water Management* 95: 1253–1260.

Kilgour, J. (1990). The Mitchell Cotts tea combine. *Journal of Agricultural Engineering Research* 45: 313–321.

Kilgour, J. and Burley, J. (1991). Tea mechanization in Uganda. *Agricultural Engineer* 46 (2): 52–55.

Kimambo, A. (2000). *Clone Inventory in Tanzania*. Unpublished report, Tea Research Institute of Tanzania, Dar es Salaam.

Kimambo, E. (2005). Experiment N9. Responses of clonal tea to fertiliser and irrigation. In *Tea Research Institute of Tanzania Annual Reports for 1996/97, 1997/98, 1998/99, 1999/2000, 2000/01, 2001/02, 2002/03, 2003/04*. Dar es Salaam.

Kingdom-Ward, F. (1950). Does wild tea exist? *Nature* 165: 297–299.

Klümper, W. and Qaim, M. (2014). A meta-analysis of the impacts of genetically modified crops. *PLoS ONE* 9: e111629.

Kodomari, S. (1988). Status of tea production in Japan. In *Recent Developments in Tea Production: Proceedings of the International Symposium* (ed. Tsai-Fua Chiu and Chie-Huang Wang). Taiwan Tea Experiment Station, 139–144.

Kottawa-Arachchi, J.D., Gunasekara, M.T.K., Ranatunga, M.A.B. *et al.* (2014). Biochemical characteristics of tea (*Camellia* L. spp.) germplasm accessions in Sri Lanka: correlation between black tea quality parameters and organoleptic evaluation. *International Journal of Tea Science* 10 (1): 3–13.

Kumar, R., Bora, D.K., Singh, A.K. and Bora, B. (2011). Seasonal and clonal variations in shoot extension rates and population density of Darjeeling tea clones (*Camellia sinensis* L.). *Two and a Bud* 58: 74–79.

Laderach, P. and Eitzinger, A. (2011). *Future climate scenarios for Kenya's tea growing areas*. Final report to the International Centre for Tropical Agriculture (CIAT), Colombia.

Landon, J.R. (ed.) (1984). *Booker Tropical Soil Manual*. London: Booker Agriculture International Ltd.

Laycock, D.H. (1961). Yield and the spacing of tea. *Tropical Agriculture (Trinidad)* 38: 195–204.

Laycock, D.H. and Wood, R.A. (1963). Some observations of soil moisture under tea in Malawi. 1. The effect of pruning mature tea. *Tropical Agriculture (Trinidad)* 41: 277–291.

Lea, V.J. (1998). Molecular analysis of genetic diversity and taxonomy of tea, *Camellia sinensis*. Thesis, University of Cambridge.

Lebedev, G.V. (1961). The tea bush under irrigation. *Izd. Akad. Nauk. SSSR Mosk.*

Lebedev, G.V. (1962). A new irrigation regime for agricultural crops. *Fiziologiya Rast.*

Lee, J. I., Ung-Kyu, L. and Kyung, J.C. (1988). Status of tea production and utilization in Korea. In *Recent Developments in Tea Production: Proceedings of the International Symposium* (ed. Tsai-Fua Chiu and Chie-Huang Wang). Taiwan Tea Experiment Station, 103–109.

Leyser, O. (2014). Moving beyond the GM debate. *PLoS Biol* 12: e1001887.

Li, S., Wu, X., Xue, H. *et al.* (2011). Quantifying carbon storage for tea plantations in China. *Agriculture, Ecosystems and Environment* 141: 390–398.

Liang, C., Ming-Zhe, Y., Xin-Chao, W. and Ya-Jun, Y. (2012). Tea genetic resources in China. *International Journal of Tea Science* 8 (2): 55–64.

Libby, W. (1982). What is the safe number of clones per plantation? In *Resistance to Disease and Pests in Forest Trees* (ed. H. Heybroek, B. Stephan and K. von Weisenberg). Proceedings of the

3rd International Workshop on the Genetics of Host-Parasite Interactions in Forestry, Wageningen, the Netherlands, 342–360.

Lockwood, G. (1999). Tea selection in Tanzania. Unpublished paper presented at a meeting of the Tea Research Institute of Tanzania, Njombe, July 1999.

Lopez, S.J., Thomas, J., Pius, P.K., Kumar, R.R. and Muraleedharan, N. (2005). A reliable technique to identify superior quality clones from tea germplasm. *Food Chemistry* 91: 771–778.

Lubang'a, N.M. (2014). Inheritance of catechin and caffeine in Kenyan tea (*Camellia sinensis* (L.) O. Kuntze). Thesis, University of Eldoret, Kenya.

Magambo, M.J.S. and Cannell, M.G.R. (1981). Dry matter production and partitioning in relation to yield of tea. *Experimental Agriculture* 17: 33–38.

Magambo, M.J.S. and Kimani-Waithaka (1988). The effect of plucking on dry matter production and partitioning in young clonal tea plants (*Camellia sinensis* L.). *East African Agriculture and Forestry Journal* 53 (4): 181–184.

Maghanga, J.K., KItuyi, J.L., Kisinyo, P.O. and Ng'etich, W.K. (2013). Impact of nitrogen fertilizer applications on surface water nitrate levels within a Kenyan tea plantation. *Journal of Chemistry* 2013: 196516. http://dx.doi.org/10.1155/2013/196516.

Mair, V.H. and Hoh, E. (2009). *The True History of Tea*. London: Thames and Hudson.

Majumder, B.A., Bera, B. and Rajan, A. (2012). Tea statistics: global scenario. *International Journal of Tea Science* 8: 121–124.

Makola, L.M., Kamunya, S.M., Muoki, R.C., CheruIyot, E.K. and Korir, R.K. (2013). The effect of genotype by environment interactions on yield and quality of tea (*Camellia sinensis* (L.) O. Kuntze) in Kenya. *Tea* 34: 31–42.

Malenga, N.E.A. (2001). Review of experiments on tea pruning practices at TRF (CA). *Quarterly Newsletter, Tea Research Foundation of Central Africa* 141: 9–14.

Mann, H.H. (1935). *Tea Soils*. Technical Communication No. 32. Harpenden, UK: Imperial Bureau of Soil Science.

Martin, P. (1999). Mechanical harvesting of tea in the central and southern African region. *Tea & Coffee Trade Journal* 1 September 1999.

Martin, P.J., Malenga, N.E.A., Mphangwe, N.I.K., Nyirenda, H.E. and Rattan, P.S. (1997). A review of current recommendations for pruning tea. *Quarterly Newsletter, Tea Research Foundation of Central Africa* 126: 21–33.

Matthews, R.B. and Stephens, W. (1998). CUPPA_TEA: a simulation model describing seasonal yield variation and potential production of tea. 1. Shoot development and extension. *Experimental Agriculture* 34: 345–367.

McCubbin, P.D. and Steenkamp, J.W. (1998). Field establishment using rooted cuttings from Speedling trays. *Quarterly Newsletter, Tea Research Foundation of Central Africa* 131: 13–15.

McCulloch, J.S.G., Pereira, H.C., Kerfoot, O. and Goodchild, N.A. (1965). Effect of shade trees on tea yields. *Agricultural Meteorology* 2: 385–399.

McDowell, I. and Owuor, P. (1992). The taste of tea. *New Scientist* 11: 30–33.

Mendham, D.S., White, D.A., Battaglia, M. *et al.* (2011). Soil water depletion and replenishment during first and early second rotation *Eucalyptus globulus* plantations with deep soil profiles. *Agricultural and Forest Meteorology* 151: 1568–1579.

Ming, J. (1999). Taiwan tea industry. In *Global Advances in Tea Science* (ed. N.K. Jain). Houston, Texas: Stadium Press, II. 39–47.

Mitei, Z. (2011). Growing sustainable tea on Kenyan smallholder farms. *International Journal of Agricultural Sustainability* 9 (1): 59–66.

Miura, N. (2003). The Japanese tea market. *International Journal of Tea Science* 2 (3): 8–16.

Mizambwa, F.C.S. (1997). Experiment N9. Responses of clonal tea to fertiliser and irrigation. In *Ngwazi Tea Research Unit Annual Report for 1997*, 3–4.

Mizambwa, F.C.S. (2002). Response of composite tea plants to drought and irrigation in the Southern Highlands of Tanzania. PhD thesis, Cranfield University, UK.

Mizambwa, F.C.S. (2004). Experiment N10. Responses of clones to drought and irrigation. In *Tea Research Institute of Tanzania Annual Report for 2002/03*, 9–11.

Mohanpuria, P., Kumar, V., Ahuja, P.S. and Yadav, S.K. (2011). Producing low-caffeine tea through post-transcriptional silencing of caffeine synthase mRNA. *Plant Molecular Biology* 76: 523–534.

Mohotti, A.J. and Lawlor, D.W. (2002). Diurnal variation in photosynthesis and photoinhibition in tea: effects of irradiance and nitrogen supply during growth in the field. *Journal of Experimental Botany* 53: 313–322.

Möller, M. and Weatherhead, E.K. (2007). Evaluating drip irrigation in commercial tea production in Tanzania. *Irrigation and Drainage Systems* 21: 17–34.

Mondal, T.K., Bhattacharya, A., Ahuja, P.S. and Chang, P.K. (2001). Transgenic tea [*Camellia sinensis* (L.) O. Kuntze cv. Kangra Jat] plants obtained by *Agrobacterium*-mediated transformation of somatic embryos. *Plant Cell Reports* 20: 712–720.

Monteith, J.L. (1972). Solar radiation and productivity in tropical ecosystems. *Journal of Applied Ecology* 9: 747–766.

Morita, A. and Konishi, S. (1989). Relationship between vesicular-arbuscular-mycorrhizal infection and soil phosphorus concentration in tea fields. *Soil Science and Plant Nutrition* 35 (1): 139–143.

Mouli, B.C., Marimuthu, S. and Sharma, V.S. (2012). Zinc: the master micronutrient for tea. *International Journal of Tea Science* 8 (1): 91–96.

Mphangwe, N.K. and Nyirenda, H.E. (2000). Residual effects of nursery manipulation on field performance of two tea clones. *Quarterly Newsletter, Tea Research Foundation of Central Africa* 137: 6–11.

Mphangwe, N.K. and Nyirenda, H.E. (2001a). Procedures for the propagation of plucking shoots and composites in Speedling trays. *Quarterly Newsletter, Tea Research Foundation of Central Africa* 142: 12–15.

Mphangwe, N.K. and Nyirenda, H.E. (2001b). Maximising crop and total value of some TRF clones by plucking at optimum shoot age. *Quarterly Newsletter, Tea Research Foundation of Central Africa* 143: 18–24.

Mphangwe, N.I.K., Vorster, J., Steyn, J.M. *et al.* (2013). Screening of tea (*Camellia sinensis*) for trait-associated molecular markers. *Applied Biochemistry and Biotechnology* 171: 437–449.

Msomba, S.W., Kamau, D.M., Uwimana, M.A., Muhoza, C. and Owuor, P.O. (2014). Effect of location of production, nitrogenous fertilizer rates and plucking intervals on tea clone TRFK 6/8 tea in East Africa: 1. *International Journal of Tea Science* 10 (3/4): 14–24.

Mukumbarezah, C.N. (2001). Comparison of potential and actual yields of tea (*Camellia sinensis*) in central and southern Africa. *Quarterly Newsletter, Tea Research Foundation of Central Africa* 143: 12–18.

Muoki, R.C., Wachira, F.N., Pathak, R.S. and Kamunya, S.M. (2007). Assessment of the mating system of *Camellia sinensis* in biclonal seed orchards based on PCR markers. *Journal of Horticultural Science and Biotechnology* 82: 733–738.

Murty, R.S.R. and Sharma, V.S. (1986). Canopy architecture in tea. *Journal of Plantation Crops* 14 (2): 119–125.

Murty, R.S.R. and Sharma, V.S. (1989). Rationalisation of plucking intervals in tea. ii. Forecasting systems. *UPASI Science Technical Bulletin* 43: 6–15.

Mwakha, E. (1983). Hail reports from Kericho and Nandi tea areas. *Tea* 4 (2): 8–10.

Mwakha, E. (1985). Rehabilitation of moribund tea plants: a review. *Tea* 4 (1): 44–51.

Mwakha, E. (1986). Clonal tea response to frequency and height of mechanical harvesting. *Tea* 7 (2): 48–57.

Mwakha, E. (1990). Response of seedling tea to height and frequency of mechanical harvesting in Kenya highlands. *Tea* 11 (1): 8–12.

Nemec-Bochm, R.I., Cash, S.B., Anderson, B.T. *et al.* (2014). Climate change, the monsoon and tea yields in China. Paper presented at the Annual Meeting of the Agricultural and Applied Economics Association, Minneapolis, July 2014.

Netto, L.A., Jayaram, K.M. and Puthur, J.T. (2010). Clonal variation of tea (*Camellia sinensis* (L.) O. Kuntze) in countering water deficiency. *Physiology and Molecular Biology of Plants* 16: 359–367.

Ng'etich, W.K. (1996). Pruning in tea: a review focusing on experiments in East Africa. *Tea* 17 (1): 41–46.

Ng'etich, W.K. and Othieno, C.O. (1993). Performance of clonal tea plants on rehabilitated moribund tea soils. *Tea* 14: 96–106.

Ng'etich, W.K. and Stephens, W. (2001). Responses of tea to environment in Kenya. 1. Genotype × environment interactions for total dry matter production and yield. *Experimental Agriculture* 37: 333–342.

Ng'etich, W.K. and Wachira, E.N. (2003). Variations in leaf anatomy and gas exchange in tea clones with different ploidy. *Journal of Horticultural Science Biotechnology* 78: 172–176.

Nijs, I., Olyslaegers, G., Kockelbergh, F. *et al.* (2000). Detecting sensitivity to extreme climatic conditions in tea: an ecophysiological vs. an energy balance approach. In *Topics in Ecology: Structure and Function of Plants in Ecosystems* (ed. R. Ceulemans, J. Bagaert, G. Deckmyn and J. Nijs.). Wilrijk: University of Antwerp, 253–263.

Nixon, D.J. (1995). Irrigation of tea in southern Tanzania: a seminar review. *Ngwazi Tea Research Unit Quarterly Report* 19: 3–23.

Nixon, D.J. (1996a). The effects of a modified water regime on the yield and yield distribution of fully irrigated clone 6/8. *Ngwazi Tea Research Unit Quarterly Report* 24: 3–9.

Nixon, D.J. (1996b). Mechanical harvesting of tea in the Southern Highlands: a summary of findings and conclusions. *Ngwazi Tea Research Unit Quarterly Report* 25: 8–26.

Nixon, D.J. (1996c). The effects of age and pruning on the development of drought resistance in maturing clones. *Ngwazi Tea Research Unit Quarterly Report* 26: 3–15.

Nixon, D.J. and Carr, M.K.V. (1995). The effects of irrigation frequency on yield and yield distribution. *Ngwazi Tea Research Unit Quarterly Report* 22: 3–15.

Nixon, D.J. and Sanga, B.N.K. (1995). Dry weight and root distribution of unirrigated mature tea. *Ngwazi Tea Research Unit Quarterly Report* 21: 18–23.

Nixon, D.J., Burgess, P.J., Sanga, B.N.K. and Carr, M.K.V. (2001). A comparison of the responses of mature and young tea to drought. *Experimental Agriculture* 37: 391–402.

Njuguna, C.K. (1977). Clonal spacing and yields. *Tea in East Africa* 17 (1): 9–12.

Njuguna, C.K. (1989). Yield performance of TRFK new released clones from the breeding programme. *Tea* 10: 64–72.

Njuguna, C.K. (1993a). Suitability of some Kenyan tea clones for replanting and effect of soil rehabilitation on replanted tea. *Tea* 14 (1): 7–12.

Njuguna, C.K. (1993b). 1994 TRFK released clones. *Tea* 14 (2): 69–70.

Nyabundi, K.W. (2009). Nursery performance of different tea shoot types. *Tea* 30 (2): 24–29.

Nyasulu, S.K.N. (2001). Shear harvesting: its effect on yield quality and long-term health of the tea bush. *Quarterly Newsletter, Tea Research Foundation of Central Africa* 141: 15–21.

Nyirenda, H.E. (1989). Effectiveness of assessment of vigour and productivity in young vs old bushes and mature clones of tea (*Camellia sinensis*). *Annals of Applied Biology* 115: 327–332.

Nyirenda, H.E. (1997). Replanting and how to realize maximum benefit from clones and composite plants. *Quarterly Newsletter, Tea Research Foundation of Central Africa* 125: 14–17.

Nyirenda, H.E. (2001). Studies of shoot growth. *Quarterly Newsletter, Tea Research Foundation of Central Africa* 141: 24–27.

Nyirenda, H.E. and Mphangwe, N.I.K. (1998a). The potential of plucking shoots for large-scale clonal production. *Quarterly Newsletter, Tea Research Foundation of Central Africa* 131: 10–12.

Nyirenda, H.E. and Mphangwe, N.I.K. (1998b). Matching of TRFCA clones for mixing in the field. *Quarterly Newsletter, Tea Research Foundation of Central Africa* 132: 8–9.

Nyirenda, H.E. and Mphangwe, N.I.K. (2000a). Production of composite plants: success rate and productivity. *Quarterly Newsletter, Tea Research Foundation of Central Africa* 137: 15–16.

Nyirenda, H.E. and Mphangwe, N.I.K. (2000b). Production of composite plants using plucking shoots. *Quarterly Newsletter, Tea Research Foundation of Central Africa* 139: 4–5.

Nyirenda, H.E. and Mphangwe, N.I.K. (2001). Procedures for the propagation of plucking shoots and composites in Speedling trays. *Quarterly Newsletter, Tea Research Foundation of Central Africa* 142: 12–14.

Nyirenda, H.E. and Ridpath, V.E.T. (1984). Criteria for clonal selection in tea (*Camellia sinensis*). I. The relation between mother bushes and their clonal derivatives for bush area, shoot number and yield. *Experimental Agriculture* 20: 339–343.

Obanda, M., Owuor, P.O. and Njuguna, C.K. (1992). The impact of clonal variation of total polyphenols content and polyphenol oxidase activity of fresh tea shoots on plain black tea quality parameters. *Tea* 13: 129–133.

Obanda, M., Owuor, P.O. and Wekesa, H. (2002). The influence of clone and plucking standards on formation of black tea chemical and sensory parameters. *Tea* 23: 81–90.

Odhiambo, H.O. and Othieno, C.O. (1992). Effect of height and month of pruning on yields of tea. *Tea* 13 (1): 18–21.

Ogino, A., Tanaka, J., Taniguchi, F., Yamamoto, M.P. and Yamada, K. (2009). Detection and characterization of caffeine-less tea plants originated from interspecific hybridization. *Breeding Science* 59: 277–283.

Olyslaegers, G., Nijs, I., Roebben, J. *et al.* (2002). Morphological and physiological indicators of tolerance to atmospheric stress in two sensitive and two tolerant tea clones in South Africa. *Experimental Agriculture* 38: 397–410.

Othieno, C.O. (1972). Hydrological investigations. In *Annual Report for 1972*. Kericho, Kenya: Tea Research Institute of East Africa, 28–29.

Othieno, C.O. (1975). Surface run-off and soil erosion in fields of young tea. *Tropical Agriculture (Trinidad)* 52: 299–308.

Othieno, C.O. (1977). *Annual Report 1976*. Kericho: Tea Research Institute of East Africa.

Othieno, C.O. (1978a). Supplementary irrigation of young clonal tea in Kenya. I. Survival, growth and yield. *Experimental Agriculture* 14: 229–238.

Othieno, C.O. (1978b). Supplementary irrigation of young clonal tea in Kenya. II. Internal water status. *Experimental Agriculture* 14: 309–316.

Othieno, C.O. (1983). Effect of weather on recovery from pruning and yield of tea in Kenya. *Journal of Plantation Crops* (supplement): 41–52.

Othieno, C.O. (1991). Tea production in Kenya and scientific research contributions to its success. In *World Tea: Proceedings of the International Symposium on Tea Science*, Shizuoka, Japan, 1991, 74–88.

Othieno, C.O. (1992). Soils. In *Tea: Cultivation to Consumption* (ed. K.C. Willson and M.N. Clifford). London: Chapman and Hall, 137–172.

Othieno, C.O. and Ahn, P.M. (1980). Effects of soil mulches on soil temperature and growth of tea plants in Kenya. *Experimental Agriculture* 16: 287–294.

Othieno, C.O. and Ng'etich, W.K. (1992). Studies on the use of shade in tea plantations in Kenya. II. Effects on yields and their components. *Tea* 13 (2): 100–110.

Othieno, C.O., Stephens, W. and Carr, M.K.V. (1992). Yield variability at the Tea Research Foundation of Kenya. *Agricultural and Forest Meteorology* 61: 237–252.

Owuor, P.O. (1992). Comparison of gas chromatographic volatile profiling methods for assessing the flavour quality of Kenyan black teas. *Journal of the Science of Food and Agriculture* 59: 189–197.

Owuor, P.O. (1994). Effects of lung pruning and nitrogen fertilizer on black tea quality. *Tea* 15 (1): 4–7.

Owuor, P.O. (1996). Problems of replanting tea or other crops on old tea lands: possible role of allelochemicals. A review. *Tea* 17 (1): 34–40.

Owuor, P.O. and Chavanji, A.M. (1986). Caffeine contents of clonal tea: seasonal variations and effects of plucking standards under Kenyan conditions. *Food Chemistry* 20: 225–233.

Owuor, P.O. and Wanyoko, J.K. (1983). Fertilizer use advisory service: a reminder to farmers. *Tea* 4: 3–7.

Owuor, P.O., Reeves, S.G. and Wanyoko, J.K. (1986). Correlation of theaflavins content and valuations of Kenyan black teas. *Journal of the Science of Food and Agriculture* 37: 507–513.

Owuor, P.O., Takeo, T., Horita, H., Tsushida, T. and Murai, T. (1987). Differentiation of clonal teas by Terpene index. *Journal of the Science of Food and Agriculture* 40: 341–345.

Owuor, P.O., Obaga, S.O. and Othieno, C.O. (1990). The effects of altitude on the chemical composition of black tea. *Journal of the Science of Food and Agriculture* 50 (1): 9–17.

Owuor, P.O., Othieno, C.O., Robinson, J.M. and Baker, D.M. (1991). Changes in the quality parameters of seedling tea due to height and frequency of mechanical harvesting, *Science, Food and Agriculture* 55: 241–249.

Owuor, P.O., Othieno, C.O., Odihambo, H.O. and Ng'etich, W. K. (1997a). Effect of fertilizer levels and plucking intervals of clonal tea (*Camellia sinensis* L. O. Kuntze). *Tropical Agriculture* 74: 184–191.

Owuor, P.O., Wachira, F.N. and Ng'etich, W.K. (1997b). Influence of region of production on relative plain tea quality parameters in Kenya. *Food Chemistry* 119: 1168–1174.

Owuor, P.O., Obanda, M., Apostolides, M. *et al.* (2006). The relationship between the chemical plain black tea quality parameters and black tea colour, brightness and sensory evaluation. *Food Chemistry* 97: 644–653.

Owuor, P.O., Kavoi, M.M., Wachira, F.N., Ogola, S.O. and Jain, N.K. (2007). Sustainability of smallholder tea growing in Kenya. *International Journal of Tea Science* 6 (1): 1–23.

Owuor, P.O., Othieno, C.O., Kamau, D.M., Wanyoko, J.A. and Ng'etich, W. K. (2008). Effects of long term fertilizer use on a high yielding clone S15/10: yields. *International Journal of Tea Science* 7 (1/2): 19–31.

Owuor, P.O., Kamau, D.M. and Jondiko, E.O. (2010a). The influence of geographical area of production and nitrogenous fertilizer on yields and quality parameters of clonal tea. *Journal of Food, Agriculture and Environment* 8: 682–690.

Owuor P.O., Wachira, F.N. and Ng'etich, W.K. (2010b). Influence of region of production on relative clonal plain tea quality parameters in Kenya. *Food Chemistry* 119: 1168–1174.

Owuor, P.O., Kamau, D.M., Kamunya, S.M. *et al.* (2011). Effects of genotype, environment and management on yields and quality of black tea. In *Genetics, Biofuels and Local Farming Systems* (ed. E. Lichtfouse). Sustainable Agriculture Reviews 7. Dordrecht: Springer, 277–307.

Owuor, P.O., Othieno, C.O., Kamau, D.M. and Wanyoko, J.K. (2012). Effects of long-term fertilizer use on a high-yielding tea clone AHPS15/10: soil pH, mature leaf nitrogen, mature leaf and soil phosphorus and potassium. *International Journal of Tea Science* 8 (1): 15–51.

Oyamo, J. (1992). The golden clone in a golden field. *Tea* 13: 1.

Pallemulla, D., Shanmugarajah, V. and Kathiravepillai, A. (1992). Effect of grafting fresh cuttings on yield and drought resistance in tea. *Sri Lankan Journal of Tea Science* 51 (2): 45–50.

Pareham, W.E. (2000). The rubber/tea agroforestry system of South China: a short review. https://walterparham.files.wordpress.com/2009/08/the-rubber-tea-agroforestry-system3.pdf (accessed August 2017).

Paul, S., Sharma, D.K., Sharma, K.L. and Misra, S.K. (1994). Genetic variability for yield and quality in tea (*Camellia sinensis*) clones. *Indian Journal of Agricultural Science* 64: 314.

Paul, S., Wachira, F.N., Powell, W. and Waugh, R. (1997). Diversity and genetic differentiation among populations of Kenyan tea (*Camellia sinensis* (L.) O. Kuntze) revealed by AFLP markers. *Theoretical and Applied Genetics* 94: 255–263.

Petinov, N.S. (1962). Physiological basis for the high effectiveness of intermittent sprinkler irrigation in tea plantations. *Izd. Akad. Nauk. SSSR Ser. Boil.* 5717–5728.

Pettigrew, J. and Richardson, B. (2008). *Tea Classified: a Tea Lover's Companion*, London: The National Trust.

Philip, M.S. (1994). *Measuring Trees and Forests.* Wallingford: CABI.

Phukan, I.K., Bhagat, R.M., Saikia, B.P. and Barthakur, B.S. (2011). Soil properties under rehabilitation and non-rehabilitation conditions in tea soils of Assam. *Two and a Bud* 58: 141–149.

Pool, P.A. (1982). Genetic analysis of some morphological, physiological and biochemical characters associated with yield and quality in tea (*Camellia sinensis* L.). *Tropical Agriculture, Trinidad* 59: 9–13.

Pool, P.A. and Nyirenda, H.E. (1981). Effects of rootstocks on the components of yield in tea (*Camellia sinensis* L.). *Journal of Horticultural Science* 56: 121–123.

Pratt, J.N. and Walcott, S.M. (2012). Tea in the USA. *International Journal of Tea Science* 8 (3): 61–74.

Pretty, J., Smith, G., Goulding, K.W.T. *et al.* (2008). Multi-year assessment of Unilever's progress towards agricultural sustainability: indicators, methodology, and pilot farm results. *International Journal of Agricultural Sustainability* 6: 37–62.

Raina, S.N., Ahuja, P.S., Sharma, R.K. *et al.* (2012). Genetic structure and diversity of India hybrid tea. *Genetic Resources and Crop Evolution* 59: 1527–1541.

Ranganathan, V. (1976). Tea soils of South India with special reference to nutritional and manuring problems. Paper presented at the UPASI-FAI Workshop on Fertilizer Requirements of Tea Industry in South India, Madras, November 1976.

Ranganathan, V. and Natesan, S. (1985). Potassium nutrition of tea. In *Potassium in Agriculture: Proceedings of the International Symposium.* ASA-CSSA-SSSA. Madison, Wisconsin, USA, 981–1022.

Ranganathan, V. and Natesan, S. (1987). Manuring of tea: revised recommendations. In *Handbook of Tea Culture*. India: UPASI Tea Research Institute, Section 11, 1–27.

Rao, T.N. and Sharma, P.K. (2001). Sulphur nutrition of tea: global review. *International Journal of Tea Science* 1: 20–27.

Ripley, E.A. (1967). Effects of shade and shelter on the microclimate of tea. *East African Agricultural and Forestry Journal* 33 (1): 67–80.

Roberts, E.A.H. (1950). The phenolic substances of manufactured tea. II. Their origin as enzymic oxidation products in fermentation. *Journal of the Science of Food and Agriculture* 9: 212–216.

Roberts, E.A.H. (1962). Economic importance of flavonoid substances: tea fermentation. In *The Chemistry of Flavonoid Compounds* (ed. T.A. Geissman). Oxford: Pergamon, 468–512.

Robertson, A. (1992). The chemistry and biochemistry of black tea production: the non-volatiles. In *Tea: Cultivation to Consumption* (ed. K.C. Willson and M.N. Clifford). London: Chapman and Hall, 555–601.

Robinson, J.M. and Owuor, P.O. (1992). Tea aroma. In *Tea: Cultivation to Consumption* (ed. K.C. Willson and M.N. Clifford). London: Chapman and Hall, 602–647.

Rono, J.K., Kenduyiwa, J. and Corley, R.H.V. (1993). Clone × environment interactions in tea. In *Proceedings, 1991 International Society of Oil Palm Breeders Workshop on Genotype–Environment Interactions in Perennial Tree Crops*. Kuala Lumpur: Palm Oil Research Institute of Malaysia, 105–109.

Saikia, D.N. and Sarma, J. (2011). Effect of continuous and intermittent shear plucking on yield and quality of plucked shoots in tea. *Two and a Bud* 58: 98–102.

Samson, R., Vandenberghe, J., Vanassche, F. *et al.* (2000). Detecting sensitivity to extreme climatic conditions in tea: sap flow as a potential indicator of drought sensitivity between clones. In *Topics in Ecology: Structure and Function of Plants in Ecosystems* (ed. R. Ceulemans, J. Bagaert, G. Deckmyn and J. Nijs.) Wilrijk: University of Antwerp, 267–278.

Sanderson, G.W. (1963). The chloroform test: a study of its suitability as a means of rapidly evaluating fermenting properties of clones. *Tea Quarterly* 34: 193–196.

Sanga, B.N.K. and Kigalu, J.M. (2005). Experiment M5: soil and water conservation in young tea. In *Annual Report*. Dar es Salaam, Tanzania: Tea Research Institute of Tanzania, 57–59.

Sarmah, P.C. and Bezbaruah, H.P. (1984). Triploid breeding in tea. *Two and a Bud* 31: 55–59.

Satyanarayana, N. and Ilango, R.V.J. (1993). Rooting of crop shoots in tea. *Planters' Chronicle* 88: 225–226.

Satyanarayana, N. and Sharma, V.S. (1984). Role of maintenance foliage in tea (*Camellia* spp.). In *Proceedings of the Fourth Annual Symposium on Plantation Crops: Genetics, Plant Breeding and Horticulture* (ed. S. Vishveshwara). Bangalore, India, 67–72.

Satyanarayana, N. and Sreedhar, Ch. (1995). Nursery grafting in tea. In *UPASI Handbook of Tea Culture*, Section 3, 3.

Satyanarayana, N., Cox, S. and Sharma, V.S. (1991). Field performance of grafts made of fresh tea clonal cuttings. *Planters' Chronicle* 86: 85–93.

Schönau, A.P.G. (1983). Silvicultural considerations for high productivity of *Eucalyptus grandis*. Paper presented at International Symposium on Strategies and Designs for Afforestation, Reforestation and Tree Planting, Wageningen, the Netherlands.

Sealy, J. (1958). *A Revision of the Genus Camellia*. London: Royal Horticultural Society.

Sebastiampillai, A.R. and Solomon, H.R. (1976). Short-term yield in the immature stage as an indicator of the yield potential of tea (*Camellia sinensis*). *Tea Quarterly* 46: 16–25.

Seurei, P. (1996). Tea improvement in Kenya: a review. *Tea* 17 (2): 76–81.

Sharma, V.S. (1982). Vegetative propagation of tea: a review. *Proceedings, 5th Symposium on Plantation Crops* 1–15.

Sharma, V.S. (2011). *A Manual of Tea Cultivation.* Hyderabad: International Society of Tea Science, 51–62.

Sharma, V.S. (2012). Tea (*Camellia* sp.) germplasm status in India. *International Journal of Tea Science* 8 (1): 3–8.

Sharma, V.S. and Murty, R.S.R. (1989). Certain factors influencing recovery from pruning in South India. *Tea* 10 (1): 32–41.

Sharma, V.S. and Satyanarayana, N. (1993). Harvesting in tea. *International Symposium on Tea Science and Human Health, Calcutta, India*, 21–34.

Sharma, V.S. and Swaminathan, P. (1976). Contrarities in tea cultural practice in S. India and Assam. *The Planters' Chronicle* 77 (4): 131–133.

Sharma, V.S., Satyanarayana, N. and Cox, S. (1990). Training of young tea by two-stage tipping. Paper presented at the Ninth Symposium on Plantation Crops, Bangalore, India, December 1990.

Shaxon, T.F. (1968). Some long-term effects of shade trees on yield of tea in Malawi. In *Proceedings of the 4th Specialist Meeting in Applied Meteorology, Nairobi, Kenya.*

Shi, C.Y., Yang, H., Wei, C.L. *et al.* (2011). Deep sequencing of the *Camellia sinensis* transcriptome revealed candidate genes for major metabolic pathways of tea-specific compounds. *BMC Genomics* 12: 131.

Sikka, A.K., Samra, J.S., Sharda, V.N., Samraj, P. and Lakshmanan, V. (2003). Low flow and high flow responses to converting natural grassland into bluegum (*Eucalyptus globulus*) in Nilgiris watersheds of South India. *Journal of Hydrology* 270: 12–26.

Simbua, E.F. and Nyanga, A.W. (2001). Smallholder tea extension activities in Rungwe District. In *Annual Report, Tea Research Institute of Tanzania 2000/2001, Dar es Salaam*, 66–69, and subsequent reports.

Simmonds, N.W. (1985). Two-stage selection strategy in plant breeding. *Heredity* 55: 393–399.

Simmonds, N.W. (1996). Family selection in plant breeding. *Eyphytica* 90: 201–208.

Singh, I.D. (1980). Nonconventional approaches in the breeding of tea in North East India. *Two and a Bud* 27: 3–6.

Singh, I.D. (2006). Selection and testing field performance of candidate tea clones in India. *International Journal of Tea Science* 5 (1): 1–12.

Singh, I.D., Chakraborty, S. and Prodhan, S.K. (1993). Genetic variability, heritability and genetic advance of yield in tea (*Camellia sinensis* L.). *Two and a Bud* 40 (1): 38–40.

Sitienei, K. and Kamau, D.M. (2013). Towards sustainable agriculture: the case of tea nutrient budgeting in the small-holder sub-sector in Kenya. *Tea* 34 (2): 68–71.

Sitienei, K., Home, P.G., Kamau, D.M. and Wanyoko, J.K. (2013). Nitrogen and potassium dynamics in tea cultivation as influenced by fertilizer type and application rates. *American Journal of Plant Sciences* 4: 59–65.

Sivapalan, P. (1991). Current status of tea production in Sri Lanka. In *World Tea: Proceedings of the International Symposium on Tea Science*, Shizuoka, Japan, 1991, 89–96.

Smith, B.G. (1998). Tea planting materials. *Tropical Agricultural Newsletter* (March): 22–24.

Smith, B.G., Burgess, P.J. and Carr, M.K.V. (1993a). Responses of tea (*Camellia sinensis*) clones to drought. II. Stomatal conductance, photosynthesis and water potential. *Aspects of Applied Biology, Physiology of Varieties*, 34: 359–368.

Smith, B.G., Stephens, W., Burgess, P.J. and Carr, M.K.V. (1993b). Effects of light, temperature, irrigation and fertilizer on photosynthetic rate in tea (*Camellia sinensis*). *Experimental Agriculture* 29: 291–306.

Smith, B.G., Burgess, P.J. and Carr, M.K.V. (1994). Effects of clone and irrigation on the stomatal conductance and photosynthetic rate of tea (*Camellia sinensis*). *Experimental Agriculture* 30: 1–16.

Soares, J.V. and Almeida, A.C. (2001). Modelling the water balance and soil water fluxes in a fast growing *Eucalyptus* plantation in Brazil. *Journal of Hydrology* 253: 130–147.

Squire, G.R. (1976). Xylem water potential and yield of tea (*Camellia sinensis* L.) clones in Malawi. *Experimental Agriculture* 12: 289–297.

Squire, G.R. (1977). Seasonal changes in photosynthesis in tea (*Camellia sinensis* L.). *Journal of Applied Ecology* 14: 303–316.

Squire, G.R. (1979). Weather, physiology and seasonality of tea (*Camellia sinensis*) yields in Malawi. *Experimental Agriculture* 15: 321–330.

Squire, G.R. and Callender, B.A. (1981). Tea plantations. In *Water Deficits and Plant Growth* (ed. T.T. Kozlowski), vol. 6. New York: Academic Press, 471–510.

Squirrell, J. (1995). A comparison of tea harvesting techniques in the Southern Highlands of Tanzania. BSc dissertation, Cranfield University, Silsoe, UK.

Stape, J.L., Binkley, D. and Ryan, M.G. (2004). *Eucalyptus* production and the supply, use and efficiency of use of water, light and nitrogen across a geographic gradient in Brazil. *Forest Ecology and Management* 19: 317–331.

Stephens, W. and Carr, M.K.V. (1990). Seasonal and clonal differences in shoot extension rates and numbers in tea (*Camellia sinensis*). *Experimental Agriculture* 26: 83–98.

Stephens, W. and Carr, M.K.V. (1991a). Responses of tea (*Camellia sinensis*) to irrigation and fertilizer. I. Yield. *Experimental Agriculture* 27: 177–191.

Stephens, W. and Carr, M.K.V. (1991b). Responses of tea (*Camellia sinensis*) to irrigation and fertilizer. II. Water use. *Experimental Agriculture* 27: 193–210.

Stephens, W. and Carr, M.K.V. (1993). Responses of tea (*Camellia sinensis*) to irrigation and fertilizer. III. Shoot extension and development. *Experimental Agriculture* 29: 323–339.

Stephens, W. and Carr, M.K.V. (1994). Responses of tea (*Camellia sinensis*) to irrigation and fertilizer. IV. Shoot population density, size and mass. *Experimental Agriculture* 30: 189–205.

Stephens, W., Othieno, C.O. and Carr, M.K.V. (1992). Climate and weather variability at the Tea Research Foundation of Kenya. *Agricultural and Forest Meteorology* 61: 219–235.

Stephens, W., Burgess, P.J. and Carr, M.K.V. (1994). Yield and water use of tea in southern Tanzania. *Aspects of Applied Biology*, 38: 223–230.

Storey, H.H. and Leach, E. (1933). A sulphur deficiency of the tea bush. *Annals of Applied Biology* 20: 24–25.

Sullivan, G. (2011). Tea and sympathy in Bangladesh. In *The Making of a One-Handed Economist*. Sussex, UK: Book Guild Publishing, 123–137.

Sumberg, J. and Thompson, J. (eds.) (2012). *Contested Agronomy: Agricultural Research in a Changing World*. Oxford: Earthscan, Routledge.

Suryanarayanan, S. and Hegde, D.G. (1993). Shear harvesting and plucker productivity. In *Proceedings of the Sixth Joint Area Scientific Symposium (JASS-VI). United Planters' Association of Southern India Bulletin* 46: 99–102.

Tanton, T.W. (1982a). Environmental factors affecting the yields of tea (*Camellia sinensis*). I. Effects of air temperature. *Experimental Agriculture* 18: 47–52.

Tanton, T.W. (1982b). Environmental factors affecting the yields of tea (*Camellia sinensis*). II. Effects of soil temperature, day length and dry air. *Experimental Agriculture* 18: 53–63.

Tanton, T.W. (1992). Tea crop physiology. In *Tea: Cultivation to Consumption* (ed. K.C. Willson and M.N. Clifford). London: Chapman and Hall, 173–199.

Taylor, S., Baker, D., Owuor, P.O. *et al.* (1992). A model for predicting black tea quality from the carotenoid and chlorophyll composition of fresh green tea leaf. *Journal of the Science of Food and Agriculture* 58: 185–191.

Tea Research Association (2012). Irrigation. Tocklai: TRA. http://www.tocklai.net/activities/tea-cultivation/irrigation (accessed June 2017).

Templer, J.C. (1970). Mountainous cultures. *Tea: Journal of the Tea Boards of East Africa* 10 (4): 19–21.

Templer, J.C. (1971). Grafting tea. *Tea: Journal of the Tea Boards of East Africa* 11 (4): 39–45.

Templer, J.C. (1978). Tea plucking studies. In *Annual Report 1977*. Kericho: Tea Research Institute of East Africa.

Thompson, J. and Sumberg, J. (2012). *Nullius in verba*: contestation, pathways and political agronomy. In *Contested Agronomy: Agricultural Research in a Changing World* (ed. J. Sumberg and J. Thompson). Oxford: Earthscan, Routledge, 204–211.

Tolhurst, J.A.H. (1961). Report of the Agricultural Chemist. In *Annual Report of the Tea Research Institute of Ceylon, 1961*, 50–58.

Tolhurst, J.A.H. (1969). *Annual Report of the Tea Research Institute of East Africa 1968/69*. Kericho, Kenya, 51–52.

Tolhurst, J.A.H. (1971). Leaf nutrient content in relation to season and irrigation. In *Water and the Tea Plant* (ed. M.K.V. Carr and S. Carr). Kericho, Kenya: Tea Research Institute of East Africa, 195–212.

TRA (2012). *Report of the Tea Research Association, Tocklai, Assam, India*.

TRFCA (1990). *Tea Planter's Handbook*. Mulanje, Malawi: Tea Research Foundation of Central Africa.

TRFCA (2000). *Clonal Catalogue*. Mulanje, Malawi: Tea Research Foundation of Central Africa.

TRFK (1986). *Tea Growers' Handbook*, 4th edn. Kericho, Kenya: Tea Research Foundation of Kenya.

TRFK (2011). *Annual Technical Report*. Kericho, Kenya: Tea Research Foundation of Kenya.

TRISL (2001). *Bringing Young Tea into Bearing*. Advisory Circular PA1. Tea Research Institute of Sri Lanka.

TRIT (2000). Experiment N10. Responses of clones to drought and irrigation. In *Tea Research Institute of Tanzania Annual Report, 1997/98 and 1998/99*, 10–13.

Tubbs, F.R. (1938). Report of the plant physiologist. In *Annual Report of Tea Research Institute of Ceylon for 1937*, 44–62.

Tunstall, A.C. (1931). A note on the propagation of tea by green shoot cuttings. *Quarterly Report of Indian Tea Association*, 49–51.

Tuwei, G., Kaptich, F.K.K., Langat, M.C., Chomboi, K.C. and Corley, R.H.V. (2008a). Effects of grafting on tea. 1. Growth, yield and quality. *Experimental Agriculture* 44: 521–535.

Tuwei, G., Kaptich, F.K.K., Langat, M.C., Smith, R.G. and Corley, R.H.V. (2008b). Effects of grafting on tea. 2. Drought tolerance. *Experimental Agriculture* 44: 537–546.

Tuwei, G.K. (2008). MRTM1: a high yielding and drought tolerant tea clone from Unilever Tea Kenya. *Tea* 29 (2): 3–7.

Van der Laan, P. and Verdooren, L.R. (1990). A review with some applications of statistical selection procedures for selecting the best variety. *Euphytica* 51: 67–75.

Vanli, H. (1991). Tea production and consumption in Turkey. In *World Tea: Proceedings of the International Symposium on Tea Science*, Shizuoka, Japan, 1991, 97–104.

Varadan, K.M. (1996). Irrigation scheduling for tea in South India. *Journal of Plantation Crops* 24 (2): 97–100.

Venkatesan, S. (2006). Notes and amendments to the recommendations on manuring of tea in south India. Reprinted (with modifications) from *Planters' Chronicle* 102 (4,5): 4–16.

Venkatesan, S. and Senthurpandian, V.K. (2006). Comparison of enzyme activity with depth under tea plantations and forested sites in south India. *Geoderma* 137: 212–216.

Verma, D.P. and Palani, N. (1997). *Manuring of Tea in South India (revised recommendations)*. Valparai, India: UPASI Tea Research Institute.

Verma, D.P. and Pund, S.A. (2014). Potassium – the key nutrient for tea grown in latosols of South India: glimpses of seven decades of potassium research. *International Journal of Tea Science* 10 (1/2): 34–56.

Visser, T. (1969). Tea – *Camellia sinensis* (L.) O. Kuntze. In *Outlines of Perennial Crop Breeding in the Tropics* (ed. F.P. Ferwerda and F. de Wit). Wageningen: Wageningen University, 459–493.

Wachira, F.N. (1994a). Breeding and clonal selection in relation to black tea quality: a review. *Tea* 15: 56–66.

Wachira, F.N. (1994b). Triploidy in tea (*Camellia sinensis*): effect on yield and yield attributes. *Journal of Horticultural Science* 69: 53–60.

Wachira, F.N. and Kamunya, S.M. (2004). Preliminary observations on inheritance of yield and some physiological traits in tea. *Tea* 25: 24–31.

Wachira, F.N. and Kamunya, S.K. (2005). Pseudo-self-incompatibility in some tea clones (*Camellia sinensis* (L.) O. Kuntze). *Journal of Horticultural Science and Biotechnology.* 80: 716–720.

Wachira, F.N. and Ng'etich, W.K. (1999). Dry-matter production and partition in diploid, triploid and tetraploid tea. *Journal of Horticultural Science and Biotechnology* 74: 507–512.

Wachira, F.N., Ng'etich, W., Omolo, J. and Mamati, G. (2002). Genotype × environment interactions for tea yields. *Euphytica* 127: 289–296.

Wanyoko, J.K., Othieno, C.O., Mwakha, E. and Cheruiyot, D. (1996). Effects of types and rates of nitrogen fertilizer on soil. 2. Extractable phosphorus and potassium contents under seedling tea in Nandi Hills, Kenya. *Tea* 17 (1): 12–19.

Weatherstone, J. (1992). Historical introduction. In *Tea: Cultivation to Consumption* (ed. K.C. Willson and M.N. Clifford). London: Chapman and Hall, 1–23.

Wekumbura, W.G.C., Mohotti, A.J., Vidhaha Arachchi, L.P. and Mohotti K.M. (2010) Comparison of assimilate partitioning in organically and conventionally grown tea. *Journal of Agricultural Sciences* 5 (1): 32–41.

Wellensiek, S.J. (1934). Research on quantitative tea selection.1: The Pajoeng reform and seed garden at Tjinj (Dutch with English summary). *Arch. Thee. Cult.* 8–37.

Whittle, A. (1997). Clonal tea in southern Africa. *Tea and Coffee Trade Journal* (December), 62–66.

Whittle, A. (1999). Country profile: tea industry in central Africa and South Africa. In *Global Advances in Tea Science* (ed. N.K. Jain). India: Aravali Books International,

Wickramaratne, M.R.T. (1981a). Genotype–environment interactions in tea (*Camellia sinensis* L.) and their implications in tea breeding and selection. *Journal of Agricultural Science (Cambridge)* 96: 471–478.

Wickramaratne, M.R.T. (1981b). Variation in some leaf characteristics in tea (*Camellia sinensis* L.) and their use in the identification of clones. *Tea Quarterly* 50 (4): 183–198.

Wickramasinghe, K.N. and Krishnapillai, S. (1986). Fertilizer use. In *Handbook on Tea* (ed. P. Sivapalan, S. Kulasegaram and A. Kathiravetpillai). Talawakelle: Tea Research Institute of Sri Lanka, 63–77.

Wight, W. (1932). The structure and growth of the tea bush. The tea leaf. *Quarterly Journal of the Indian Tea Association* IV: 200–217.

Wight, W. (1938). Recent advances in the classification and selection of tea. In *Proceedings of Second Tocklai Annual Conference, Assam*, 36.

Wight, W. (1956). Commercial selection and breeding of tea in India. *World Crops* (8): 263–268.

Wight, W. (1958). Selection policy in tea estates of North East India. *Annual Report of Tocklai Station* 1957: 58–68.

Wight, W. (1959a). Nomenclature and classification of the tea plant. *Nature* 183: 1726.

Wight, W. (1959b). The shade tree traditions in tea gardens of northern India. *Report of Tocklai Experimental Station* 1958: 77–150.

Wight, W. (1962). Tea classification revised. *Current Science* 81: 298–299.

Wight, W. and Gilchrist R.C.J.H. (1961). Concerning the quality and morphology of tea. *Report of the Indian Tea Association, Tocklai Experimental Station* 1959: 69–86.

Wilkie, A.S. (1995). TRF regional shear plucking seminar at Ratelshoek Estate Zimbabwe: a report. *Quarterly Newsletter, Tea Research Foundation of Central Africa* 117: 15–20.

Willatt, S.T. (1970). A comparative study of the development of young tea under irrigation. I. Establishment in the field. *Tropical Agriculture (Trinidad)* 47: 243–249.

Willat, S.T. (1971). A comparative study of the development of young tea under irrigation. II. Continued growth in the field. *Tropical Agriculture (Trinidad)* 48: 271–277.

Willatt, S.T. (1973). Moisture use by irrigated tea in southern Malawi. *Ecological Studies* 4: 331–338.

Willson, K.C. (1969). The mineral nutrition of tea. *Potash Review* 27: 1–17.

Willson, K.C. (1970). Foliar analysis of tea. *Experimental Agriculture* 6: 263–265.

Willson, K.C. (1972). Paraquat, mulch and the mineral nutrition of tea. *Outlook on Agriculture* 7 (2): 74–77.

Willson, K.C., Hainsworth, E., Green, M.J. and O'Shea, P.B.T. (1975). Studies on the mineral nutrition of tea. III. Phosphate. *Plant and Soil* 43: 259–278.

Winter, E.J. and Bora, M. (1940). *The Species of Shade Trees Used on Tea Gardens*. Memorandum No. 12, Tocklai Experimental Station, Indian Tea Association.

Yu, S., Wang, S., Wei, P. *et al.* (1991). A study on Paulownia-tea intercropping system microclimate modification and economic benefits. In *Agroforestry Systems in China* (ed. Z. Zhu, M. Cal, S. Wang and Y. Jiang). Canada: Chinese Academy of Sciences, China and International Development Research Centre, 150–161.

Yu, Y. (2012). Agrotechnology of tea in China. CD enclosed with *International Journal of Tea Science* 8 (2). (Reproduced from *Global Advances in Tea Science*, 1999.)

Zhang, D., Kuhr, S. and Engelhardt, U.H. (1992). Influence of catechins and theaflavins on the astringent taste of black tea brews. *Zeitschrift für Lebensmittel-Untersuchung und Forschung* 195: 108–111.

Zheng, C., Zhao, L., Wang, Y. *et al.* (2015). Integrated RNA-Seq and sRNA-Seq analysis identifies chilling and freezing responsive key molecular players and pathways in tea plant (*Camellia sinensis*). *PLoS ONE* 10: e0125031.

Further Reading

Azam-Ali, S.N. and Squire, G.R. (2002). *Principles of Tropical Agronomy.* Oxford: CABI.

Baffes, J. (2005). Reforming Tanzania's tea sector: a story of success? *Development Southern Africa* 22: 589–604.

Bandyopadhyay, S.K. (1999). Rooting patterns of tea (*Camellia sinensis*) under trickle irrigation. *Annals of Agricultural Research* 20: 497–501.

Barooah, H.P. (1999). Indian tea industry. In *Global Advances in Tea Science* (ed. N. K. Jain). Houston, Texas: Stadium Press, II.15–17.

Baruah, A., Saikia, H.K. and Bera, B. (2010). Detection of close genetic relatedness in some tea genotypes of Assam and Darjeeling using RAPD marker. *Journal of Plantation Crops* 38: 11–15.

Biswas, A.K., Karmokar, P.K. and Barbora, B.C. (1996). Crop–weather relationships for forecasting weekly crop of tea. *Journal of Plantation Crops* 24 (2): 107–114.

Bore, J.K.A., Cheserek, B.C., Ng'etich, W.K. and Yegon, S.K. (2011). An analysis of the incidence of hail damage in tea areas of Kenya. *Tea* 32: 15–20.

Bungard, K. and Carr, M.K.V. (1987). *Management Training Needs in Africa.* Unpublished report to Brooke Bond Group, London. Cranfield University, UK.

Carr, M.K.V. (2012). Tea. In *Advances in Irrigation Agronomy: Plantation Crops.* Cambridge: Cambridge University Press, 222–274. (With G. Lockwood and J. Knox.)

Carr, M.K.V., Stephens, W. and Congdon, T.C.E. (1988). Tea in Tanzania. *Outlook on Agriculture* 17 (1): 18–23.

Chen, D. and Wu, H. (2012). Black tea in China. *International Journal of Tea Science* 8 (2): 35–46.

Dale, M.O. (1971). Mechanical harvesting. *Quarterly Newsletter, Tea Research Foundation of Central Africa.*

De Costa, W.A.J.M. and Surenthran, P. (2005). Resource competition in contour hedgerow intercropping systems involving different shrub species with mature and young tea on sloping highlands in Sri Lanka. *Journal of Agricultural Science (Cambridge)* 143: 395–405.

Dutta, R., Stein, A., Smaling, E.M.A., Bhagat, R.M. and Hazarika, M. (2010). Effects of plant age and environmental and management factors on tea yield in northeast India. *Agronomy Journal* 102: 1290–1301.

Fordham, R. and Holgate, M. (1972). Estimation of leaf area of tea (*Camellia sinensis*, L.) from linear measurements. *Journal of Horticultural Science* 47: 131–135.

George, U. and Barpujari, N. (2005). Influence of some soil chemical properties on yield of tea (*Camellia sinensis*) in the Dooars region of North Bengal, India. *International Journal of Tea Science* 4 (3/4): 27–33.

Gerami, M. (2003). Marketing tea in the Middle East. *International Journal of Tea Science* 2 (3): 12–16.

Guo Ya-ling (2012). Oolong tea in China. *International Journal of Tea Science* 8 (2): 23–34.

Hanks, R.J., Keller, J., Ramussen, V.P. and Wilson, G.D. (1976). Line source sprinkler for continuously variable irrigation – crop production studies. *Soil Science Society America Journal* 44: 426–429.

Harikrishnan, B. and Sharma, V.S. (1980). Drought resistance in tea (*Camellia* L. spp.): I. Studies in cuticular resistance in tea clones. *Journal of Plantation Crops* 8 (2): 73–77.

Hossain, M. (2012). Tea growing countries in the world: Bangladesh. *International Journal of Tea Science* 8 (4): 77–80.

Kamunya, S.M. (2011). TRFK seeks plant breeders rights for a new tea variety. *Quarterly Bulletin, Tea Research Foundation of Kenya* 16 (4): 6–7.

Kimambo, E. (2001). Experiment N15. Partial pruning of S15/10. In *Tea Research Institute of Tanzania, Annual Report 2000/2001*, 32–34.

Klasra, M.A., Khawar, K.M. and Aasim, M. (2007). History of tea production and marketing in Turkey. *International Journal of Agriculture and Biology* 9 (3): 523–529.

Koehler, J. (2015). *Darjeeling: a History of the World's Greatest Tea.* London, Bloomsbury.

Kwach, B.O., Kamau, D.M., Owuor, P.O. *et al.* (2011). Effects of location of production, fertilizer rates, and plucking intervals on mature leaf nutrients of clone TRFK 6/8 in East Africa. *Tea* 32 (2): 56–68.

Kwach, B.O., Owuor, P.O., Kamau, D.M. and Msomba, S.W. (2016). Variations in the precursors of plain black tea quality parameters due to location of production and nitrogen fertilizer rates in Eastern African clonal tea leaves. *Experimental Agriculture* 52: 266–278.

Lane, P.N.J., Morris, J., Zhang, N. *et al.* (2004). Water balance of tropical eucalypt plantations in south-eastern China. *Agricultural and Forest Meteorology* 124: 253–267.

Lin, Z. (2012). Green tea in China. *International Journal of Tea Science* 8 (2): 17–22.

Maidment, R.I., Allan, R.P. and Black, E. (2015). Recent observed and simulated changes in precipitation over Africa. *Geophysics Research Letters* 42: 8155–8164.

Mbadi, P.K. and Owuor, P.O. (2008). The role of the Kenya Tea Development Agency Limited in the small-scale tea holder development in Kenya. *International Journal of Tea Science* 7 (3/4): 7–15.

Mohotti, A.J., Damayanthi, M.M.N., Anandacoomaraswamy, A. and Mohotti, K.M. (2008). Comparative dynamics of tea (Camellia sinensis L.) roots under organic and conventional management systems with special reference to water use. Paper presented at 16th IFCAM Organic World Congress, Modena, Italy.

Nagarajah, S. (1979). Differences in cuticular resistance in relation to transpiration in tea (*Camellia sinensis*). *Physiologia Plantarium* paper nos. 46 (2): 89–92.

Nagarajah, S. and Ratnsooriya, G.B. (1977). Studies with antitranspirants on tea (*Camellia sinensis* L.). *Plant and Soil* 48: 185–197.

Okano, K. (1999). Current status of Japanese tea industry. In *Global Advances in Tea Science* (ed. N. K. Jain). Houston, Texas: Stadium Press, II.27–31.

Pethiyagoda, V. and Rajendran, N.S. (1965). The determination of leaf area in tea. *Tea Quarterly* 36: 48–58.

Rajasekar, R., Cox, S. and Satyanarayana, N. (1991). Evaluation of certain morphological factors in tea (*Camellia* L. spp.) cultivars under water stress. *Journal of Plantation Crops* 18 (supplement): 83–92.

Renard, C., Flemal, J. and Barampama, D. (1979). Evaluation of the resistance of the tea bush to drought in Burundi. *Café Cacao The* 23: 175–182.

Saito, K. and Ikeda, M. (2012). The function of roots of tea plant (*Camellia sinensis*) cultured by a novel form of hydroponics and soil acidification. *American Journal of Plant Sciences* 3: 646–648.

Sarma, G. (2013). A historical background of tea in Assam. *Journal of Humanities and Social Science* 1 (4): 123–131.

Shi, Z. (2012). Dark tea in China: a type of post fermentation tea only made in China. *International Journal of Tea Science* 8 (2): 47–55.

Simmonds, N.W. (1989a). How frequent are superior genotypes in plant breeding populations? *Biological Reviews* 61: 341–365.

Simmonds, N.W. (1989b). Economic aspects of plant breeding with special reference to economic index selection. *Research and Development in Agriculture* 6: 53–62.

Sivapalan, P. and Gnanapragasam, N.C. (2008). The tea smallholder sector in Sri Lanka. *International Journal of Tea Science* 7: 17–24.

Sivapalan, P. and Smaling, E.M.A. (2010). Sustainable agriculture with a focus on tea production systems and soil fertility management. In *Proceedings of Tocklai Tea Centenary Conference 'Ensuring the Future of Tea'*, New Delhi, May 2010.

Stephens, W. and Carr, M.K.V. (1999). Responses of tea (*Camellia sinensis*) to irrigation and fertilizer IV. Shoot population density, size and mass. *Experimental Agriculture* 30: 189–205.

Tien, D.M. (1999). Vietnam tea industry. In *Global Advances in Tea Science* (ed. N. K. Jain). Houston, Texas: Stadium Press, II.48–51.

Tsai-Fua, C. (1988). Tea production and research in Taiwan. In *Recent Developments in Tea Production: Proceedings of the International Symposium* (ed. Tsai-Fua Chiu and Chie-Huang Wang). Taiwan Tea Experiment Station, 121–130.

Tsai-Fua, C. (1991). The present status of tea cultivation, manufacture and utilization in Taiwan. In *World Tea: Proceedings of the International Symposium on Tea Science*, Shizuoka, Japan, 1991, 34–44.

Van Lalyveld, L.J., Smith, B.L. and Fraser, C. (1990). Nitrogen fertilisation of tea: effect on chlorophyll and quality parameters of processed black tea. *Acta Horticulturae* 275: 283–287.

Venkatesan, S. and Ganapathy, M.N.K. (2004). Impact of nitrogen and potassium fertilizer application on quality of CTC teas. *Food Chemistry* 84 (3): 325–328.

Venkatesan, S., Murugesan, S., Ganapathy, M.N.K. and Verma, D.P. (2004). Long-term impact of nitrogen and potassium fertilisers on yield, soil nutrients and biochemical parameters of tea. *Journal of the Science of Food and Agriculture* 84: 1939–1944.

Venkatesan, S., Murugesan, S., Senthurpandian, V.K. and Ganapathy, M.N.K. (2005). Impact of sources and doses of potassium on biochemical and green leaf parameters of tea. *Food Chemistry* 90: 535–539.

Venkatesan, S., Senthurpandian, V.K., Murugesan, S., Mathuam, W. and Ganapathy, M.N.K. (2006). Quality standards of CTC black teas as influenced by sources of potassium fertilizer. *Journal of the Science of Food and Agriculture* 86: 799–803.

Venkateswaran, G. and Radhakrishnan, B. (2011). Studies on the effect of drip irrigation on mature clonal tea. *Two and a Bud* 58: 93–97.

Whitehead, D. and Beadle, C.L. (2004). Physiological regulation of productivity and water use in *Eucalyptus*: a review. *Forest Ecology and Management* 193: 113–140.

Wilbowo, Z.S. (1987). Fertilization scheme for tea in Indonesia. Paper presented at Regional Tea (Scientific) Symposium in Sri Lanka, October 1987.

Williams, E.N.D. (1971). Investigations into certain aspects of water stress in tea. In *Water and the Tea Plant* (ed. M.K.V. Carr and S. Carr). Kericho, Kenya: Tea Research Institute of East Africa, 79–87.

Index

Locators in **bold** refer to tables; those in *italic* to figures